Product Design for the Environment

A Life Cycle Approach

Product Design for the Environment

A Life Cycle Approach

Fabio Giudice

Guido La Rosa

Antonino Risitano

CRC Press
Taylor & Francis Group
Boca Raton London New York

CRC Press is an imprint of the
Taylor & Francis Group, an **informa** business
A TAYLOR & FRANCIS BOOK

CRC Press
Taylor & Francis Group
6000 Broken Sound Parkway NW, Suite 300
Boca Raton, FL 33487-2742

First issued in paperback 2019

© 2006 by Taylor & Francis Group, LLC
CRC Press is an imprint of Taylor & Francis Group, an Informa business

No claim to original U.S. Government works

ISBN-13: 978-0-8493-2722-3 (hbk)
ISBN-13: 978-0-367-39134-8 (pbk)

Library of Congress Cataloging-in-Publication Data

Catalog record is available from the Library of Congress

Visit the Taylor & Francis Web site at
http://www.taylorandfrancis.com

and the CRC Press Web site at
http://www.crcpress.com

The Tao works to use the excess, and gives to that which is depleted.
The way of man is to take from the depleted,
and give to those who already have an excess.
La Via del Cielo toglie il sovrappiù e aggiunge ciò che manca.
La Via degli uomini, al contrario, non è così:
essi tolgono dove c'è mancanza per offrirlo dove c'è un sovrappiù.
Tao Tê Ching
VI-III BC

Design, if it is to be ecologically responsible and socially responsive,
must be revolutionary and radical in the truest sense.
It must dedicate itself to nature's principle of least effort [...]
That means consuming less, using things longer, recycling materials,
and probably not wasting paper printing books.
La progettazione, se vuole essere ecologicamente responsabile e socialmente
rispondente deve essere rivoluzionaria e radicale nel senso più vero.
Deve votarsi al principio del minimo sforzo adottato dalla natura [...]
Ciò significa consumare meno, usare più a lungo, riciclare i materiali,
e probabilmente non sprecare carta stampando libri (come questo).
Victor Papanek
Design for the Real World
XX AD

Contents

Part II—Methodological Statement

Part III—Methods, Tools, and Case Studies

List of Figures

List of Tables

Preface

Technological innovation prompted by the need to satisfy the changing needs of society, ever more efficiently and economically, involves complex interactions between three basic systems: the production system, the economic system, and the ecosystem.

An analysis of the relationships between these systems can provide an interesting index of the quality of technological innovation. In a high-quality innovative process, the economic system should adapt to the necessities of the production system, which in turn adapts to those of the ecosystem. Directives should, therefore, come from the ecosystem and pass through the production system to the economic system.

A lack of environmental awareness has led us to mistakenly consider ourselves to be outside the global ecosystem and, consequently, to satisfy our needs according to the sole criterion of "the greatest efficiency at the lowest cost." The resulting environmental crisis has shown how the ecosystem has been seriously degraded by the use of modern means of production, conceived without concern for either the environment or the balanced use of resources. It has also evidenced the negative effects of another closely related issue—the incompleteness of the innovator's understanding, often resulting in unforeseen side effects.

It is appropriate to note that, with regard to the problems inherent in the economic, political, and social systems, the constraints imposed by economic pressure must be challenged, and this is possible on the basis of some considerations.

Above all, the widespread idea that profit and respect for the environment are incompatible (a dangerous prejudice delaying a process of recovery that can no longer be postponed) is based on an inadequate vision of the problem. Any costs avoided by a production system in neglecting environmental issues will fall, redoubled, onto the community. Clearly, industry must respect the elementary condition of earning more than it spends, but it is crucial that profit is made while reducing environmental impacts to a minimum.

Regarding explicitly industrial activity, establishing company strategies that give due consideration to environmental issues must not be seen solely as an obligation toward the community, but also as an opportunity to produce benefits at various levels. An approach to the planning of production activities with the objective of achieving economically advantageous eco-compatible production is of primary strategic importance; the manufacturer can anticipate regulatory norms and so avoid the costs involved in adapting to them, and can also obtain substantial advantages in market competition, offering the transparency necessary to improve its relations with a public ever more sensitive to environmental issues.

The process of technological innovation that is taking shape in this way, still motivated by economic expediency, is far from the ideal where production and economic systems are subordinate to the priorities of the ecosystem. Nevertheless, it is tending in this direction and will continue to do so if prompted by ever-greater environmental awareness and by an effective regulatory structure.

The factors associated with environmental degradation, identified in the level of pollution, and in the intensity of resources consumption and the search for an industrial ecology that attains the complete equilibrium of resources typical of biological cycles, lead to those aspects of the environmental question that are the subject of this book—the scientific and technological factors at the base of product innovation.

Frame of Reference

Of the vast range of human activities, without doubt industrial activity has the most significant effect on the environment. The main environmental issues concerned can be summarized as: controlling and limiting the consumption of resources; avoiding the saturation of waste dumps; achieving maximum energy conservation in production processes; reducing as much as possible all types of emissions, whether inherent to the process or accidental; and intensifying the processes for the recovery of resources.

Increasing awareness of environmental issues has recently materialized in a move toward the optimization of production systems to ensure an elevated level of product eco-compatibility. This process has led to the development of a new methodological approach to product design, known as Design for Environment (or Green Design, Ecodesign). According to this approach, the most effective interventions guaranteeing the compatibility of an industrial product with the environment are those undertaken in the first phases of product development.

This perspective resulted in Life Cycle Design, a design intervention which considers all the phases of the product's life cycle (development, production, distribution, use, recovery, and disposal) during the entire design process, from concept definition to detailed design development. It therefore uses design methods and tools to correlate product evolution, from conception to disposal, and a wide range of design requirements.

The characteristics distinguishing this approach from other design approaches make it more suitable when the aim is to design for the environmental quality of products. One of the principal objectives of Life Cycle

Design, that of safeguarding the environment, is particularly relevant to this book, which intends to identify effective methods and tools for a product design oriented toward the environmental performance of products over their life cycles.

Objectives and Directions

Given the large number of issues involved, which clearly require a multidisciplinary approach, we choose to focus on one particular aspect in the field of product design: the need to develop design methodologies which, by optimizing the physical properties of products (architecture, geometries, systems, junctions, parts, materials), ensure an efficient product life, with full support during their useful life and different types of recovery at end-of-life. This is necessary to reduce the consumption of resources and limit all emissions involved in the various subprocesses making up the life cycle.

A design intervention oriented in this way requires the development of methodologies and mathematical models that can provide an overall vision of the problem and address product optimization. We, therefore, define a series of tools and techniques that can be used to improve the environmental performance of the life cycle phases, conferring elevated eco-compatibility on the final product while respecting the constraints imposed by conventional design criteria (functionality, safety, reliability, quality) and other company functions (production, marketing). The proposed approach regards the study of techniques for Life Cycle Design, with particular attention given to methodologies for the optimization of product life, guaranteeing the extension of a product's useful life, and the recovery of resources at the end of its life through the disassembly of components, maintenance and repair, and reuse and recycling.

The final objective is to develop a set of design tools to aid designers in making choices regarding the definition of product characteristics, integrating a series of analysis, calculation, and optimization tools in the most appropriate manner in order to allow as complete an approach as possible to the design problem. A secondary objective is to develop all these tools in strict correlation with the parameters of conventional engineering design, seeking to highlight the needs and potentials of an integrated approach to the problem.

Structure of This Book

This book is divided into an introductory chapter and three parts which present main concepts, basic design frameworks and techniques, important themes and related innovative design methods and tools, and practical applications.

Chapter 1 introduces the concepts of Sustainable Development, Industrial Ecology, and Design for Environment as defined in the literature. The life cycle theory and approach are presented and applied in Part I—Life Cycle Approach (Chapters 2 through 6), defining the main techniques (Life Cycle Design and Management, Life Cycle Assessment). This part also considers the bases of Life Cycle Cost Analysis for the full integration of the economic problems linked to product development.

Part II—Methodological Statement (Chapters 7 through 10) includes the main premises and reference models for the process of product design and development, delineating how it is possible to achieve an effective integration of environmental aspects in modern product design. In this context, particular attention is given to the environmental strategies that can help the designer achieve the requisites of eco-efficiency in the various phases of the product life cycle, and to how these strategies are closely correlated to the functional performance of the product and its components and, therefore, to some aspects of conventional engineering design. In this respect, it was considered useful to introduce phenomena of performance deterioration, together with principles of design for component durability and methods for the assessment of residual life.

In Part III—Methods, Tools, and Case Studies (Chapters 11 through 16), entirely new methods and tools are defined in relation to some issues of Life Cycle Design deserving further analysis, given their effectiveness in the design intervention. Each theme provides an introduction to the problem and some original proposals based on the authors' experience. The new concepts developed are then implemented in design practice, differentiating between different levels of intervention (materials, components, system) and demonstrating their use and effectiveness in specific case studies. In this final phase, we intend to concretize the knowledge acquired, presenting experiences that not only evidence the potential of the approach and methods proposed, but also analyze some of the problems involved in developing eco-compatible products in the company context.

Fabio Giudice
Guido La Rosa
Antonino Risitano

Catania, Italy
March 2005

Acknowledgments

The contents of this book are the fruit of several years of research activity, still in progress today, at the Department of Industrial and Mechanical Engineering at the University of Catania, Italy. Clearly, many people have contributed to this activity in various ways during its development.

In the early years, invaluable research was undertaken at CRIED—European Institute Design Research Center of Milan, and particular thanks are given for the untiring help of Matteo Ragni, designer; Prof. Amilton Arruda (Department of Design, Federal University of Pernambuco, Recife, Brazil); and Prof. Carlo Vezzoli (Research Unit on Design and Innovation for Environmental Sustainability, Milan Polytechnic).

Subsequently, prolific and stimulating collaboration was offered by CRF—Fiat Research Center of Orbassano (Turin), and our sincere thanks go to Eng. Gian Carlo Michellone, CRF's Managing Director; Eng. Giuseppe Rovera; Eng. Edoardo Rabino; and to Alessandro Levizzari (now at CRF—Bari Branch).

More recently, we have been involved in setting up an interesting research program, organized and coordinated by our research team, and currently in progress. This consists of an interuniversity program of scientific research, entitled Environmental Quality-Oriented Product Design, approved by the Italian Ministry of Education, University and Research, and involving three other Italian universities. Encountering the experience, competence, and ideas of the other partners in this project has proved extremely stimulating and has greatly enriched our knowledge and understanding. In this respect, our thanks go to the directors of the other research groups: Prof. Raffaele Balli (Department of Industrial Engineering, Perugia University); Prof. Paolo Citti (Department of Mechanics and Industrial Technologies, Florence University); Prof. Piermaria Davoli (Department of Mechanical Engineering, Milan Polytechnic); and to all those collaborating in the research.

Again regarding this same initiative, particular thanks are given to Prof. Rinaldo Michelini (Department of Mechanics and Machine Construction, Genoa University), who encouraged us to organize the program, and who has always shown great interest in our research activity.

Within our own department, we would like to thank Prof. Luigi Marletta for his interest shown in our research and for his readiness to exchange ideas; the whole Machine Construction faculty for their unfailing esteem and support; and the students and graduates who over the years have responded with enthusiasm to the themes treated in this book. Among our colleagues, special thanks go to Dr. Giuseppe Mirone and Eng. Guido Strazzeri for his help in the final drafting of some parts of the manuscript.

For their contribution to some of the studies presented in this book, particular thanks go to Dr. Giovanna Fargione and Dr. Lia Maiolino (Department of Industrial and Mechanical Engineering, Catania University), and to Eng. Rino Furnò (CRF—Fiat Research Center, Catania Branch).

Finally, we would like to thank Mike Wilkinson for the care he has taken in the translation, and Cindy Renee Carelli and Jessica Vakili of CRC Press/ Taylor & Francis, whose enthusiasm and constant support for this publishing venture cannot be overestimated.

Author Biographies

Fabio Giudice
Ph.D., Associate Researcher

Fabio Giudice is currently Associate Researcher at the University of Catania, Italy. He graduated in Mechanical Engineering at the University of Catania, obtained a Master's in Industrial Design at the Research Centre of the European Institute of Design in Milan, and a Ph.D. in Mechanical Engineering at the University of Catania. With research interest in product design and design for X, at present Dr. Giudice is developing research on design for environment, with particular interest in life cycle design, design for disassembly, cost–benefit analysis of recovery, and life cycle simulation, and has published a number of papers in this area.

Guido La Rosa
Full Professor of Design of Mechanical Structures

Guido La Rosa is a Full Professor of Design of Mechanical Structures at the University of Catania, Italy. Graduated in Electronic Engineering at the University of Pisa, Italy, he has held prior teaching and research position in the areas of mechanical engineering and biomechanics. Responsible for the program of the National Research Council (CNR) and of Ministry of the University and Scientific Research (MIUR), Prof. La Rosa is author of more than 100 papers, presented at congresses and published in national and international journals, in the field of machine design, structural and experimental mechanics, biomechanics, and design for environment.

Antonino Risitano
Full Professor of Machine Design

Antonino Risitano is a Full Professor of Machine Design at the University of Catania, Italy. Graduated in Mechanical Engineering at the Polytechnic of Torino, Italy, he has held prior teaching and research position in the areas of mechanical and aeronautical engineering. Formerly dean of the Faculty of Engineering at the University of Catania, Prof. Risitano is currently head of the Department of Industrial and Mechanical Engineering at the same university. With research interest in vibration, fatigue, strength analysis, nonconventional methods in mechanical analysis of materials, conventional and water engines, environmental protection, he has published more than 100 scientific papers.

Chapter 1

From Sustainable Development to Design for Environment

The last 40 years or so have seen a more attentive examination of the factors characterizing the processes of development in industrialized countries, evidencing the environmental risks implicit in an industrial development conditioned exclusively by economic mechanisms.

One result of our new comprehension of the limits to resources and of the risks from phenomena of pollution is the concept of sustainable development. This advocates the reconciliation of processes of development with respect for the environment, in the interests of future generations. Going as far as drawing an analogy between the processes of natural transformation and those of industry, sustainability concepts take inspiration from the teachings of nature in seeking to optimize the flows of resources characterizing the whole industrial system and the life cycles of products. From this perspective, whether directed at processes or products, the design phase is that stage in the life of systems or products with the greatest potential.

This first chapter presents an overview, trying to define, contextualize, and correlate the main concepts and approaches to environmental protection in the ambit of industrial production, considering in greater detail those held to be more important to the goals of this book.

1.1   Sustainable Development

"Sustainable development is development that meets the needs of the present without compromising the ability of future generations to meet their own needs" (WCED, 1987). With this definition of sustainable development, in 1987 the World Commission on Environment and Development (WCED) mapped out what is now widely recognized as the guiding objective of the current process of economic and technological development—to ensure that the use of environmental resources to satisfy present demands is managed in a way that they are not left so damaged or impoverished they cannot be used by future generations.

1

This new hypothesis of developmental orientation, which was first formulated in the early 1980s (Brown, 1981) and then used in the WCED Brundtland Report (from the name of the commission chairperson), was also intended as a response to the worrying conclusions reached in 1972 by a group of scientists and experts in a project known as the Club of Roma. Collating the provisional results of a mathematical model of the world development system, based on the interaction between several key factors (population growth, food production, industrialization, resources depletion, pollution), the Club of Roma study concluded that economic and industrial growth would come to a stop in the near future due to the exiguity of natural resources, with a consequent decline in the population level and in the industrial system. The study said, "If the actual line of development continues unchanged in these five principal sectors, humanity is destined to reach the natural limits of development within the next 100 years" (Meadows, 1972).

Although the Brundtland Report is based on models of economic–environmental interaction analogous to those formulated in the Club of Roma report, it reached the conclusion that growth could continue as long as it took different forms and directions in line with the concept of sustainable development outlined above.

1.1.1 Key Factors in Sustainable Development and the Role of Environmental Protection

Although some contents of the Brundtland Report may be open to criticism today (primarily some of the development scenarios foreseen, which time has shown were not wholly accurate), it nevertheless deserves to be considered an important milestone in the process of raising the issue of environmental protection, as shown by some of its analyses and long-term suggestions. In particular, the report's explicit call for a change in direction with regard to industrial activity, promoting the concept of "producing more with less," is still undoubtedly relevant today. In the same way, the development strategies indicated for achieving this change in direction are now fully assimilated into the precepts of environmental protection:

- Establish environmental goals, regulations, incentives, and standards
- Make more effective use of economic instruments
- Broaden environmental assessment
- Encourage action by industry
- Increase capacity to deal with industrial hazards
- Strengthen international efforts to help developing countries

In July 1992, delegates from more than 170 countries met in Rio de Janeiro for the United Nations Conference on Environment and Development (UNCED).

They recognized common orientations for sustainable world development, defining the most appropriate actions to ensure the necessary harmonization between environmental protection and economic development, and ratified the statement, "In order to achieve Sustainable Development, environmental protection shall constitute an integral part of the development process and cannot be considered in isolation from it" (UNCED, 1992). This conclusion, which openly asserts the key role of environmental protection in the context of a completely sustainable development process, is the result of a definitive maturing of the concept of sustainability itself.

In fact, it initially originates from an essentially economic approach, according to which a productive process is sustainable if it allows the maximum flow of generable profit to be obtained, maintaining at least the same reserve of capital that produced it. This concept of sustainability has often been associated with another—sociocultural orientation—which instead promotes the importance of maintaining the stability of social and cultural systems and of making them evolve toward a condition of greater equity, with particular regard to the problems of the elimination of poverty and the promotion of civil rights for future generations.

Several authors have, however, observed that human activities are part of a socioeconomic subsystem integrated into the ecosphere (Odum, 1975), going on to emphasize that economic growth that is uncontrolled and based essentially on the intensive consumption of resources is not sustainable in the long term, precisely because the ecosphere is finite and contains limited resources (Goodland et al., 1991). According to this view, the consequences of impoverishing environmental resources must be taken into great account in decisional processes, since the environment supports human society through three important typologies of service (Munasinghe, 1993):

- Providing raw materials and all other primary resources that support human activities
- Constituting a dumping reservoir to absorb, and sometimes recycle, waste from activities
- Performing other irreplaceable functions for the life of mankind (e.g., stabilization of climatic conditions on the global scale, or defense from UV rays by the ozone layer)

These considerations have led to the development of a third concept of sustainability from an ecological viewpoint, which integrates the economic and sociocultural aspects considered above. This new concept is based on the survival capacity of biological and physical systems, which are crucial to the stability of the entire ecosystem and to the protection of biological diversity.

After more than 30 years since the publication of the Club of Roma report, it now seems that a complete vision of the problem has finally matured, where sustainable development is considered a process involving all three key factors: economic, sociocultural, and environmental. This vision, as ratified by the Rio Declaration in 1992, is based on a full understanding of the need to promote behavior that respects environmental issues. This understanding has, for some years now, been the driving force behind initiatives at different levels and involving diverse actors (governments, manufacturers, and consumers), as evidenced also by the great proliferation of approaches and tools oriented toward achieving environmental sustainability (Robèrt et al., 2002).

1.1.2   Role of Science and Technology

In seeking a schematic vision, the sociocultural, economic, and environmental elements representing the principal factors involved in the process of sustainable development, can be imagined as ideally placed at the vertices of an equilateral triangle (Munasinghe, 1993). The graphical representation in Figure 1.1 evidences the interactions between these key elements which must be harmonized for sustainable development. The center of the triangle, the point of equilibrium between the three factors, represents the condition where sustainable development is fully achieved, while all other points represent conditions where some elements have a different weight in defining the direction of development. The points on the edges of the triangle, for example, represent conditions where only two factors are considered, excluding the third.

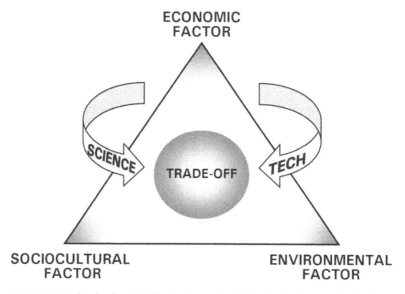

FIGURE 1.1   Triangle of sustainable development and the role of science and technology.

In this view, scientists (and, therefore, the activity of scientific and technological research they embody) are placed in direct relation to each of the key factors and must provide information and tools to reinforce choices aimed at the best equilibrium (Munasinghe and Shearer, 1995) to achieve the desired condition of balanced, sustainable development.

After several decades of debate, therefore, the competencies of the actors involved in industrial development seem to have been delineated (O'Brien, 1999): industry must acquire the knowledge and capacity to assume its responsibilities in the development of sustainable production systems, while government has the responsibility of creating those socioeconomic conditions that allow companies to perform this task without losing competitiveness. In this scenario, research and development policies assume fundamental roles.

1.2 Industrial Ecology

Although the premises at the base of the sustainable development concept are of irreplaceable ethical value, clearly evidencing the obligation to take new routes to development which guarantee future generations the resources required to meet their needs, they themselves do not give clear indications of how to undertake this task in concrete terms. The introduction of the concept now known as Industrial Ecology can contribute to a better understanding and application of the system-based approach required to achieve the condition of sustainability (Korhonen, 2004a).

The observation of natural cycles, characterized by the physical–chemical transformation of organic and inorganic compounds, together with concern for the environmental problems associated with industrial activity, immediately suggested that inspiration could be taken from the efficiency of ecological systems, as noted in even the earliest studies on design theory (Asimow, 1962). In these systems, in fact, there are no waste products; the cycles transforming resources are closed because each component of the system provides the means of sustenance for other components. A second consideration highlighted the analogy between industrial and natural systems: both system typologies are characterized by processes of transforming resources (materials, energy), which are the object of the optimization required to resolve environmental problems (Allenby and Cooper, 1994).

This same analogy also highlights the essential difference between natural and economic–industrial systems; the former are cyclic systems where materials circulate and transform continuously, without generating waste, while the latter are linear systems where resources are used and transformed into products and waste. This linear characteristic of economic–industrial systems, making the consumption of resources and the generation of wastes inevitable, is in its present state subjected to a dangerous pressure pushing it down a slippery slope. The current system of industrial production can be considered to resemble an organism that ingests resources taken from nature, subjects them to processes of transformation, produces objects, and expels waste. The characteristic that renders this organism devastating for the environment is its insatiable physiological bulimia (Regge and Pallante, 1996); the indicator of its well-being is the quantity of goods produced. Consequently, the greater the increase in resources it ingests and transforms, and the greater the increase in objects it produces and waste it discards, the better is its state of health, even though the latter does not coincide with that of mankind and the biosphere.

Reflecting on the system of industrial production in this manner is highly effective in favoring the dissemination of a new understanding of the need to conceive models of activity in terms of an industrial ecosystem, where the consumption of materials and energy is optimized, the production of waste is minimized, and the discarded materials from a generic process become the raw materials of another process (Frosch and Gallopoulos, 1989).

The industrial ecosystem must, therefore, hold to the ideal closed model best represented by biological ecosystems—plants synthesize the food of herbivores which, in turn, fuel the carnivores whose organic remains and waste go to feed further generations of plants. Analogously, in an ideal industrial ecosystem, a piece of steel can be used in a can one year, in an automobile the next, and in the structure of a building in later years.

These considerations led to the evolution of the concept of Industrial Metabolism (Ayres, 1989), based on the affinity between the biosphere and the economic–industrial system in the transformation of material resources, into that of Industrial Ecology (Allenby, 1992; Jelinski et al., 1992). In recent years this has rapidly evolved into a systematic study, based on a holistic approach, of the processes making up the whole life cycle of artifacts, from production to retirement (i.e., from the transformation of resources to their disposal).

1.2.1   Scope and Evolution of the Ecological Metaphor

The transposition of the organizational principles of ecological systems into economic–industrial systems underlying the concepts of industrial metabolism and industrial ecology derives from the perception of fundamental analogies between the two system typologies. In both cases, in fact, it is possible to identify some common characteristics:

- Cyclical structure of the subject's life (conception, birth, development, maturity, the end of life)
- Functions of metabolic type (ingestion of resources, transformation, growth of systems)

- Capacity to reuse and to recycle resources (potentially zero waste, in terms of the system)

As will become clear below, over the last decade these analogies have become a major focus of interest for researchers, initiating a new perspective oriented toward an effective approach to the question of sustainability in relation to industrial activities, expressly based on this kind of ecological metaphor. The effective potential of this approach is, however, still under discussion, as evidenced by several recent critical analyses, some originating within the very circle of researchers who, in the early 1990s, contributed to the dissemination of this new perspective. Attention has been drawn to some profound dissimilarities between ecosystems and industrial systems, which could make the ecological metaphor misleading for a clear comprehension of the sustainability of industrial activities (Ayres, 2004). These dissonances are mainly related to the economic dimension of industrial systems—the differing roles that some objects have in the two systems (e.g., resources and products); the fact that other important factors are not common to both systems (labor, market, and money exist only in the economic system); and the different meanings assumed by important processes (e.g., growth and evolution).

Another interesting criticism concerns the way in which industrial ecology, in general, proposes using the study of natural models for the planning of industrial systems (i.e., learning from the former to improve the latter; this is known as the "eco-mimicry approach"). That is, in a prescriptive way, with the object of establishing solutions and making suggestions for the modification and improvement of industrial systems based on the organizational dynamics of natural systems. Some authors propose a substantial re-elaboration of this approach, preferring an "ecology as constraint" approach, limited to using the knowledge derived from the natural sciences and ecology to trace the boundaries that industrial activities must not cross if they are to avoid compromising the environmental equilibrium (Harte et al., 2001). Although other authors do not exclude the prescriptive scope of the ecological metaphor, above all in relation to the physical dimension of systems (flows of material and energy) (Korhonen, 2004b), its scientific foundations are still the subject of debate (Ehrenfeld, 2004).

Despite these critical analyses of the theoretical basis of the ecological metaphor, there is no indication that its development has been arrested. One of the most recent and particularly interesting interpretations is one based on an understanding of the organizing principles continuously developed by living systems to sustain what some authors have called "the web of life" (Capra, 1996). This reading, which gives an important contribution to the very concept of sustainable development (generalized and raised to the level of "sustainability of life"), is based on a system-oriented vision of the life of natural systems, whose evolution is governed by organizing principles.

These can be identified in some fundamental principles of ecology which human societies can aspire to in the organization of their own activities in a sustainable manner:

- The various systems are distinct and correlated through a complex network with a structure of nested boxes, but their boundaries have the sole function of demarcating identities, not of preventing communication and the exchange of resources.
- The processes of transformation can be considered cyclical, given that single organisms draw on flows of material and energy and produce waste, but the entire network to which they belong does not produce waste.
- The exchange of resources between systems is sustained by forms of cooperation.
- The stability of systems and their capacity to recover is greater as the complexity of the networks linking them together increases, where complexity implies richness and variety.
- Each network of systems is in a constant state of dynamic equilibrium, where no variable arrives at the extremes of its range of values, but all fluctuate close to their optimal values.

When transferred to the organization of human activities, these wide-ranging principles take on clear ethical and sociocultural connotations, but they also delineate a clear trajectory to follow in the development and planning of industrial activities. According to this perspective, such activities can be understood in terms of complex networks, subsystems, and organisms corresponding to industrial sectors, assemblies of production activities, and single production activities and products. This trajectory provides a clearer vision of the strategy required to achieve sustainability in human activities, on the basis of some fundamental and ineluctable premises (Capra, 2002):

- Existing systems that interpret human activities do not need to be rebuilt from zero but, rather, must be remodeled to reflect natural ecosystems and the organizing principles that express their intrinsic capacity to sustain life.
- The sustainability of a system is not a static condition, but a dynamic operational process in continuous interaction with other systems.

1.2.2 Definition of Industrial Ecology

The radically new perspective based on the considerations discussed above leads to a profound change in the way we relate to natural systems. Accepting the validity of the ecological metaphor and its prescriptive possibilities, this change could signal the beginning of a new industrial era, no longer based on the idea of what can be extracted from nature but, rather, on what can be learned from nature (Benyus, 1997). This new viewpoint finds its complete formulation in the concept (and in the now-affirmed discipline correlated to it) known as industrial ecology (Allenby, 1992; Jelinski et al., 1992; Graedel et al., 1993; Frosch, 1994; Socolow et al., 1994; Graedel and Allenby, 1995; Ayres and Ayres, 1996, 2002; Allenby, 1998).

Given that it is still under debate, it is not yet possible to propose an unequivocal definition of industrial ecology, although some proposals are beginning to emerge (Seager and Theis, 2002). Most of the definitions formulated to date do, however, contain common elements, with differing emphases, which can provide an overview of the concept (Garner and Keoleian, 1995):

- Study of the flows and transformations of materials and energy
- Change in the conception of transformation processes from linear (open) to cyclical (closed)
- Holistic vision of the interaction between industrial and ecological systems
- Harmonization between industrial systems and ecological systems
- Emulation, in the structuring of industrial systems and in their organizational principles, of the more efficient and sustainable natural systems

One of the first definitions formulated appropriately highlights how industrial ecology interprets, and aims to achieve, the condition of sustainable development based on a systems-oriented perspective. According to B.R. Allenby, industrial ecology "consists of a systems view of human economic activity and its interrelationship with fundamental biological, chemical, and physical systems with the goal of establishing and maintaining the human species at levels that can be sustained indefinitely, given continued economic, cultural, and technological evolution" (Allenby, 1992). Another well-known definition takes up essentially the same premises but clarifies how the condition of sustainability can be achieved through industrial ecology: "It is a system view in which one seeks to optimize the total materials cycle […]. Factors to be optimized include resources, energy, and capital" (Graedel and Allenby, 1995).

Of the numerous definitions found in the literature, the following is perhaps the most in tune with the contents of this book, since it highlights that aspect of industrial ecology most associated with the design activity and applies the life cycle concept to material flows. According to L.W. Jelinski et al., "Industrial ecology is a new approach to the industrial design of products and processes and the implementation of sustainable manufacturing strategies [...]. It seeks to optimize the total materials cycle from virgin material to finished material, to component, to product, to waste products, and to ultimate disposal" (Jelinski et al., 1992).

1.2.3   Objectives and Approaches of Industrial Ecology

On the basis of the main concepts reported above, the objectives that industrial ecology sets for itself can be summarized in these important points:

- Development of conceptual structures for the understanding and evaluation of the impacts of industrial systems on the environment, and for the implementation of strategies targeted at reducing the impacts of products and processes
- Conversion of the linear structure of industrial systems (where raw materials are usually transformed, used, and dumped) to a cyclical structure (where the outgoing flows of resources are used as input by other processes of transformation)
- Harmonization between the processes making up the life cycle of products, between different interacting life cycles, between the system of life cycles, and the environment

With these objectives, industrial ecology proposes the application of an integrated approach to the management of environmental impacts correlated to the use of all the resources in play (energy, materials, economic capital) in the context of industrial ecosystems. To optimize resource use, R.A. Frosch and M. Uenohara said, "[M]anagers need a better understanding of the metabolism (use and transformation) of materials and energy in industrial ecosystems, better information about potential waste sources and uses, and improved mechanisms (markets, incentives, and regulatory structures) that encourage systems optimization of materials and energy use" (Frosch and Uenohara, 1994).

1.2.4   Typologies of Cycles in Nature and Translation into Industrial Ecosystems

Biological ecology, defined as the scientific study of the interactions determining the distribution and aggregation of organisms, can suggest interesting ideas for the analysis of the interactions between industrial processes and natural systems. On the basis of some results obtained in ecology, it has been possible to hypothesize a past process of the evolution and adaptation of biological systems in response to changes in the availability of resources (Jelinski et al., 1992; Graedel et al., 1993).

According to a general scheme shown in Figure 1.2, a biological system consisting of n components interacts with the external environment through inflows of resources (R) and outflows of waste (W). Furthermore, each component C_i can potentially interact with each of the others through unidirectional or bidirectional flows (F_{ij} is the interaction flow between components C_i and C_j).

For a primitive biological system, such as that which might have existed in the first stages of life on Earth, the quantity of resources is so enormous in relation to the number of living examples that they have no impact on the availability of resources. In a system of this kind (a type I ecosystem, which can be considered linear), each biological system can be understood to consist of a single component organism (n = 1), which, respectively, consumes and produces unlimited quantities of resources and waste (R and W unlimited), since the flow of resources is independent of any other factor.

The multiplication of the first forms of life, in relation to which the resources become limited, leads to the evolution of the now-inefficient linear flow system into a recursive cycle system, whose domain extends to more component organisms (n > 1). The flows within the domain are numerous, and the flows into and out of the domain are limited (R and W limited). A system organized in this way, called a type II ecosystem, is more efficient than the previous but clearly is not sustainable over the long term because of the one-way direction of the wastage outflow. For a system to be biologically compatible it must be completely cyclical (i.e., such that it is not possible to make a distinction between resources and waste, since the waste from one activity is the resources for another). This is the type III ecosystem model, analogous to the preceding model but so efficient that the disposal outflows are annulled

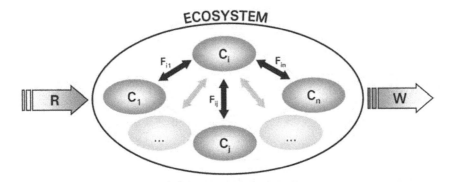

FIGURE 1.2   General scheme for a biological ecosystem.

(W = 0), being fed solely by inflow resources in the form of energy (R = Energy) necessary for the internal processes of transformation.

Although there is a clear need to develop industrial systems in line with models as close as possible to those of type III, at present the use of resources continues to be dissipative, for the most part, following conventional models where materials are degraded and disposed of in the course of a single use in a way analogous to their use in type I ecosystems (Ayres, 1989). The aim of industrial ecology is to favor the evolution of industrial systems from the linear model of type I to the semicyclic model of type II, and subsequently to the ideal model of type III, by seeking a fuller understanding of the complex relations between processes and products and an efficient optimization of the set of factors involved.

1.2.5   Efficiency of Industrial Ecosystems and Determining Factors

The semicyclical industrial system (type II) can be represented by the interaction between four main activities (Graedel and Allenby, 1995): extraction of materials, working of materials, use of product, and management of waste. Therefore, according to the scheme of Figure 1.3, the phases undergone by the material resources are extraction, working, collecting after product use, and recycling. This scheme is clearly derived from the general model shown in Figure 1.2. The intended ideal is to move toward a system of the type III model, to develop a cyclical model for the flows of materials between the various activities in a way that limits as much as possible the consumption of external resources and the impact that the overall industrial system has on the external supporting system.

Having identified the flows of resources in play, as shown in Figure 1.3, it is possible to define some particularly interesting indicators of the ecological

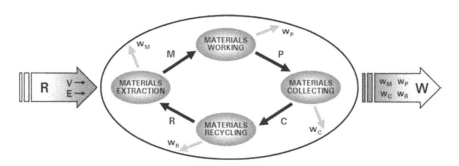

FIGURE 1.3   Model of semicyclical industrial system. V, virgin material; E, energy; M, processed material; P, product; C, collected material; R, recycled material; w_M, w_P, w_C, w_R, waste flows relative to the various phases; W, overall waste flow.

efficiency of activities (Graedel and Allenby, 1995). These indicators are valuable in understanding the environmental question from a systems perspective, in that they make it possible to evidence the factors which can influence the eco-efficiency of industrial systems. These efficiency indicators represent the performance from each phase traversed by the material resources (extraction, working, collecting, recycling), and therefore assume unitary value in the ideal case where the corresponding disposal flows are null and the resources are recovered in their entirety.

The efficiency of the whole system, which is expressed by the product of the efficiency indicators relative to each phase, assumes unitary value in the case of a perfect industrial ecology system where the inflows and outflows are null and the system maintains itself solely through an external energy input, fulfilling the type III ecosystem model. With reference to the figure, this condition occurs when the system does not need an input of virgin material ($V = 0$), and all the disposal flows are null ($w_M = w_P = w_C = w_R = 0$, and therefore $W = 0$).

It is important to highlight the determining factors for the efficiency of each phase of transformation of the resources:

- The efficiency of the extraction of materials depends on the technology used in the process, and on the specific requirements of the manufacturer who will process the materials.
- The efficiency of working the materials depends on the technology used, and on the design of the production processes and of the products and their implementation.
- The efficiency of collecting used products and materials depends on the design of the products, on government policies, and on the market for recycled materials.
- The efficiency of recycling materials depends on the design of the recycling processes and products, on the technology used, on government policies, on the market for recycled materials, and on regulations imposed on the consumer and producer.

Among these factors determining the efficiency of the system, shown in relation to each phase, it is of particular interest here to note the importance of an effective and appropriate design of products and processes.

1.3 Design in the Context of the Environmental Question

In this era of mass production, when all activities tend to be planned in detail, design becomes a powerful instrument by which mankind forges the world we live in. The scope of this instrument clearly also extends to the management of environmental problems. Having examined the concepts of sustainable development and industrial ecology, it is possible to consider the interpretation of these concepts in the context of the design of industrial processes and products. Design, essentially consisting of molding material and energy flows for the purposes of satisfying the needs of humankind, ultimately becomes a process of transformation when the needs generating it are contextualized in the patterns and flows of natural systems, assimilating the organizing principles of the ecosphere.

Those factors influencing the indicators of the environmental efficiency of industrial systems, noted at the end of the previous section, allow the identification of the contexts most appropriate for a design intervention directed at environmental protection. In particular, they emphasize the impact of process design on the efficiency of working and recycling materials, and that of product design on the efficiency of working, re-collecting, and recycling materials. Due to its great potential, therefore, design becomes one of the most influential factors in the development of sustainable production systems and products.

A full understanding of the indefeasible environmental needs, and of the potential and responsibility that the vast typology of design interventions has toward the environmental question, has been slow to arrive. Although attention was first drawn to the necessary influence of socioecological systems on technical design in the early 1960s (Asimow, 1962), the transition from a "design for needs" to a "design for environment" first began in the early 1970s (Madge, 1993). In fact, this period saw the first ideas which, starting out from different points of view, decisively raised the environmental question and underlined its revolutionary effects on the structure of conventional design.

In the context of the culture of design, it should be noted that there was already an explicit reference to the potential of using biological systems as worthy models for systems developed by mankind (an idea that anticipates the basic concepts of industrial ecology), and to the consequent opportunities of reusing, repairing and recycling artifacts (Papanek, 1971). Starting out from different viewpoints (economic and social in character), other authors highlighted similar aspects, with a clear emphasis on the design phase. They stressed the importance of moving toward the optimization of production systems, promoting the principle of obtaining the maximum well-being with the least possible consumption of resources (Schumacher, 1973) and the need to spread a correct perception of the environmental question among consumers, fundamental to promoting an industrial production directed at limiting the obsolescence of products and at encouraging their recycling (O'Riordan, 1976).

These first incentives to revising conventional design paradigms were consolidated over the following decade. This second phase is perceived as a turning point from an industrial to a postindustrial design (Cross, 1981). Taking up some concepts already discussed regarding the nonsustainability of development exclusively oriented toward economic expansion, the new paradigms of design must draw inspiration from alternative models of development ("sane humane ecological future," Robertson, 1980). Against specialized industrial products with limited functionality and of short duration, postindustrial design contrasts multifunctional products, repairable and durable, taking the form of a design that is socially responsive and eco-sustainable. Conventional product requirements regarding functionality and cost are integrated with new requisites; energy efficiency, duration, and recyclability all appeal to consumers sensitive to environmental issues (Elkington et al., 1988). At the same time, it is emphasized that this extension in product requirements must not be seen as a disadvantage by the manufacturer, as environmentally compatible products can be not only economically competitive but also innovative and particularly attractive for the consumer (Elkington and Burke, 1987). At the beginning of the 1990s it was thus possible to have an overall vision of the effects of environmental issues on the design activity, extending to the most diverse areas, and clarified by the results of the first experiences (Mackenzie, 1991).

1.4 Design for Environment

The first consideration of the technical aspects associated with the practice of a design action directed at reducing the environmental impact of products appeared in the first half of the 1980s (Overby, 1979; Lund, 1984). In the early 1990s, these first experiences were followed by a phase of greater understanding of new needs to safeguard resources, which consolidated in a wide diffusion of new ideas and experiences developed with the clear objective of integrating environmental demands in traditional design procedures (Overby, 1990; Navin-Chandra, 1991; OTA, 1992). In this way a new approach to the design intervention was born, known as Design for Environment (DFE), Green Design (GD), Environmentally Conscious Design (ECD), and EcoDesign (Ashley, 1993; Allenby, 1994; Dowie, 1994; Fiksel, 1996; Billatos and Basaly, 1997; Zhang et al., 1997; Brezet and van Hemel, 1997; Graedel and Allenby, 1998), characterized by the priority objective of, already in the design phase, minimizing the impact of products on the environment.

1.4.1 DFE: Definition and Approach

Some reflections on the relation between technology and the environment evidence the fundamental principles that can serve as the basis for the development of more effective methodologies for their complete integration (Allenby, 1994):

- The methodologies must be wide-ranging and based on systemic statements.

- The approach must be multidisciplinary (i.e., it must include technical, legal, economic, and political aspects).
- Technology plays a key role in the solution of environmental problems; it is therefore necessary to encourage the evolution of process and product technologies oriented toward environmental protection.
- All the economic factors involved in the process of developing industrial products must assimilate a system of environmental constraints as extensive as possible.
- Development policies and legislative pressures must encourage the development of programs of research and experimentation that support the innovations required by these changes.

The implementation of these principles in industrial practice, at the various levels noted, requires the direct involvement of the product design and development process, as a vector of dissemination and integration of the new environmental needs. Design for Environment (DFE) originated precisely to play this strategic role. Its definition, which at least initially was not clearly univocal, has evolved over the last decade. First presented in a reductive manner as a design approach directed at the reduction of industrial waste and the optimization of the use of materials (OTA, 1992), DFE subsequently acquired a more appropriate dimension. Maintaining the necessary attention on the management of waste and resources, and integrating it in a systems vision clearly inspired by the principles of industrial ecology, it can be understood more completely as "a design process that must be considered for conserving and reusing the earth's scarce resources; where energy and material consumption is optimized, minimal waste is generated and output waste streams from any process can be used as the raw materials (inputs) of another" (Billatos and Basaly, 1997).

Ultimately, DFE can be defined as a methodology directed at the systematic reduction or elimination of the environmental impacts implicated in the whole life cycle of a product, from the extraction of raw materials to disposal. This methodology is based on evaluating the potential impacts throughout the entire course of the design process. In addition to its specific primary objective and its orientation toward the life cycle, DFE is characterized by two other aspects, in particular (Figure 1.4):

- The dual level of intervention, regarding both products and processes
- The proactive action of intervention, based on the presupposition of the greater efficacy of intervening early in the product development process (i.e., in the early design phases)

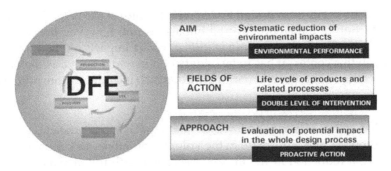

FIGURE 1.4 **Objectives and characteristics of DFE.**

Bringing these concepts into the dimension of design practice, a similar concept known as Green Engineering Design suggests a clear, two-part approach (Navin-Chandra, 1991):

- Evaluation of designs to assess their environmental compatibility, using a spectrum of indices and measures ("green indicators")
- Analysis of the relationship between design decisions and green indicators

In the description of this approach, particular attention is given to the use of indicators quantifying the environmental benefit of design choices. These make up part of environmental metrics, which can generally be considered the algorithmic interpretations of levels of performance within an environmental criterion (i.e., an attribute of the product found to be significant in determining the environmental performance of alternative product designs) (Veroutis and Fava, 1996). As will be seen in subsequent chapters, environmental metrics have an important role in the design of the life cycle of products.

As already noted by other authors, any type of design intervention directed at environmental protection cannot be separated from the requirements of product functionality, performance, reliability, quality, and cost (Navin-Chandra, 1991; Weinbrecht et al., 1993; Johnson and Gay, 1995). Having guaranteed these obligatory requirements, environmental demands must evolve from simple constraints to new opportunities and incentives to innovation.

1.4.2 Approaches to Optimal Environmental Performance

The central theme unifying the various studies of DFE can be identified in the common objective of reducing the environmental impact of a product over its entire life cycle, from design to disposal (Coulter et al., 1995). The concept of "reduction of the environmental impact" is not, however, limited to the simple quantification and minimization of direct impacts on the ecosystem. Rather, in this context it has to be understood in wider terms, as the optimization of the environmental performance, which includes a more articulated range of aspects:

- Reduction of scrap and waste, allowing a more efficient use of resources and a decrease in the volumes of refuse, and, more generally, a reduction in the impact associated with the management of waste materials
- Optimal management of materials, consisting of the correct use of materials on the basis of the performance required, in their recovery at the end of the product's life and in the reduction of toxic or polluting materials
- Optimization of production processes, consisting of the planning of processes that are energetically efficient and result in limited emissions
- Improvement of the product, with particular regard to its behavior during the phase of use, to reduce the consumption of resources or the need for additional resources during its operation

With these premises, it appears clear how DFE also becomes a bridge connecting two traditionally separate functions—production development and environmental management. The aim of DFE is, therefore, that of bringing these two functions into close contact and giving prominence to those problems of a product's life cycle, which are often ignored.

These objectives of improving the environmental performance of a product can be given form by following two different approaches to the problem, according to the circumstance they address. In the case of intervention on a preexisting production cycle, a first evaluation of the negative impact of the product's life cycle is followed by research into the interventions most effective in reducing that impact, principally through an increase in the level of technology implemented. Instead, for an intervention in the context of the development of a new product it is possible to move directly toward the complete sustainability of the life cycle, with the ideal objective of obtaining a product manufactured, used, and retired without giving rise to significant impacts on the environment. This can be achieved through a design intervention which has precisely the characteristics of DFE, where the most significant benefits can only be obtained by taking into account the entire life cycle of a product. As well as the specific production phase, the life cycle includes the procurement of materials upstream, and successively the use, disposal, and possible recovery of resources downstream, as simplified in Figure 1.5. In

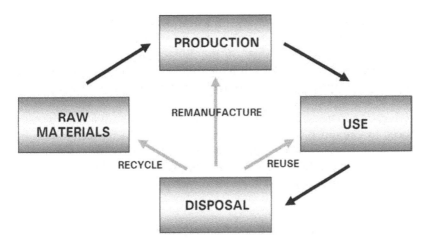

FIGURE 1.5   Main phases of product life cycle and flows of resources.

agreement with the basic concepts of industrial ecology, with the aim of optimizing the use of resources and of reducing waste, it is then necessary to attempt to produce a system characterized by closed cycles (through reuse, remanufacturing, and recycling) in a way that allows resources in play to reenter the cycle.

1.4.3   Area of Intervention of DFE

Although our attention here will be focused on the theme of product design and development, it is should again be emphasized that, in general, the field of intervention of DFE is considered to cover the design of both products and processes. For this reason a distinction is frequently made between "environmentally conscious product design" and "environmentally conscious process design" (the latter also being known as "environmentally conscious manufacturing") (Zhang et al., 1997).

Generally, those who address the design of processes do not deal with the design of products, and vice versa. The industry–environment interaction is therefore mainly treated by two distinct groups of designers who must direct their efforts in the same direction and harmonize the results obtained in their different areas of competence. In addition, the influence of process design on the environmental question takes different forms and gives different results than those in product design.

Processes have an importance extending well beyond a specific application, since the same processes can be part of the life cycles of many different products. They perform the transformation of raw materials into products and, therefore, define the flows of solids, liquids, and gases, as well as energy flows, and are largely responsible for the outflows from the industrial ecosystem they are part of. From this point of view, their optimization has greater strategic importance than that of the life cycle of a specific product. Furthermore, once a process is implemented in a production system, it generally cannot undergo significant modification in the short term without heavy costs.

Product design has greater margins for flexibility. This advantage is counterbalanced by the much wider field of investigation, which inevitably includes problems of process. In fact, an exhaustive analysis of the environmental performance of a product within its life cycle also includes the choice and planning of the processes required to perform each phase of the life cycle itself.

1.4.4   Implementation of DFE and General Guidelines

Whether the subject of environmental improvement is a product, a process, or each single flow of resources, DFE is implemented in design practice through three successive phases (Allenby, 1994):

- Scoping, which consists of defining the target of the intervention (product, process, resource flow), identifying possible alternatives, and determining the depth of analysis
- Data gathering, which consists of acquiring and evaluating the more significant environmental data
- Data translation, which consists of transforming the results from the preliminary analysis data into tools (from simple guidelines and design procedures to more sophisticated software systems assisting the design team in applying environmental data in the design process)

In practice, the second and third phases are implemented using two instrument typologies:

- Tools aiding the analysis of the life cycle (Life Cycle Assessment, Life Cycle Cost Analysis), allowing the acquisition, elaboration, and interpretation of environmental data
- Tools aiding the design or redesign (Product and Process Design, Design for Use, Design for End-of-Life, or, more generally, Design for X)

These tools, and the issues correlated with them (evaluation of environmental impact of products and processes, choice of materials and processes, disassembly of the product or subsystems, extension and optimization of the useful life, and recovery at end-of-life through reuse of components and recycling of materials) are the subject of subsequent chapters. However, it should be noted here that these tools are based on a wide-ranging series of suggestions and guidelines for the designer (OTA, 1992; Fiksel, 1996; Billatos and Basaly, 1997) that can be summarized as follows:

- Reducing the use of materials, using recycled and recyclable materials, and reducing toxic or polluting materials
- Maximizing the number of replaceable or recyclable components
- Reducing emissions and waste in production processes
- Increasing energy efficiency in phases of production and use
- Increasing reliability and maintainability of the system
- Facilitating the exploitation of materials and recovery of resources by planning the disassembly of components
- Extending the product's useful life
- Planning strategies for the recovery of resources at end-of-life, facilitating reuse, remanufacturing and recycling, and reducing waste
- Controlling and limiting the economic costs incurred by design interventions aimed at improving the environmental performance of the product
- Respecting current legal constraints and evaluating future regulations in preparation

Applying these guidelines in relation to the main phases of the product's life cycle (Figure 1.5), it is possible to obtain useful information and to explore the whole set of opportunities for an eco-efficient intervention in the product design and development process.

1.4.5 Orientation of DFE Evolution

The guidelines outlined above to give an idea of the complexity of implementing DFE, and the main characteristics of DFE summarized in this overview, are directly reflected in several issues of strategic importance for the dissemination of this new approach to industrial and engineering design:

- Integration between design of products, processes, and systems for the selection of materials
- Development of durable materials, with low impacts in their production and operation

- Improvement of methods, tools, and procedures for the evaluation of environmental impacts and risks, in relation to the costs and benefits of corrective interventions

- Development of models for the evaluation of integration between the consumer's requirements and the use, disposal, and recycling of the product

- Development of techniques to predict the effects of legislation on the whole life cycle of the product

Each of these wide-ranging questions, already identified some 10 years ago (NSF, 1995), fully reflects the problems still confronting us today in the ambit of the evolutionary process of design for the environment. However, it should be noted that environmental protection cannot be achieved through an approach addressing purely technical problems, the more so if localized in a single actor—the "producer." The resolution of the environmental question must be put into the context of the entire society, considered as a complex system with several main actors: government, manufacturers, recyclers, and consumers (Sun et al., 2003). Since it is the interaction between these actors that finally determines the environmental performance of a product over its entire life cycle, the development of the issues raised must be accompanied by an in-depth study of the mechanisms of the entire complex system.

1.5   Concepts, Tools, and Approaches to the Environmental Question: Overview

The aim of this first chapter is to present a panorama of the main concepts, tools, and approaches to environmental protection in the context of industrial production, examining in greater detail those held to be of major importance. To complete this panorama and provide an overview of the concepts introduced, it is worth considering a graphical representation proposed by other authors (Coulter et al., 1995), which defines their range, competence, and fields of application in relation to two factors:

- Time scale (horizontal axis), quantifying the time range and extending, ideally, from the life cycle of a product to the entire evolutionary process of a civilization

- Spatial scale (vertical axis), showing the entities involved and extending from the life cycle of a product to the entire society

This approach results in the graph shown in Figure 1.6, which identifies the range and highlights the spatial and temporal relations between:

- Interventions to control and prevent pollution phenomena (Pollution Control and Prevention—PC&P)
- Methods and tools for the design and development of eco-sustainable products (Design for Environment and other similar approaches described in Section 1.4)
- System-based approaches to the environmental question (Industrial Ecology)
- Underlying concepts of the environmental debate (Sustainable Development)

The term "pollution prevention" is understood to mean all the techniques of "source reduction" (i.e., of reducing the harmful or polluting substances which enter the waste flows or are released into the environment, and of reducing their harmfulness to the health of living organisms and the environment). Also included are all the other interventions that reduce or eliminate the generation of polluting substances by increasing the efficiency of the use of virgin materials, energy, or other resources, or through the conservation of resources (EPA, 1997).

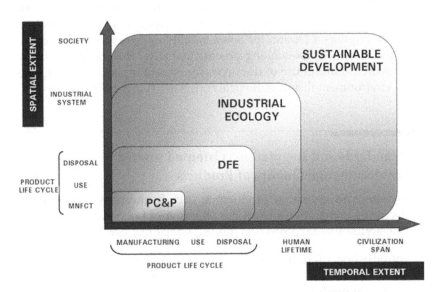

FIGURE 1.6 Overview and relations between concepts, tools, and approaches to the environmental question (Adapted from Coulter, S., Bras, B., and Foley, C., A lexicon of green engineering terms, in *Proceedings of ICED `95 10th International Conference on Engineering Design*, Prague, Czech Republic, 1995, 3. With permission).

In practice, the pollution prevention approach translates into reducing or eliminating the use of production processes that create pollution, and is achieved in two ways: through the redesign of processes to eliminate the generation of polluting byproducts and emissions, and through the redesign of products to eliminate the need to use polluting processes. Thus, it differs from the "pollution control" approach in that it is directed at avoiding the creation of pollutants and waste, rather than applying containment and reclamation measures after they have been generated (Wilhelm et al., 1993). It is, however, clear that in either case the techniques of pollution control and prevention mainly devolve on the manufacturing phase only, as shown in the figure.

Conversely, Design for Environment and other similar approaches (Green Engineering Design, Environmentally Conscious Design, and Environmentally Conscious Manufacturing) extend to cover the product's entire life cycle, in both spatial and temporal terms. In the same graph, the concepts of industrial ecology and sustainable development extend to cover a much greater spatial and time span, as can be readily deduced from their respective features described above.

This overview, which defines, contextualizes, and correlates the main approaches to the environmental question, can finally be summed up in the concept of "sustainable production," achieved when products are designed, produced, distributed, used, and disposed of with minimal environmental and health damages, and with minimal use of resources (Alting and Jorgensen, 1993). Highlighting the strategic role which design may play, sustainable production requires a Design for Sustainability approach, which should be thought of as "a decision-making process that aims at achieving maximum benefits with minimum use of resources, by integrating all economic, social, human, environmental, and ecological concerns" (Ling, 1997).

1.6 Standards and Regulations Oriented toward Environmental Quality of Products

With reference to industrial production, diverse motivating factors drive manufacturing companies toward the adoption of policies and instruments aimed at environmental protection (Fiksel, 1996; Bras, 1997):

- Introduction of standards for the management of environmental systems, and the promotion of certification of the environmental quality of products

- Legislation oriented toward extending the manufacturer's responsibility beyond the commercialization of products, going as far as imposing the management of the end-of-life phase with a strong drive toward recovery and recycling

• Increasing attention on the part of consumers with regard to environmental protection

In this last section we introduce some of the most significant regulatory initiatives for incorporating environmental considerations into product development. These initiatives intend to reinforce and redirect environmental policies, with the ultimate aim of evoking the primacy of ecological production in companies, integrating environmental requirements into product standards.

1.6.1 Environmental Standards and Product Certification

Perhaps the most effective means of promoting the principles of environmental quality in the context of design intervention is the creation of standards that encourage a preventive approach to environmental problems.

Standards have the great advantage of promoting the importance of design in industrial production and product development. It is now normal practice for the designer to have knowledge of and to refer to the relevant national and international norms. Thus, the inclusion of environmental parameters within these standards guarantees an immediate response. Various national and international regulatory bodies have initiated research into instruments of standardization aimed at improving the environmental impact of industrial production. The International Organization for Standardization, through its ISO 14000 series, has first begun to standardize the implementation of Environmental Management Systems, and subsequently has considered the instruments and procedures that serve to introduce environmental variables into company management (Environmental Audit, Environmental Labeling, Life Cycle Assessment) (Culley, 1998). Of the standards of the 14000 series, we note those expressly addressing product life cycle and DFE:

• ISO 14040 series—Reference standards for Life Cycle Assessment (ISO 14040, 1997), give guidelines on the principles and conduct of environmental analysis of product's life cycle (see Chapter 4)

• ISO 14062—Reference technical report for integrating environmental aspects into product design and development (ISO 14062, 2002), gives guidelines for improvement of environmental performance of product development processes (see Chapter 8)

Environmental certification is another instrument able to introduce innovations in products and production technologies, and to stimulate their promotion through mechanisms of market competition. With regard to product certification, different procedures have been developed for the assigning of environmental quality labels worldwide. Germany introduced the Blauer Engel program in 1977, making it the first country to implement a national ecolabeling program. Subsequently, Sweden, Norway, Finland, Denmark, and Iceland introduced the Nordic Swan ecolabel, and the Netherlands adopted its national ecolabel Stichting Milieukeur. In 1992 the European Union issued Regulation 880/92, the community system for assigning a seal of ecological quality for some product typologies, subsequently defined in specific directives. The regulation was revised in 2000 (EC 1980/2000). Similar initiatives on environmental labeling spread elsewhere (Green Seal and Energy Star in the United States, Environmental Choice in Canada, EcoMark in Japan, Ecomark in India, etc.) (EPA, 1998).

The principal objectives behind these regulations can be summarized as follows:

- To create a mechanism of voluntary adhesion to promote the market presence of more environmentally friendly products
- To indicate to the consumer the more environmentally favorable products among those present on the market

Also, the ISO 14000 standards have specific areas regarding the ecological labeling of products. The ISO 14020 series (ISO 14020, 2000) addresses a range of different approaches to environmental labels and declarations, including self-declared environmental claims, ecolabels, and a possible scheme of environmental declarations for products.

1.6.2   Extension of Manufacturer Responsibility

An underlying principle of more recent environmental regulations is that of extended producer responsibility (EPR), an environmental policy approach in which a producer's responsibility for a product is extended to the postconsumer stage of the product's life cycle, focusing on product–systems instead of production facilities (Davis et al., 1997). This can be translated into restrictions at different levels, from the obligation of producers to take on the costs of disposal and, in some cases, the organization of recalling their products after use, to the explicit demand for product requisites such as disassemblability and recyclability.

The principle of EPR relies for its implementation upon the life cycle concept to identify opportunities to prevent pollution and reduce resource use in each stage of the product life cycle, through changes in process technology and product design. Therefore, the ultimate aim of this type of regulatory action is to stimulate the redesign of some categories of products in order to obtain a reduction in their environmental impact, encouraging producers to adopt an integrated approach for the development and management of eco-compatible products, and introducing an extended vision of the problem to cover the entire life cycle of products. This extended view, typical of the DFE approach, includes a wide range of aspects (energy consumption, use of materials, component duration, reuse of components, and recycling of materials).

Since its introduction, EPR has become a characteristic of regulations regarding various production sectors, constituting a valid stimulus for a process of innovation directed at environmental sustainability. In the following section, with meaningful examples, we briefly report the current situation in the European Union regarding EPR implementation in two important industrial sectors: vehicles, and electrical and electronic equipment.

1.6.2.1 The European Example

Extending the limit of producer responsibility beyond the point of sale, EPR favors a life cycle–oriented approach to product design and development. The Green Paper on Integrated Product Policy (IPP), adopted by the European Commission in 2001 (EC-IPP, 2001), strengthened and amended by the Communication from the Commission to the Council and the European Parliament on IPP in 2003, clearly confirms this assumption. It launched a broad debate on how to achieve a new growth paradigm through wealth creation and competitiveness on the basis of so-called greener products, and proposed strategies intended to reinforce environmental policies, with the aim of integrating environmental requirements into product standards. In this context, IPP avers "the concept of producer responsibility relates to the integration of costs occurring once the product has been sold into the price of new products. This encourages prevention at the design stage and allows consumers to bring back end-of-life products free of charge" (EC-IPP, 2001).

The concept of producer responsibility for the disposal of products at the end of their useful life has recently been integrated into the European Commission's Directive on End-of-Life Vehicles and the Directive on Waste Electrical and Electronic Equipment. The implications of these directives have great impact on product design because they encourage alternatives to hazardous substances, and design for disassembly and recycling (Kumar and Fullenkamp, 2005).

Directive 2000/53/EC on End-of-Life Vehicles (ELV) established prevention measures for waste from scrapped vehicles, its collection and treatment to promote reuse and recycling, and restrictions on the use of dangerous substances in new vehicles (EC 2000/53, 2000). The salient points of the ELV directive are:

- Prevention of waste from vehicles and improvement in the environmental performance of all of the operators involved in the life cycle of vehicles

- Fundamental principles that waste should be reused and recovered, and that preventive measures should be applied from the conception and design phases of vehicle in order to facilitate recycling and to avoid the disposal of hazardous waste

- Precise targets for rate of recovery and recycling (by January 1, 2006, the reuse and recovery shall be increased to a minimum of 85%, and reuse and recycling shall be increased to a minimum of 80%, by an average weight per vehicle and year; by January 1, 2015, these limits should be raised to 95% and 85%, respectively)

Similarly, Directive 2002/96/EC on Waste Electrical and Electronic Equipment (WEEE), amended by Directive 2003/108/EC, is aimed at increasing the recovery flows of various categories of products (household appliances, IT and telecommunications equipment, lighting equipment, electrical and electronic tools, medical devices, and monitoring and control instruments), extending producer responsibility to cover collection and recycling (EC 2002/96, 2003). As of August, 2005, the following obligations exist:

- Producers must finance the collection, treatment, recovery, and environmentally sound disposal of WEEE from private households and from other users, and meet precise targets for rate of recovery and the recyclable fraction to be reached by December 2006.

- Distributors must be prepared to take in old, similar equipment for waste treatment on a one-to-one basis when supplying new electrical or electronic products.

- All products falling under the Directive and put on the European Union market must be marked by producers with the symbol of a crossed-out wheeled bin.

- Member states must encourage the design and production of electrical and electronic equipment that facilitates disassembly, reuse, and recycling of WEEE, components, and materials.

1.7   Summary

In the wide panorama of human activities having a significant effect on the environment, industrial processes are undoubtedly crucial, and the increased attention paid to the environmental consequences of these processes has led to the evolution of a new vision of productive systems. Starting from the concepts of sustainable development and industrial ecology, these are interpreted in the context of the design of industrial processes and products.

Design, which essentially consists of molding flows of material and energy in order to satisfy the needs of humankind, thus becomes a process of transformation wherein the needs generating it are contextualized in the patterns and flows of natural systems, assimilating the organizing principles of the ecosphere. Design for Environment (DFE) interprets this new role; it can be defined as a methodology based on constantly evaluating the potential impacts implicated in the entire life cycle of a product, throughout the whole course of the design process, and directed at the systematic reduction or elimination of these impacts.

The overview presented in this chapter defines, contextualizes, and correlates the main approaches to the environmental question; it can be summed up in the concepts of sustainable production and Design for Sustainability (DFS). The former is an approach to production wherein products are designed, produced, distributed, used, and disposed of with a minimal use of resources and minimal damage to the health of humans and the ecosphere. The latter consists of a decision-making process directed at achieving maximum benefits with minimum use of resources, by integrating economic, social, and ecological concerns.

1.8   References

Allenby, B.R., Achieving sustainable development through industrial ecology, *International Environmental Affairs*, 4(1), 56–68, 1992.

Allenby, B.R., Integrating environment and technology: Design for environment, in *The Greening of Industrial Ecosystems*, Allenby, B.R. and Richards, D.J., Eds., National Academy Press, Washington, DC, 137–148, 1994.

Allenby, B.R., *Industrial Ecology: Policy Framework and Implementation*, Prentice Hall, Englewood Cliffs, NJ, 1998.

Allenby, B.R. and Cooper, W.E., Understanding industrial ecology from a biological systems perspective, *Total Quality Environmental Management*, 3(3), 343–354, 1994.

Alting, L. and Jorgensen, J., The life cycle concept as a basis for sustainable industrial production, *Annals of CIRP*, 42(1), 163–167, 1993.

Ashley, S., Designing for the environment, *Mechanical Engineering*, 15(3), 53–55, 1993.

Asimow, M., *Introduction to Design*, Prentice Hall, Englewood Cliffs, NJ, 1962.

Ayres, R.U. and Ayres, L.W., *A Handbook of Industrial Ecology*, Edward Elgar, Cheltenham, UK, 2002.

Ayres, R.U., Industrial metabolism, in *Technology and Environment*, Ausubel, J.H. and Sladovich, H.E., Eds., National Academy Press, Washington, DC, 1989, 23–49.

Ayres, R.U., On the life cycle metaphor: Where ecology and economics diverge, *Ecological Economics*, 48, 425–438, 2004.

Ayres, R.U. and Ayres, L.W., *Industrial Ecology: Towards Closing the Materials Cycle*, Edward Elgar, Cheltenham, UK, 1996.

Benyus, J.M., *Biomimicry*, Morrow, New York, 1997.

Billatos, S.B. and Basaly, N.A., *Green Technology and Design for the Environment*, Taylor & Francis, Washington, DC, 1997.

Bras, B., Incorporating environmental issues in product design and realization, *Industry and Environment*, 20(1–2), 7–13, 1997.

Brezet, H. and van Hemel, C., *Ecodesign: A Promising Approach to Sustainable Production and Consumption*, UNEP United Nations Environment Programme, Paris, 1997.

Brown, L.T., *Building a Sustainable Society*, Norton, New York, 1981.

Capra, F., *The Web of Life*, Doubleday, New York, 1996.

Capra, F., *The Hidden Connections: A Science for Sustainable Living*, Doubleday, New York, 2002.

Coulter, S., Bras, B., and Foley, C., A lexicon of green engineering terms, in *Proceedings of ICED 95 10th International Conference on Engineering Design*, Prague, Czech Republic, 1995, 1–7.

Cross, N., The coming of post-industrial design, *Design Studies*, 2(1), 6–7, 1981.

Culley, W.C., *Environmental and Quality Systems Integration*, Lewis Publishers, Boca Raton, FL, 1998.

Davis, G.A. et al., Extended Product Responsibility: A New Principle for Product-Oriented Pollution Prevention, U.S. Environmental Protection Agency, Office of Solid Waste, Washington, DC, 1997.

Dowie, T., Green design, *World Class Design to Manufacture*, 1(4), 32–38, 1994.

EC-IPP, Green Paper on Integrated Product Policy, COM(2001) 68, 7/2/2001, Commission of the European Communities, Brussels, 2001.

EC 1980/2000, Regulation (EC) No. 1980/2000 of the European Parliament and of the Council on Revised Community Eco-Label Award Scheme, Official Journal of the European Communities, L 237, 21/9/2000, 1–12, 2000.

EC 2000/53, Directive 2000/53/EC of the European Parliament and of the Council on End-of-Life Vehicles (ELV), Official Journal of the European Communities, L 269, 21/10/2000, 34–42, 2000.

EC 2002/96, Directive 2002/96/EC of the European Parliament and of the Council on Waste Electrical and Electronic Equipment (WEEE), Official Journal of the European Communities, L 37, 13/2/2003, 24–38, 2003.

Ehrenfeld, J., Industrial ecology: A new field or only a metaphor?, *Journal of Cleaner Production*, 12, 825–831, 2004.

Elkington, J. and Burke, T., *The Green Capitalists: How Industry Can Make Money and Protect the Environment*, Victor Gollancz, London, 1987.

Elkington, J., Burke, T., and Hailes, J., *Green Pages: The Business of Saving the World*, Routledge, London, 1988.

EPA, Pollution Prevention 1997: A National Progress Report, EPA 742-R-97–00, U.S. Environmental Protection Agency, Office of Pollution Prevention and Toxics, Washington, DC, 1997.

EPA, Environmental Labeling: Issues, Policies, and Practices Worldwide, EPA 742-R-98–009, U.S. Environmental Protection Agency, Office of Prevention, Pesticides and Toxic Substances, Washington, DC, 1998.

Fiksel, J., *Design for the Environment: Creating Eco-Efficient Products and Processes*, McGraw Hill, New York, 1996.

Frosch, R.A., Industrial ecology: Minimizing the impact of industrial waste, *Physics Today*, 47(11), 63–68, 1994.

Frosch, R.A. and Gallopoulos, N.E., Strategies for manufacturing, *Scientific American*, 261(3), 94–102, 1989.

Frosch, R.A. and Uenohara, M., Chairmen's overview, in *Industrial Ecology, U.S.–Japan Perspectives*, Richards, D.J. and Fullerton, A.B., Eds., National Academy Press, Washington, DC, 1994, 1–6.

Garner, A. and Keoleian, G.A., Industrial Ecology: An Introduction, National Pollution Prevention Center for Higher Education, University of Michigan, Ann Arbor, MI, 1995.

Goodland, R., Daly, H., and El-Serafy, S., Environmentally Sustainable Economic Development Building on Brundtland, ENV 46, Environment Department, The World Bank, Washington, DC, 1991.

Graedel, T.E., Allenby, B.R., and Linhart P., Implementing industrial ecology, *IEEE Technology and Society Magazine*, 12(1), 18–26, 1993.

Graedel, T.E. and Allenby, B.R., *Industrial Ecology*, Prentice Hall, Englewood Cliffs, NJ, 1995.

Graedel, T.E. and Allenby, B.R., *Design for Environment*, Prentice Hall, Upper Saddle River, NJ, 1998.

Harte, J. et al., Business as a living system: The value of industrial ecology (A roundtable discussion), *California Management Review*, 43(3), 16–25, 2001.

ISO 14020, Environmental Labels and Declarations—General Principles, ISO 14020:2000(E), International Organization for Standardization, Geneva, 2000.

ISO 14040, Environmental Management—Life Cycle Assessment—Principles and Framework, ISO 14040:1997(E), International Organization for Standardization, Geneva,1997.

ISO 14062, Environmental Management—Integrating Environmental Aspects into Product Design and Development, ISO/TR 14062:2002(E), International Organization for Standardization, Geneva, 2002.

Jelinski, L.W. et al., Industrial ecology: Concepts and approaches, in *Proceedings of National Academy of Sciences, Colloquium on Industrial Ecology*, Washington, DC, 89(3), 1992, 793–797.

Johnson, E.F. and Gay, A., A practical, customer-oriented DFE methodology, in *Proceedings of IEEE International Symposium on Electronics and the Environment*, Orlando, FL, 1995, 47–50.

Korhonen, J., Industrial ecology in the strategic sustainable development model: Strategic applications of industrial ecology, *Journal of Cleaner Production*, 12, 809–823, 2004a.

Korhonen, J., Theory of industrial ecology, *Progress in Industrial Ecology*, 1, 61–88, 2004b.

Kumar, S. and Fullenkamp, J., Analysis of European Union environmental directives and producer responsibility requirements, *International Journal of Services and Standards*, 1(3), 379–398, 2005.

Ling, J.T., Next stop: Designing for sustainability, in *Pollution Prevention 1997: A National Progress Report*, EPA 742-R-97–00, U.S. Environmental Protection Agency, Office of Pollution Prevention and Toxics, Washington, DC, 1997, 239–241.

Lund, R.T., Remanufacturing, *Technology Review*, 87(2), 18–29, 1984.

Mackenzie, D., *Green Design: Design for the Environment*, Laurence King, London, 1991.

Madge, P., Design, ecology, technology: A historiographical review, *Journal of Design History*, 6(3), 149–166, 1993.

Meadows, D.H., *The Limits to Growth: A Report for the Club of Rome's Project on the Predicament of Mankind*, 2nd ed., Universe Books, New York, 1972.

Munasinghe, M., *Environmental Economics and Sustainable Development*, The World Bank, Washington, DC, 1993.

Munasinghe, M. and Shearer, W., An introduction to the definition and measurement of biogeophysical sustainability, in *Defining and Measuring Sustainability: The Biogeophysical Foundations*, Munasinghe, M. and Shearer, W., Eds., The World Bank, Washington, DC, 1995, xvii-xxxii.

Navin-Chandra, D., Design for environmentability, in *Proceedings of ASME Conference on Design Theory and Methodology*, Miami, FL, 1991, DE-31, 119–125.

NSF, Environmentally Conscious Manufacturing, National Science Foundation, Program guideline, Arlington, VA, 1995.

O'Brien, C., Sustainable production: A new paradigm for a new millennium, *International Journal of Production Economics*, 60–61, 1–7, 1999.

Odum, E.P., *Ecology: The Link Between the Natural and Social Sciences*, 2nd ed., Holt-Saunders, New York, 1975.

O'Riordan, T., *Environmentalism*, Pion, London, 1976.

OTA, Green Products by Design: Choices for a Cleaner Environment, Report OTA-E-541, Office of the Technology Assessment, Congress of the United States, Washington, DC, 1992.

Overby, C., Product design for recyclability and life extension, in *Proceedings of American Society of Engineering Education Annual Conference*, Baton Rouge, LA, 1979, 181–196.

Overby, C., Design for the entire life-cycle: A new paradigm?, in *Proceedings of American Society of Engineering Education Annual Conference*, Toronto, 1990, 552–563.

Papanek, V., *Design for the Real World: Human Ecology and Social Change*, Pantheon Books, New York, 1971.

Regge, T. and Pallante, M., *Scienza e Ambiente*, Bollati Boringhieri, Turin, Italy, 1996.

Robèrt, K.-H. et al., Strategic sustainable development: Selection, design and synergies of applied tools, *Journal of Cleaner Production*, 10, 197–214, 2002.

Robertson, J., *The Sane Alternative: A Choice of Futures*, 2nd ed., River Basin Publishing Co., St. Paul, MN, 1980.

Schumacher, E.F., *Small Is Beautiful: A Study of Economics as if People Mattered*, Blond & Briggs, London, 1973.

Seager, T.P. and Theis, T.L., A uniform definition and quantitative basis for industrial ecology, *Journal of Cleaner Production*, 10, 225–235, 2002.

Socolow, R.H. et al., *Industrial Ecology and Global Change*, Cambridge University Press, Cambridge, UK, 1994.

Sun, J. et al., Design for environment: Methodologies, tools, and implementation, *Journal of Integrated Design and Process Science*, 7(1), 59–75, 2003.

UNCED, Rio Declaration on Environment and Development, United Nations Conference on Environment and Development, Rio de Janeiro, Brazil, 1992.

Veroutis, A.D. and Fava, J.A., Framework for the development of metrics for design for environment assessment of products, in *Proceedings of IEEE International Symposium on Electronics and the Environment*, Dallas, TX, 1996, 13–18.

WCED, *Our Common Future*, World Commission on Environment and Development, Oxford University Press, New York, 1987.

Weinbrecht, E. et al., Industrial waste reduction program, *International Journal of Environmentally Conscious Design and Manufacturing*, 2(1), 47–55, 1993.

Wilhelm, M.R. et al., Selection of waste management technologies to implement manufacturing pollution prevention strategies, *International Journal of Environmentally Conscious Design and Manufacturing*, 2(2), 41–47, 1993.

Zhang, H.C. et al., Environmentally conscious design and manufacturing: A state of the art survey, *Journal of Manufacturing Systems*, 16(5), 352–371, 1997.

Part I

Life Cycle Approach

Chapter 2

Life Cycle Approach and the Product–System Concept and Modeling

The most important benefits of Design for Environment (DFE) can only be obtained when the entire life cycle of a product is already taken into consideration at the design stage. Only a systematic vision of the product over its life cycle can, in fact, ensure that the design activity not only identifies the product's environmental criticalities, but also reduces them effectively and avoids simply transferring impacts from one arena to another.

In this chapter a holistic vision of the product and its life cycle is presented, where the latter is no longer thought of as a series of independent processes expressed exclusively by their technological aspects, but rather as a complex product–life cycle system set in its environmental, economic, and sociotechnological context.

2.1 Life Cycle Concept and Theory

Originally conceived in the context of studies on biological systems, the concept of "life cycle" has become widely used as a model for the interpretation and analysis of phenomena characterized by processes of change. It is applied in many wide-ranging fields, from social sciences to processes of technological innovation. This second case, in particular, represents one of the more interesting examples of the metaphor of biological evolution used in the management of industrial activities (Abernathy and Utterback, 1978). Beginning from this type of experience, the application of Life Cycle Theory to the development of industrial products has become a key factor in the management of technological innovation, where it is recognized as an effective instrument of analysis and a useful aid to decision making.

2.1.1 Life Cycle Theory: General Concepts

With regard to the study and understanding of the processes of development and evolution of organizational structures, management science has adopted different concepts and theories typical of other disciplines, used to explain processes of change in the context of social, physical, and biological sciences. These theories differ substantially in terms of the model by which they represent the sequence of events (event progression), and in the mechanism by which they generate and guide change (generating force) (van de Ven and Poole, 1995). The Life Cycle Theory is one of the most widely used. It is based on the metaphor of the phenomena of organic growth typical of evolutionary biology, and has two salient characteristics:

- Event progression is linear and irreversible (i.e., characterized by a unitary sequence wherein each intermediate phase is a necessary precursor of the subsequent phase).

- Generating force consists of a predefined program, inherent in the entity that evolves, which is regulated by the environment in which the entity is conceived and develops (nature, in the case of biological systems; society, the market, and institutions in the case of manufacturing organizations).

Regarding the first characteristic (event progression), Life Cycle Theory presumes that the progression of change events in a life cycle model is "a unitary sequence (it follows a single sequence of stages or phases), which is cumulative (characteristics acquired in earlier stages are retained in later stages) and conjunctive (the stages are related such that they derive from a common underlying process)" (van de Ven and Poole, 1995). According to this viewpoint, each phase of the cycle contributes to the development of the final product and must be undertaken following a preestablished order, since its contribution is required for the completion of successive phases.

Considering the second characteristic (generating force), according to Life Cycle Theory "the developing entity has within it an underlying form, logic, program, or code that regulates the process of change and moves the entity from a given point of departure toward a subsequent end that is prefigured in the present state" (van de Ven and Poole, 1995). This characteristic, which defines the mechanism generating and guiding change, further clarifies the relation between the entity's internal evolutionary factor and the environment in which it evolves: "External environmental events and processes can influence how the entity expresses itself, but they are always mediated by the immanent logic, rules, or programs that govern the entity's development" (van de Ven and Poole, 1995).

With these premises, Life Cycle Theory can, in principle, be applied to any system that undergoes a series of changes over the course of its existence. The entire life of the system takes the name "life cycle," and the various phases following one after the other in the evolutionary process are called "life cycle phases" or "stages."

2.1.2 Life Cycle Theory in the Management of Product Development

At present, the use of Life Cycle Theory as an aid to decision making is fully accepted in the managerial context, above all with regard to some strategic management issues in industrial production—the management of the organizational structures of production activities; market analysis and predictions based on the evolution of technologies; and the development of new products and their introduction into the market. At the base of this acceptance of the life cycle concept as an analytical model for such widely varying phenomena, there is the understanding that both production activities and technologies, and products themselves, theoretically develop following an evolutionary path passing through different phases.

With regard to products, this evolutionary perspective is now well-rooted in the field of marketing (Massey, 1999). In the context of the management of products in relation to market dynamics, in fact, the life cycle is understood as the period during which the product is on the market. This period consists of four successive phases: introduction, growth, maturity, and decline. In this context, the objective of Life Cycle Theory is to describe the behavior of the product from development to retirement, to optimize the value of and the potential for profit in each phase of the cycle (Ryan and Riggs, 1996). With this aim, life cycle becomes a representation of the product's market history and each phase is characterized by the trend of the sales volumes and profit performance (Cunningham, 1969), so as to guide the decisional choices of management regarding possible intervention strategies (marketing actions, pricing, service strategies, product substitution, etc.).

In the same limited field of marketing, the breadth of the potential offered by the life cycle approach has been clearly described by D.M. Gardner: "[P]roduct life cycle is an almost inexhaustible concept because it touches on nearly every facet of marketing and drives many elements of corporate strategy, finance and production" (Gardner, 1987). Likewise, the conceptual premises of Life Cycle Theory summarized above evidence the potential for its use in the management of other aspects of a product, as well. Considering, then, the product as a single entity that includes both the abstract dimension (need, concept, and project) and the concrete, physical dimension (finished product), its life cycle can be understood as a preestablished sequence of evolutionary phases wherein each phase is necessary for the execution of subsequent phases, and each provides a different contribution to the development of the final product. This is in full agreement with the concept of event progression, one of the fundamental principles of Life Cycle Theory.

With clear reference to the management of product design and development, and as shown in Figure 2.1, the evolutionary sequence includes all the phases from product conception and design to manufacturing and distribution, and potentially can be extended to also consider the phases of use and

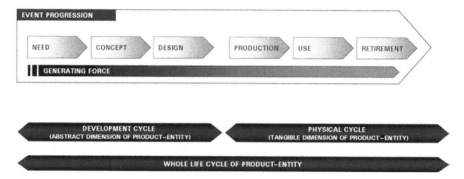

FIGURE 2.1   Life Cycle Theory: Product–entity application.

disposal. The entire life cycle represented by this sequence is composed of two parts:

- Development cycle—Indicates the first part of the life cycle of the product–entity, understood in its abstract dimension. This part includes all the conventional process of product design and development, through which the need is translated into the finished design.

- Physical cycle—Indicates the subsequent part of the life cycle of the product–entity, understood here in its tangible dimension as a finished product. This part includes all the phases the product passes through during its physical life.

In this context, moreover, the need underlying the product concept and the design requisites interpret, respectively, the roles of generating factor and internal evolutionary factor of the product–entity. This follows the second fundamental principle of Life Cycle Theory, that of generating force. In particular, design requirements are translated into product properties that ideally can condition its behavior over the entire life cycle, and can therefore guide its evolution in relation to the different environments in which the product–entity evolves (not only the market, but the entire economic system, ecosystem, and society).

The application of Life Cycle Theory to the management of product development, in the sense described above, and the concept of product life cycle corresponding to it, are summed up in the concept of product–system, introduced in the following section, fully interpreting the requirements of DFE.

2.2   Life Cycle and the Product–System Concept

As noted previously, the most significant benefits of DFE can only be obtained if the product's entire life cycle, including other phases together with those specific to development and production, is already considered at the design stage.

Products must be designed and developed in relation to all these phases, in accordance with a design intervention based on a life cycle approach, understood as a systematic approach "from the cradle to the grave," the only approach able to provide a complete environmental profile of products (Alting and Jorgensen, 1993; Keoleian and Menerey, 1993). Only a systematic view can in fact guarantee that the design intervention manages to both identify the environmental criticalities of the product and reduce them efficiently, without simply moving the impacts from one phase of the life cycle to another.

As noted in the previous section, the concept of product life cycle has different meanings in different contexts. Excluding the strictly marketing context (where it is understood to mean the phases of introduction, growth, maturity, and decline with regard to a product's performance on the market), the term life cycle can be used in the management of product development to mean the entire set of phases from need recognition and design development to production. This usage can go so far as to include any possible support services for the product, but does not usually take into consideration the phases of retirement and disposal.

This limited view of the life cycle has its origins in a statement of the problem conditioned by the competencies and direct interests of different actors involved in the life of manufactured goods. This leads to a fragmentation of the life cycle according to the main actors: the manufacturer (design, production and distribution); the consumer (use); and a third actor, defined on the basis of the product typology (retirement and disposal). It is clear, therefore, that the managerial concept of life cycle springs from the interests of the manufacturer and does not usually include those phases subsequent to the distribution of the product.

Given that the environmental performance of a product over its entire life cycle is influenced by interaction between all the actors involved, an effective approach to the environmental problem must be considered in the context of the entire society, understood as a complex system of actors including government, manufacturers, consumers, and recyclers (Sun et al., 2003). This system is also characterized by complex dynamics, since the various actors interact through the application of reciprocal pressures dependent on political, economic, and cultural factors (Young et al., 1997).

Therefore, from a more complete perspective (not limited by the point of view of a specific actor), the life cycle of a product must include both its abstract and physical dimensions and extend the latter to include the phase of product retirement and disposal. This aspect fully interprets the life cycle approach which, in contrast to the limited view of the environmental question held by the single actor "manufacturer," imposes a sort of "social planner's view" (Heiskanen, 2002).

In general terms, therefore, the life cycle of a product can be considered well-represented by the event progression shown in Figure 2.1, or, in a similar manner, by the main phases of need recognition, design development, production, distribution, use, and disposal, as has already been suggested by other authors (Alting, 1993; Jovane et al., 1993).

The concepts underlying industrial ecology (Section 1.2) require that the actions of the system of all actors are placed in the context of the global ecosystem, which includes the biosphere (i.e., all living organisms) and the geosphere (all lands and waters). On these premises, environmental analysis is oriented toward a view of the life cycle of a product associated with its physical reality (physical dimension of product–entity, Figure 2.1), focusing on the interaction between the environment and all the processes involved in the product's life, from inception to disposal.

From this perspective, the product becomes "a transient embodiment of material and energy occurring in the course of material and energy process flows of the industrial system" (Frosch, 1994), and the life cycle is understood as a set of activities, or processes of transformation, each requiring an input of flows of resources (quantities of materials and energy) and generating an output of flows of byproducts and emissions. This vision is in perfect harmony with the analogy between industrial and natural systems at the basis of industrial ecology, according to which both system typologies are characterized by cycles of transformation of resources.

For a complete analysis aimed at the evaluation and reduction of a product's environmental impact, it is therefore necessary to take into account not only the manufacturing phases of production and machining, but also the phases of preproduction of materials and those of use, recovery, and disposal. Furthermore, all these phases must not be considered in relation to the specific actors involved, but rather in relation to the whole environment–system, taking a wider view and sidestepping direct responsibilities.

These considerations can be summarized in a holistic vision of the product and its life cycle, wherein the latter is no longer thought of as a series of independent processes expressed exclusively by their technological aspects, but rather as a complex product–life cycle system set in its environmental and sociotechnological context (Zust and Caduff, 1997). It is then possible to speak of a product–system. In its most complete sense, the product–system includes the product (understood as integral with its life cycle) within the environmental, social, and technological context in which the life cycle evolves (Figure 2.2).

2.3   Product–System and Environmental Impact

From the specific viewpoint of environmental analysis, the product–system is characterized by flows of resources transformed through the various

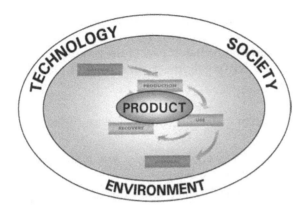

FIGURE 2.2 Schematic representation of a product–system.

processes constituting the physical life cycle. The environmental impact of this product–system is the result of life cycle processes that exchange substances, materials, and energy with the ecosphere. The different effects produced can be summarized in three main typologies (Guinée et al., 1993):

- Depletion—The impoverishment of resources, imputable to all the resources taken from the ecosphere and used as input in the product–system (e.g., depletion of mineral and fossil fuel reserves as a result of their extraction and transformation into construction materials and energy)
- Pollution—All the various phenomena of emission and waste, caused by the output of the product–system into the ecosphere (e.g., dispersion of toxic materials or phenomena caused by thermal and chemical emissions such as acidification, eutrophication, and global warming)
- Disturbances—All the phenomena of variation in environmental structures due to the interaction of the product–system with the ecosphere (e.g., degradation of soil, water, and air)

Some of these impacts have a local effect while others act at the regional, continental, or global level. This distinction is important because the effects of these impacts on the environment can vary in different geographical contexts due, for example, to differing climatic conditions or soil typologies.

Ultimately, to undertake the environmental evaluation of a product is "to define and quantify the service provided by the product, to identify and quantify the environmental exchanges caused by the way in which the service is provided, and to ascribe these exchanges and their potential impacts to service" (Wenzel et al., 1997).

Ascribing the environmental impact of the product–system to the flows of exchange with the ecosphere, the main factors of life cycle impact can be summarized as:

- Consumption of material resources and saturation of waste disposal sites
- Consumption of energy resources and loss of energy content of products dumped as waste
- Combined direct and indirect emissions of the entire product–system

With regard to the first aspect, the quantification of the impact can be made only on the basis of an analysis of the distribution of the volumes of material in play over the entire life cycle. The energy and emission aspects, on the other hand, require a more complete approach that takes into account the energy and emission contents of the resources and of the final products.

In an elementary production process such as that shown in Figure 2.3, each typology of resource introduced (materials and energy) is characterized in terms of both energy and emission content, and a distinction is made between direct and indirect emissions. The energy and emission content of a material resource are, respectively, understood as:

- The energy cost (i.e., the energy expended to produce the material)
- All the emissions correlated with its production

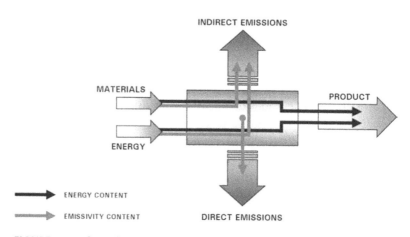

FIGURE 2.3   Scheme for the definition of a product's environmental impact.

The energy and emission content of an energy resource are, respectively, understood as:

- The sum of energy expended to produce this energy resource in the form in which it is used in the process
- The sum of emissions correlated with its production

Regarding the distinction between direct and indirect emissions, these are understood as, respectively:

- The sum of characteristic emissions of the process itself (dependent on the materials, the type of process, and on the product of this process)
- The sum of the emissions correlated with the production of the resources used by the process, therefore corresponding to the emission content of the resources

With this structure, again referring to Figure 2.3, it is possible to say that:

- The sum of the direct and indirect emissions quantifies the total emissivity that can be associated with the process and, therefore, with the final product.
- The sum of the energy contents of the materials and of the energy introduced quantifies the energy content of the final product, and expresses the consumption of energy resources associable with it and with the activity that generated it.

Following this scheme, the practical quantification of a product's energy and emission impacts comes down to obtaining the following information:

- The quantity of material and energy resources introduced
- The energy cost per unit weight of each material used
- The energy cost per unit of energy used by the process (i.e., the quantity of energy needed to produce the unit of energy in the form used by the process)
- The emissivity associable with the production of the unit weight of each material used
- The emissivity associable with the production of the unit of energy
- The direct emissivity associable with the process per unit of final product (this can also encompass the waste per unit of final product)

The structure proposed above represents a conceptual schematization with which it is possible to define in detail the environmental impact of an elementary process. In theory, it is easily extended to the product–system by considering all the processes that make up its life cycle. To complete the picture of the environmental impact of a product, it is worth examining in greater detail the two concepts of energy and emissivity content.

2.3.1 Environmental Aspects of the Consumption of Energy Resources

The energy content or cost of a material resource is understood to be the total quantity of energy which must be consumed in order to obtain the unit quantity of material. This quantity can be considered in two different ways:

- At a first level of analysis, it can mean the quantity of energy expended in the production processes of the material, in the form used by these processes.
- At a deeper level of analysis, it is intended to mean the quantity of primary energy expended to produce the energy used by the processes of producing the material.

It is clear, therefore, that an accurate evaluation requires the distinction between primary energy and energy in the form used by the processes of transformation. In this sense, the analogous concept of energy content or cost of an energy resource is clear: it indicates the quantity of primary energy expended to produce this energy resource in the form in which it is employed. The need for this distinction is due to some aspects of energy transformation. According to the principle of entropy (Second Law of Thermodynamics), all natural and artificial processes are irreversible because of inevitable dissipation effects, measured, in fact, by entropy production. Because of the irreversibility manifested in any real process, a part of the energy powering a system is returned as energy that can no longer be converted into usable forms. The sum of these nonconvertible portions of energy and the remaining portion that can still be converted equals the total energy entering the system. From these considerations, it is clear that powering any type of process requires not generic energy but convertible energy, called exergy. When speaking of energy content or cost, therefore, it is necessary to establish whether one is referring to energy or exergy.

These considerations are also valid for the processes of energy production. If the efficiency of a conventional thermoelectric power plant is 35% to 40% (meaning that 60% to 65% of the energy supplied by the combustible is dissipated into the environment), it is evident how important it is to make a distinction between the quantity of usable energy powering a process (exergy) and the total amount of energy that must be expended to power the same process, when also taking into account the production of this energy in the form used.

The energy dissipated due to the irreversibility of transformation processes, called anergy, can be identified with all the forms of thermal waste released into the environment. This factor, in conjunction with the thermal emissions of domestic heating, industrial activities, and motor vehicles, results in the formation of a layer of warm air lying over more densely industrialized areas (a local phenomenon termed a "heat island") and has an indirect effect on the global phenomenon known as the greenhouse effect.

Energy consumption, therefore, entails an environmental impact due to the impoverishment of resources and an impact due to the chemical emissions of the combustion processes at the base of the production of the energy consumed, and also produces an impact due to the thermal emission of these processes. Thus, it is possible to make a distinction between the chemical and thermal emissivity of energy resources.

2.3.2 Emission Phenomena and Environmental Effects

The distinction between the chemical and thermal emissivity of energy resources extends to all the forms of emissivity involved in an elementary process such as that shown in Figure 2.3. It can be said, therefore, that:

- The direct emissivity of a process consists of all the chemical and thermal emissions characteristic of that process.
- The indirect emissivity of a process consists of all the chemical and thermal emissions correlated to the production of the resources used by the process, corresponding therefore to the emission content of the resources.
- The emission content of an energy resource consists of all the chemical and thermal emissions correlated with its production.
- The emission content of a material resource consists of all the chemical and thermal emissions correlated with its production (considering also the chemical and thermal emissions associable with the energy used in its production).

It is clear that direct and indirect emissivity so defined provide a solely quantitative indication of the emission phenomenon, without an evaluation of the effects of the different forms of emission (chemical and thermal) or of the different substances emitted. In order to obtain significant indications regarding this type of environmental impact, it is necessary to apply evaluation processes that elaborate the quantitative data in relation to some factors:

- The scale of the evaluation (local, regional, global)
- The type of environmental damage to be investigated

When the emission per unit of product is determined quantitatively based on these factors, these quantities are usually translated into a unit equivalent that can characterize the damage caused to the environment by the quantity of substances emitted. Some examples of environmental effects (on global and regional scales) and their more significant unit equivalents are:

- Greenhouse effect—CO_2-equiv (global scale)
- Hole in ozone layer—CFC11-equiv (global scale)
- Acid rain— SO_2-equiv (regional scale)
- Toxicity—H_2SO_4-equiv (regional scale)

The same distribution of emissions must then be repeatedly weighed by varying the scale of evaluation (global or regional) to determine which particular environmental effect is deemed appropriate to investigate. These questions are the specific object of study in Life Cycle Impact Assessment, which will be discussed in detail in Chapter 4.

2.4   Life Cycle Modeling

The need for a preliminary evaluation of the capacity of a product, process, or system to achieve its intended functionality requires the use of models of different typologies and varying complexity (input–output, dynamic, stochastic, etc.). In such models, the object to be represented is reduced to an abstraction, simplifying the functional mechanisms and limiting the information in play, with the aim of simulating its behavior and estimating a wide range of attributes (performance level, quality, reliability, cost, etc.).

In the present context, the product must be understood as a product–system characterized by flows of resources transformed through the various processes making up the life cycle and by interactions with the ecosphere. The model of the life cycle must then be a fundamentally physical model and must represent a system with accurately predefined boundaries. Everything that falls outside these boundaries constitutes the environment in which the system operates, and with which it interacts through flows of resources, energy, and information. The product–system can be further broken down into subsystems and elementary activities which interact according to a functional structure such that the functionality of the original system is achieved. With this general structure, "the task for life cycle modeling is to construct an appropriate framework in which the system architecture (hierarchy) and structure (connections) can be first represented and then evaluated consistently and rigorously" (Tipnis, 1998).

2.4.1 Approach to Environmental Performance

The often-noted complexity of the environmental question indicates that a complete evaluation of a product's performance requires a holistic vision of the product–system (i.e., that the product is understood as integral with all the phases of its life cycle, in relation to the environmental, social, and technological context). Only with this holistic approach is it possible to reveal the effects of choices made in the design and production planning phases.

Therefore, only an adequate modeling of a product's life cycle can constitute a valid instrument for prediction and planning. On the other hand, the modeling of a system generally tends to reduce its complexity, with a consequent loss of information. Such simplification becomes necessary in the case of environmental evaluations because of the elevated complexity of the real systems. This aspect is clearly demonstrated in the ISO 14040 series of standards (ISO 14040, 1997), which explicitly treat product–system modeling for the purposes of evaluating environmental impacts, suggesting some fundamental stratagems in the construction of the model:

- Breaking down the product–system into subsystems, in line with a perspective oriented toward the functionality of the system to be modeled
- Defining elementary units (unit processes) that perform specific functions and necessitate resource flows in input and produce flows in output

With this clearly physical–technical-based approach, the behavior of the model can be described and simulated using mathematical models of limited complexity, in that they refer to the analysis of a system with static, linear behavior. The complexity would be markedly greater if the life cycle were treated from a sociotechnological perspective, since in this case the system would be characterized by dynamic, nonlinear behavior.

Such considerations justify the choice of the physical–technological viewpoint in modeling the life cycle of the product–system, as is generally proposed in the literature (Graedel et al., 1993; Vigon et al., 1993; Keoleian and Menerey, 1993; Billatos and Basaly, 1997; Hundal, 2002).

2.4.2 Modeling by Elementary Function or Activity

In modeling the life cycle with these premises, the entire product–system is subdivided into elementary functions (Zust and Caduff, 1997; Hundal, 2002), also represented by activity models (Navin-Chandra, 1991) which summarize the elementary processes characterizing the main phases of the cycle.

In general terms, modeling by activity (Activity Modeling) consists of defining a set of single activities that make up a complex system. These activities can be the transformation, handling, generation, use, or disposal of material resources, energy, data, or information (Tipnis, 1998). Appropriate activity modeling first requires a clear definition of the primary objective that is to be attained using the model, and of the initial viewpoint from which the model will be developed. In fact, both of these factors are necessary in defining the boundaries of the system to be modeled and in structuring the model, which must be broken down into subsystems, sequences, operating units, and processes in relation to the aims and the viewpoint.

Having defined these factors on the basis of the environmental requisites, it is possible to apply activity modeling to the product–system in its life cycle. The reference activity model is therefore of the type shown in Figure 2.4, characterized by input flows of physical resources, by output flows, and by a possible input flow of information when there is a margin of choice in how the activity is performed.

For the input flows, given that they are physical resources and can consist of materials and forms of energy, it is possible to distinguish between resources produced by preceding activities and resources coming directly from the ecosphere. For the output flows, consisting of products of the activity, it is possible to distinguish between true main products, secondary byproducts, and various types of emissions into the ecosphere. Having defined the reference activity model, the product's life cycle is translated into a system model by the following procedure (Zust and Caduff, 1997):

- Define the boundaries of the system
- Identify the elementary processes and functionalities

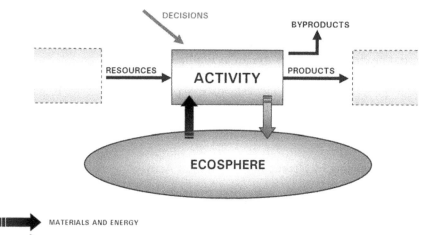

FIGURE 2.4   Reference activity model.

- Identify and quantify the connections between elementary activities
- Evaluate any possible changes in the activities and connections over time

Having developed the system model by activity, it is possible to perform simulations of the life cycle and to interpret the results.

2.4.3   Typologies of Activity Models

The reference activity model of Figure 2.4 can be read in different ways according to what kind of environmental evaluation is to be undertaken. As noted above, in terms of the physical–chemical exchanges of technological processes with the ecosphere, a product's environmental impact can be principally expressed in terms of:

- Consumption of material resources and saturation of waste disposal sites
- Consumption of energy resources and loss of energy content of products disposed of as waste
- Combined direct and indirect emissions of the entire product–system

In the first case, a quantification of the impact can be based only on an analysis of the distribution of the volumes of materials in play, in the context of the entire life cycle. On the other hand, the energy and emission aspects require the more complete approach proposed in Section 2.3, considering the energy and emission contents of the resources and final products. For a complete environmental analysis, therefore, the reference activity model can be read as in Figure 2.3. With the activity model represented in this way, it is possible to evaluate all the main environmental aspects, given that it identifies not only flows of materials but also those of energy and emissions in both their explicit and implicit forms.

In the case where the aim is, instead, to develop a life cycle model that supports only the analysis of the material resources in play, a more simplified reading of reference activity is possible, such as that represented in Figure 2.5.

This representation takes only the flows of material into account, considering as input the resources fueling the activity, and as output the product of the activity and any possible discards and waste. Regarding the input resources, however, it is necessary to make a distinction between:

- Primary or virgin resources, coming directly from the ecosphere
- Secondary or recycled resources

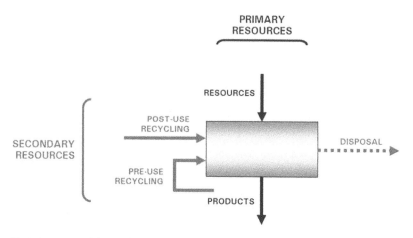

FIGURE 2.5   Activity model: Flows of material resources.

The latter can, in turn, be divided into:

- Preconsumption secondary resources (i.e., originating from discards and waste generated by the activity itself)
- Postconsumption secondary resources (i.e., originating from recycling the product after use and retirement)

2.5   Product Life Cycle: Reference Model

The various considerations noted in the preceding sections, particularly those concerning the concept of product life cycle, the appropriateness and the modality of considering the physical life cycle in environmental analysis, and the basic principles of modeling for elementary activities, are interpreted by the general life cycle model introduced below which, here, can be considered the reference model.

2.5.1   Main Phases of the Life Cycle

All the processes of transformation of resources involved in the product's entire physical life cycle can be grouped according to the following main phases (Manzini and Vezzoli, 1998; Sánchez, 1998):

- Preproduction, where materials and semifinished pieces are prepared for the production of components
- Production, involving the transformation of materials, production of components, product assembly, and finishing

- Distribution, comprising the packing and transport of the finished product
- Use (as well as the use of the product for its intended function) also includes any possible servicing operations
- Retirement (corresponding to the end of the product's useful life) can consist of various options, from product reuse to disposal as waste, depending on the possible recovery levels.

Each of these phases interacts with the ecosphere, since it is fueled by input flows of material and energy and produces not only byproducts or intermediate products that fuel the successive phase, but also emissions and waste (Figure 2.6).

2.5.1.1 Preproduction
The first phase, preproduction, consists of the production of materials and semifinished pieces required for the subsequent manufacture of components. In turn, the production of each finished material is divided into two main activities, as shown in Figure 2.7, where the activity modeling used is the simplified form for the representation of material resources only (Figure 2.5):

- Extraction—Extraction and collection of virgin materials
- Processing—Separation and refining of virgin materials, and subsequent physical and chemical processing necessary to obtain the finished materials

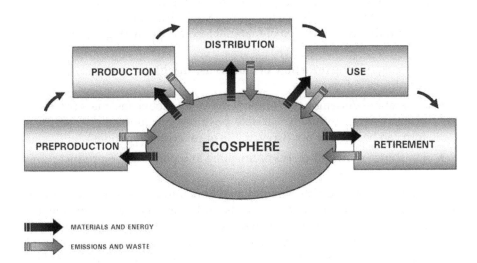

FIGURE 2.6 Main phases of physical life cycle and interaction with ecosphere.

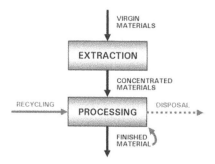

FIGURE 2.7 Phases of material production.

As shown in Figure 2.7, the resources employed in these processes are divided into primary resources, taken directly from the ecosphere, and secondary resources, coming both from discards and waste generated during the processing phase itself (preconsumption secondary resources), and from recycled materials obtained from the used product (postconsumption secondary resources). The production processes of the materials (in particular, the processing phase) also generate discards and waste that cannot be recovered and are therefore destined for disposal.

2.5.1.2   Production

The finished materials and semifinished pieces are used in the successive production phase, where it is possible to distinguish between three main activities, as in the model shown in Figure 2.8, which again refers to the simplified form for the representation of material resources only (Figure 2.5):

- Forming—Transformation of materials, machining processes, and forming of components
- Assembly—Assembly of components using mobile fasteners (union of elements in mutually variable positions), or junctions which in turn can be fixed or not (irreversible or reversible junctions)
- Finishing—Final processes of finishing and painting the product

The resources used are also differentiated here into primary and secondary. The latter can come from the discards and waste generated during the production processes themselves, particularly during the phase of component machining and forming (preconsumption secondary resources), or from the recovery of the used product (postconsumption secondary resources). The production phase, in particular the process of forming,

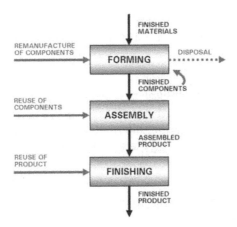

FIGURE 2.8 Phases of product manufacture.

generates discards and waste that cannot be recovered and are, therefore, destined for disposal as waste.

Finally, it should be noted that the manufacture of products usually requires a wide variety of materials. In a very broad, systematic view, it is possible to make a distinction between direct and indirect materials (Keoleian and Menerey, 1993); the former are those which, appropriately transformed and worked, constitute the final product, while the latter are those constituting the plants and equipment necessary for the manufacture of the product. This observation highlights the importance of defining a precise domain on which to perform the environmental analysis of industrial activities.

2.5.1.3 Distribution

Having manufactured the finished product, it must be distributed to be sold. The distribution phase entails packing (packaging) and transport (shipping). The resources necessary for this phase are principally those expended to obtain a packaging that will guarantee that the product is integral and functional when it reaches the user, and those resources relevant to the consumption associated with transport. Also in this case, however, in a broader systematic view, it is possible to take into account the use of resources for the production of the means of transport themselves and of the structures required for storing the product.

2.5.1.4 Use

The product is used for a certain period of time or, in some cases, is consumed. A product's phase of use often involves the consumption of material and energy resources for its operation, and produces waste and emissions.

Furthermore, during their use products can require servicing interventions such as maintenance, repair or substitution of worn components, and the upgrading of obsolete parts.

2.5.1.5   Retirement and Disposal

Once the product has been used, it reaches the phase of retirement, which can be structured according to various alternatives. Depending on the opportuneness and potential for recovering the resources employed, it is possible to:

- Regain the original functionality of the product, reusing it whole
- Reuse some components, either directly or after they have been reconditioned
- Exploit the resources used through processes of recycling materials or of energy recovery
- Eliminate all or part of the product in waste disposal sites

The first three options feed the flows of recovered resources, providing post-consumption secondary resources for use in the phases of preproduction and production.

2.5.2   Flows of Material Resources and Recovery Levels

By developing each main phase according to the different primary activities it encompasses, it is possible to obtain a vision of a product's entire physical life cycle and of the resource flows that characterize it, such as that of Figure 2.9, where the flows of material resources are shown according to the simplified activity model of Figure 2.5.

As noted above, the first phase of preproduction consists of the production of materials and semifinished pieces required for the subsequent production of components. Preproduction, therefore, includes the production phases of all the materials which will go to make up the final product. Once the product is manufactured, distributed, and used, it arrives at the final phase of retirement and disposal.

Dividing all these phases according to their primary activities, Figure 2.9 provides an overview of all the waste flows generated during the cycle, which in the model proposed are principally due to the phases of processing the various materials and of forming the components, together with, naturally, the disposal of the product.

Figure 2.9 also offers a complete picture of all the alternatives to disposing of the product as waste at the end of its life. It also shows how the recovery flows can be distributed within the same life cycle that generated them, providing the postconsumption secondary resources for various activities or, alternatively,

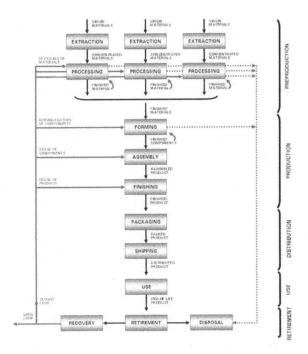

FIGURE 2.9 Complete physical life cycle of product and flows of material resources.

can be directed outside the cycle. In fact, it is necessary to make a distinction between the two different typologies of recycling flows (Vigon et al., 1993):

- Internal recycling (closed loop)—The resources recovered reenter the life cycle of the same product which generated the flows, replacing the input of virgin resources. This can occur by directly reusing the product at the end of its useful life, reusing some parts, reusing other parts after appropriate reprocessing (remanufacturing), or by recycling materials. From the viewpoint of the environmental consequences, these recovery processes lead to an increase in the expenditures and emissions for the treatment and possible transport of these volumes before they reenter the cycle. The recovery processes also lead to a decrease in the consumption of materials in general, due to the partial reduction in the input of virgin materials and a reduction in the volumes disposed of as waste.
- External recycling (open loop)—At the end of the product's life, some of its parts are directed to the production processes of other materials or products external to the cycle under examination. This can result in

recovering part of the energy content of materials to be eliminated, saving virgin materials in other production cycles, and obtaining financial benefits through the sale of materials for recycling.

2.6   Summary

For a complete analysis directed at evaluating and reducing the environmental impact of a product, it is necessary to consider, together with the phases of development and production, those phases of use, recovery, and treatment of the retired product. Furthermore, all these phases must be understood not only in relation to the specific actors involved, but also from a wider perspective, going beyond their direct competencies.

It is, therefore, possible to speak of a product–system wherein, in its most complete sense, the product is considered integral with its life cycle and within the environmental, technological, economic, and social context in which the life cycle develops. From the specific viewpoint of environmental analysis, this system is characterized by physical flows of resources transformed through the various processes making up the life cycle and by interactions with the ecosphere. The impact this product–system has on the environment is the result of life cycle processes that exchange substances, materials, and energy with the ecosphere.

2.7   References

Abernathy, W.J. and Utterback, J.M., Patterns of industrial innovation, *Technology Review,* 80(7), 40–47, 1978.

Alting, L., Life-cycle design of products: A new opportunity for manufacturing enterprises, in *Concurrent Engineering: Automation, Tools and Techniques,* Kusiak, A., Ed., John Wiley & Sons, New York, 1993, 1–17.

Alting, L. and Jorgensen, J., The life cycle concept as a basis for sustainable industrial production, *Annals of the CIRP,* 42(1), 163–167, 1993.

Billatos, S.B. and Basaly, N.A., *Green Technology and Design for the Environment,* Taylor & Francis, Washington, DC, 1997.

Cunningham, M.T., The application of product life cycles to corporate strategy: Some research findings, *European Journal of Marketing,* 3(1), 32–44, 1969.

Frosch, R.A., Manufactured products, in *Industrial Ecology, U.S.–Japan Perspectives,* Richards, D.J. and Fullerton, A.B., Eds., National Academy Press, Washington, DC, 1994, 28–36.

Gardner, D.M., Product life cycle: A critical look at the literature, in *Review of Marketing 1987,* Houston, M.J., Ed., American Marketing Association, Chicago, 1987, 162–194.

Graedel, T.E., Allenby, B.R., and Linhart, P., Implementing industrial ecology, *IEEE Technology and Society Magazine,* 12(1), 18–26, 1993.

Guinée, J.B., Udo de Haes, H.A., and Huppes, G., Quantitative life cycle assessment of products: Goal definition and inventory, *Journal of Cleaner Production*, 1(1), 3–13, 1993.

Heiskanen, E., The institutional logic of life cycle thinking, *Journal of Cleaner Production*, 10(5), 427–437, 2002.

Hundal, M.S., Introduction to design for the environment and life cycle engineering, in *Mechanical Life Cycle Handbook*, Hundal, M.S., Ed., Marcel Dekker, New York, 2002, 1–26.

ISO 14040, Environmental Management—Life Cycle Assessment—Principles and Framework, ISO 14040:1997(E), International Organization for Standardization, Geneva, 1997.

Jovane F. et al., A key issue in product life cycle: Disassembly, *Annals of the CIRP*, 42(2), 651–658, 1993.

Keoleian, G.A. and Menerey, D., Life Cycle Design Guidance Manual, EPA/600/R-92/226, U.S. Environmental Protection Agency, Office of Research and Development, Cincinnati, OH, 1993.

Massey, G.R., Product evolution: A Darwinian or a Lamarckian phenomenon?, *Journal of Product and Brand Management*, 8(4), 301–318, 1999.

Navin-Chandra, D., Design for environmentability, in *Proceedings of ASME Conference on Design Theory and Methodology*, Miami, FL, 1991, DE-31, 119–125.

Ryan, C. and Riggs, W.E., Redefining the product life cycle: The five-element product wave, *Business Horizons*, 39(5), 33–40, 1996.

Sánchez, J.M., The concept of product design life cycle, in *Handbook of Life Cycle Engineering: Concepts, Models and Technologies*, Molina, A., Sánchez, J.M., and Kusiak, A., Eds., Kluwer Academic Publisher, Dordrecht, The Netherlands, 1998, 399–412.

Sun, J. et al., Design for environment: Methodologies, tools, and implementation, *Journal of Integrated Design and Process Science*, 7(1), 59–75, 2003.

Tipnis, V.A., Evolving issues in product life cycle design: Design for sustainability, in *Handbook of Life Cycle Engineering: Concepts, Models and Technologies*, Molina, A., Sánchez, J.M., and Kusiak, A., Eds., Kluwer Academic Publisher, Dordrecht, The Netherlands, 1998, 413–459.

van de Ven, A.H. and Poole, M.S., Explaining development and change in organizations, *The Academy of Management Review*, 20(3), 510–540, 1995.

Vigon, B.W. et al., Life-Cycle Assessment: Inventory Guidelines and Principles, EPA/600/R-92/245, U.S. Environmental Protection Agency, Office of Research and Development, Cincinnati, OH, 1993.

Wenzel, H., Hauschild, M., and Alting, L., *Environmental Assessment of Products*, Vol. 1, Chapman & Hall, London, 1997.

Young, P., Byrne, G., and Cotterell, M., Manufacturing and the environment, *International Journal of Advanced Manufacturing Technology*, 13(7), 488–493, 1997.

Zust, R. and Caduff, G., Life-cycle modeling as an instrument for life-cycle engineering, *Annals of the CIRP*, 46(1), 351–354, 1997.

Chapter 3

Life Cycle Design and Management

The primary objective of design activity consists of translating an idea into a product and then incorporating the set of needs that this product must satisfy into detailed design. In the development of new products today, the designer must achieve this transformation while taking account of an ever-greater range of requisites, not solely functional (time to market, profitability, reliability, safety, recyclability), which arise in relation to the diverse life cycle phases the product must pass through. The principal difficulty inherent in a design intervention of this type lies in the fact that the most effective choices concerning a single requirement often conflict with other aspects of this wide-ranging problem.

The subject of this chapter is the extension of the life cycle approach, previously discussed, in the context of product design. Having drawn a general picture of Life Cycle Design, which fully interprets this approach, the main aspects linked to environmental protection are analyzed here, delineating how the complexity of the product design and development process necessitates an intervention at the management level of production activity.

3.1 Life Cycle Approach in Product Design

Design choices result in effects that can be propagated throughout a product's entire life cycle. They are evidenced by alterations in the behavior of the product itself with regard to the principal metrics of performance (costs, times, and quality) during the various phases of the cycle (Borg et al., 2000). This direct relation between design choices and the behavior of a product over its life cycle distinctly recalls what was said concerning the application of Life Cycle Theory to the management of product development (Section 2.1.2), with particular reference to concepts of product–entity and generating force.

According to this perspective, the life cycle of a product is understood in terms of an evolutionary sequence (event progression) of a linear type which, as evidenced in Figure 2.1, can be considered in two distinct parts. The first regards the product in its abstract dimension (the need is translated into product concept, and then into design), while the second concerns the product's physical life cycle (production, use, disposal).

Of the two parts, the first plays a determining role in the formation of the generating force (i.e., the mechanism generating and guiding the evolution of the product–entity over its life cycle). This first part, previously indicated as the development cycle, in fact consists of the process of product design and development, and de facto defines the generating force (i.e., it creates that kind of predefined program, implicit in the product itself, which will condition its behavior over its life cycle). The need which lies at the basis of the product idea interprets the role of the factor generating the product and its life cycle. Design requisites interpret the role of the product–entity's internal evolutionary factor since, through the process of design, they are translated into product properties. They condition its behavior over the entire life cycle and, therefore, guide the evolution of the product–entity in relation to the different environments in which it evolves. All this is summarized in Figure 3.1. Product design, which concerns the product's abstract dimension, creates the generating force through the definition of the needs the product must satisfy, and the realization of the properties the product acquires as a direct consequence of the design choices. Created and accumulated during the abstract part of the life cycle, the generating force is subsequently manifested in the product's physical life cycle through its properties, which condition its behavior in the phases of production, use, and disposal, directly related to the environment in which these phases develop (the sociotechnological context, the economic system, and the ecosystem).

This vision of the roles assumed by design requisites and choices, and by product properties in the relationship between product development and physical life cycle, constitute the full expression of the life cycle approach in

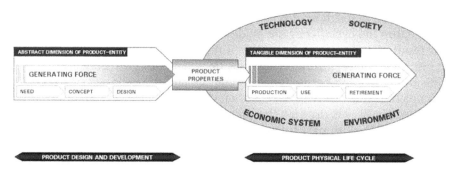

FIGURE 3.1   Life cycle approach in product design: The role of generating force.

product design. It is the basis of the decision-making design methodology known as Life Cycle Design.

3.1.1 Life Cycle Design

The original concept of Life Cycle Design can be traced back to the same limited vision of the life cycle discussed in Section 2.2, which arose from a statement conditioned by the direct competencies and interests of one actor—the manufacturer. It was, therefore, initially understood as an approach that considered "from the early product concept, the complete projected life of the product, including the product/market research, design phases, manufacturing process, qualification, reliability issues, and customer service/maintainability/supportability issues" (Keys, 1990).

The first attempts to outline a clear definition and methodological structure date back to the 1980s, when it was denoted by the expression "design for the life cycle" (Fabrycky, 1987), or "system life cycle engineering" (Keys, 1990). A more complete (and for the purposes of this book, more appropriate) orientation developed during the 1990s, interpreting the conceptual premises considered so far and incorporating their implications into product design. In particular, life cycle modeling took on a fundamental role.

A model of the product's life cycle is indispensable for evaluating the improvements obtained by choosing different materials, product configurations, and processes (Alting and Legarth, 1995). At the product design stage, modeling can be used in combination with design principles and construction guidelines to exert a direct influence on the efficiency of design solutions during their development, as well as providing an instrument to evaluate design choices (Zust and Caduff, 1997). Besides, the very concept of a product understood as integral with its life cycle and within the environmental, social, and technological context in which the cycle develops (i.e., the product–system discussed earlier) is no more than the theorization of a single system for the purposes of design (Keoleian and Menerey, 1993).

To what degree the life cycle approach is suited to the design activity is confirmed by the widely held view that the phase of product design and development must be considered the first phase of the cycle, thus becoming an integral part of the product–system (Alting, 1993; Tomiyama et al., 1997; Tipnis, 1998; Hundal, 2002).

Life Cycle Design (LCD), which is the application of the life cycle concept to the design phase of the product development process, denotes a design intervention which takes into consideration all the phases of a product's life cycle (development, production, distribution, use, maintenance, disposal, and recovery) in the context of the entire design process, from concept definition to detailed design development (Kusiak, 1993). It uses design models, methodologies, and tools to reconcile the evolution of the product, from conception to retirement, with a wide range of design requirements related to different phases of the life cycle (e.g., ease of production, functional performance, maintainability, and environmental impact) (Ishii, 1995).

From the very beginning, a term widely used in this field is Life Cycle Engineering (LCE). Although it would be more correct to apply this to an approach to the life cycle of all the functions of engineering, not solely that of design, LCE and LCD are generally used without distinction (Zhang et al., 1995). More recently, LCE has been assimilated to the wider concept of "a decision-making methodology that considers performance, environmental and cost requirements for the duration of a product" (Wanyama et al., 2003).

As a design approach, LCD is characterized by three main aspects:

- The perspective is broadened to include the entire life cycle.
- The assumption is that the most effective interventions are those made in the first phases of design.
- There is simultaneity in the operations of analysis and synthesis on the various aspects of the design problem.

This last characteristic merges LCD with design techniques oriented toward the control and compression of the times and costs of product development found in the field of Concurrent Engineering, which will be discussed later.

As proposed by other authors (Alting, 1993), the concept of LCD can be summarized by the schematic representation shown in Figure 3.2. Having identified the main phases of a product's life cycle as need recognition and design development (development cycle) and preproduction, production, distribution, use, and retirement (physical cycle), all these phases must be considered, beginning with the definition of product concept, since this represents the most effective level of intervention where the evolution of the design idea has the least economic impact. The selection of design alternatives must be guided by considering the main factors of product success, which define the design targets, in relation to all phases of the life cycle:

- Resources utilization (optimization of the materials and energy use)
- Manufacturing planning (optimization of the production processes)
- Life cycle cost (optimization of the total cost of life cycle)
- Product properties (harmonizing a wide range of required product properties, such as ease of production, functionality, safety, quality, reliability, aesthetics)
- Company policies (respect for the common company position and objectives)
- Environmental protection (control and minimization of environmental impacts)

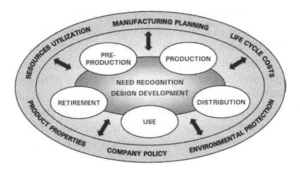

FIGURE 3.2 Life Cycle Design: Schematization of the concept. (Adapted from Alting, L., Life-cycle design of products: A new opportunity for manufacturing enterprises, in *Concurrent Engineering: Automation, Tools and Techniques,* Kusiak, A., Ed., John Wiley & Sons, New York, 1993, 4. With permission.)

These premises have led to an evolution of the very concept of engineering design. It can no longer be limited to the transformation of a need into the detailed description of a product satisfying this need, but must also guarantee that this transformation is achieved while taking into consideration a vast range of physical and functional requisites associated with the different phases of the product's life cycle (Asiedu and Gu, 1998).

The environmental issue is generally only one of the aspects examined by LCD. Nevertheless, a design approach of this kind, which simultaneously a wide range of design parameters and product development costs and relates them to specific requisites of the various phases of the entire life cycle, can be the most suitable for an effective implementation of the criteria of Design for Environment (DFE).

3.2 Life Cycle Design Oriented toward Environmental Performance of Products

The thread that ties together the various studies on design for environmental protection is the common objective of reducing the environmental impacts of products during their entire life cycle, from design to disposal (Coulter et al., 1995). From this perspective, LCD becomes a systematic "cradle to grave" approach able to "provide the most complete environmental profile of goods and services" (Keoleian and Menerey, 1993).

In the schematic representation of LCD proposed in Figure 3.2, environmental protection is seen as one of the main objectives of the design intervention (which must consider all the attributes required by a new product, such as functionality, time to market, profitability, reliability, safety, quality, etc.). In the literature, however, the emphasis is frequently placed on the environmental aspect, so that this becomes the primary criterion of LCD (Alting and Legarth, 1995; Stuart, 1998; Wanyama et al., 2003). LCD then loses its generality and becomes an approach specifically oriented toward the environmental performance of products. Using this approach, it is possible to (Cooper and Vigon, 2001):

- Determine the relations between environmental requirements and engineering requisites, and to evaluate the environmental implications of design alternatives
- Identify the opportunities for improving the environmental performance of a product throughout its entire life cycle

From this viewpoint, LCD and DFE are often assimilated into a single concept, as though environmental requirements are the sole objective of LCD. Although this is reductive and it is more correct to consider DFE as part of LCD, in terms of this book it is appropriate to emphasize the environmental aspect. The intention here is to identify effective methods and instruments for LCD directed at a product's environmental performance. This can be called Life Cycle Design for Environment (LCDFE) and defined as an approach to design based on a systematic view of the life cycle which, integrating the environmental requirements in every phase of the design process, seeks the reduction of a product's overall impact to make it as ecologically and economically sustainable as possible.

From this perspective, the general schematization of LCD in Figure 3.2 can be interpreted by highlighting the environmental aspect of each of the factors of product success, even if they are not specifically oriented toward environmental protection:

- The use of resources must be planned in a manner that takes into account the environmental efficiency of the distribution of all the resource flows in the entire life cycle.
- The production system must be organized to take into account the environmental efficiency of the production cycle, minimizing discards and waste and optimizing the energy efficiency of processes.
- The costs of the life cycle must include the environmental costs associable with the phases of production, use and disposal, and also

with environmental externalities (this will be discussedfurther in Chapter 5).

- The properties of the product must feature the environmental aspects linked to the life cycle, without neglecting the mandatory functional performance and economic aspects.
- Company policies must include environmental criteria, in line with the growing sensitivity of markets and government policies with regard to environmental protection and the issue of sustainable development.

While intensifying the effort for environmental protection, LCD must always maintain its primary objective of enabling an integrated design which takes into account a vast range of design requisites, not exclusively environmental. This is an ineluctable premise for what must constitute the final result of a truly effective intervention—the manufacture of a product whose life cycle can be considered sustainable in the widest sense (i.e., optimizing the distribution of resources, minimizing emissions and waste, maintaining adequate performance standards, and guaranteeing the product's economic sustainability).

3.2.1 Characteristics, Objectives, and Approach

Having developed the life cycle activity model (with reference to the activity model in Figure 2.4 of Chapter 2), the main objective of the design intervention can be summarized as "[D]etermine the shapes and arrangement of material that will provide the optimum flow of energy, materials, and information to fulfill the desired requirements" (Hundal, 2002). The purpose of LCDFE is the optimization of the product's environmental performance over its entire life cycle. The approach is that of operating on the product–system so that the input of resources and the impacts of all emissions and waste are reduced to a minimum in both quantitative and qualitative terms.

A full summary of the complex system of factors and requisites involved in a process of LCDFE has been presented by G.A. Keoleian and D. Menerey, who proposed the scheme shown in Figure 3.3 (Keoleian and Menerey, 1993). The fundamental objective is to develop an environmentally sustainable product, and this objective must be clear to and accepted by all members of the design team. Furthermore, the design process is significantly conditioned by a series of internal and external factors that must be appropriately defined and managed from the outset:

- Internal factors include company policies, product strategies, and the resources available during the product development process.

FIGURE 3.3 Process of Life Cycle Design oriented toward environmental performance. (Adapted from Keoleian, G.A. and Menerey, D., Life Cycle Design Guidance Manual, EPA/600/R-92/226, U.S. Environmental Protection Agency, Office of Research and Development, Cincinnati, OH, 1993, 23.)

- External factors include government policies, legislative restrictions, customer demand, and the technological level of society.

Under the influence of these factors, the process of development is divided into the following phases: analysis of needs, definition of requirements, identification of optimal solution, and implementation of the design idea in product manufacture. However, some characteristics of the LCDFE process clearly differentiate it from a conventional design process:

- The ascendancy of the analysis of design requirements and of intervention to optimize performance is extended over the entire life cycle of the product. This aspect has a strong influence on the needs analysis and definition of requirements phases.
- The assumption is that the most effective interventions are those made in the first phases of design. As shown in Figure 3.3, design strategies are applied throughout the development process, starting from the earliest phases.
- The simultaneity of interventions of analysis and synthesis of the various aspects of the design problem is evidenced by the continuous interaction between evaluation of environmental performance and design intervention in all the phases of the process, with the aim of guaranteeing a constant improvement in the final result.

The design process begins, therefore, with the needs analysis phase where the following are defined:

- Significant needs, on the basis of specific customer requirements and market demand, and on the interaction between the needs of the consumer and those of society, from the viewpoint of production sustainability.
- Scope and purpose of the project, including the boundaries of the system under examination, the analysis techniques to be used, the timing of the project, and the budget. In particular, the boundaries of the field of analysis and evaluation of results can include the entire life cycle or only parts of it, with subsequent differences in the efficacy of the design action in meeting environmental goals.

These first choices are the basis of the entire design process, since they significantly condition the set of design factors and variables identified and operated on by the designers. The next phase, requirements, is considered one of the most important and has dual aims:

- Introduce environmental requisites into the design process as early as possible.
- Integrate environmental needs into the set of design criteria, which include the conventional requirements (performance, cost, regulations).

This phase identifies the results to be obtained and aids the designers in interpreting the requisites of the final product. It also makes it possible to structure the design intervention in order to reconcile all the requirements as effectively as possible and to obtain a successful, functional, eco-compatible, and economically sustainable product. This result can only be achieved through an accurate definition of the vast range of design requirements and effective management of conflicting demands.

Proceeding with the conventional design phases (conceptual development, embodiment design, and detail design), the requirements are translated into coherent design solutions. Finally, having identified the optimal solution, the development process ends with implementation, where the product is manufactured and marketed.

Requirements are translated into solutions through the application of design strategies. As shown in Figure 3.3, these are generally distributed over the entire process and therefore represent the moment when the designer actively intervenes in product development. A wide range of design strategies can be used to achieve the environmental requirements. These strategies can be grouped in several typologies: extension of the life of product and materials, choice of materials, organization of processes, and so on. The effectiveness of the design strategies applied in product development must be quantified in an evaluation phase, using the various tools discussed in Section 3.2.3 to analyze the environmental performance of the life cycle of the product under development.

Finally, it should be noted that both the application of design strategies and the evaluation of the results must constitute the means for a process of continuous improvement of the final product, exploiting the ever-increasing knowledge base provided by research and development and the continuous monitoring of the state of the environment. A systematic process of product evolution, based on the reiteration of interventions revising the design solutions, can make it easier to achieve the environmental objectives while respecting the other design requirements and constraints (Coulter and Bras, 1997).

3.2.2   Guidelines for Design

To provide a clearer view of how the methodological premises described above operate in design practice, Table 3.1 lists some guidelines for design, summarizing the more significant trends given in the core literature (OTA, 1992; Keoleian and Menerey, 1993; Graedel and Allenby, 1995; Fiksel, 1996; Billatos and Basaly, 1997). To reflect the characteristic LCD approach, the guidelines are grouped according to the phases of the life cycle where design choices can determine the most significant improvements in environmental performance—product manufacture, product use, and retirement and disposal. The first group includes some design criteria whose ameliorative effect can be considered transversal to the various phases of the cycle. The specific environmental effects of each guideline are also shown in the table.

Some of these guidelines will be considered subsequently, when the more relevant strategies of the environmental performance approach are discussed in greater detail. However, it is worth emphasizing here that, while the formulation of design criteria has the advantage of clarifying in practical terms what problems are to be considered in the design development, these criteria rarely have a general validity. In some cases, in fact, a greater environmental benefit can require interventions that are not in harmony with prescribed design criteria (Hauschild et al., 1999). Furthermore, although the guidelines may be formulated as generically as possible, they are often conflicting (Luttropp and Karlsson, 2001). Their use in design practice can become truly efficient only if integrated into a more complete approach that also makes use of other instruments of DFE (Glazebrook et al., 2000).

3.2.3 Tools to Evaluate Environmental Performance of Products

In the description of the process of LCD directed at environmental performance summarized in Figure 3.3, the importance of the evaluation of results is highlighted, noting that this evaluation must be continuous and distributed throughout all the phases of the design process. To evaluate design alternatives and identify the one that best satisfies the environmental requirements, it is necessary to use suitable tools able to quantify the performance of the product under development.

The problem of identifying appropriate metrics to evaluate the general performance of a product over its life cycle is well-known. This stems from the observation that a converging approach to design generally produces conflicts between different aspects, making it difficult to achieve a well-balanced trade-off between the different performances that final product must display in relation to the different requisites demanded of it (Prasad, 2000).

In the specific case of evaluating environmental performance, exhaustive metrics must take into account some fundamental aspects: reduction of the consumption of nonrenewable resources, efficient management of renewable resources, and reduction of polluting emissions (Rivera-Becerra and Lin, 1999). A clear definition of "environmental metric" was proposed previously (Chapter 1, Section 1.4.1) and is restated here: Environmental metrics can generally be considered the algorithmic interpretations of levels of performance within an environmental criterion (i.e., an attribute of the product found to be significant in determining the environmental performance of alternative product designs) (Veroutis and Fava, 1996).

In current design practice, it is possible to identify various methods and tools available for use in evaluating the environmental results of a design. These fall into three main categories:

- Tools allowing a complete environmental analysis of the whole product–system, giving an evaluation of the performance in relation to the entire life cycle
- Tools allowing a broad environmental evaluation of the product on the basis of limited qualitative and quantitative information
- Tools evaluating the product's performance with respect to specific aspects, using opportunely defined metrics and indicators

The first category includes tools based on the analysis of the life cycle (Life Cycle Assessment—LCA), the most common and complete method for evaluating the environmental impacts of processes and products. Its potential with regard to LCD (or LCE) is evident: "Life Cycle Engineering is the art of designing the product life cycle through choices about product concept,

TABLE 3.1 Guidelines for Life Cycle Design

LIFE CYCLE PHASE	GUIDELINES	ENVIRONMENTAL EFFECTS
Transversal	Avoid or minimize the use of polluting, toxic, or environmentally dangerous materials	Decreasing polluting and/or toxic emissions during production and subsequent phases of life cycle
	Avoid high-energy-cost materials	Decreasing energy content of product
	Use renewable, recyclable, or recycled materials	Decreasing consumption of nonrenewable materials
	Conceive and design products in a way that reduces the use of production materials, avoid oversizing	Decreasing consumption of material resources
Product Manufacture	Design the product and production processes to minimize discards and waste	Decreasing volumes of material lost during manufacturing phases
	Recycle discards and waste from machining within the production process or externally	Decreasing consumption of virgin materials and volumes of waste from production processes
	Limit the use of highly energy–intensive machining processes	Reducing energy expenditure in product manufacture
	Optimize and integrate the distribution of energy flows over the entire production system so as to recover and use energy flows dispersed in machining processes	Increasing energy efficiency of production system
Product Use	Minimize energy consumption during use (select components with low energy consumption, ensure efficient thermal insulation)	Reducing energy consumption during useful life
	Optimize the duration of product life by increasing reliability and durability	Extending useful life of product, consequently reducing resources used in production of new products as substitutes
	Design for easy maintenance and repair (give instructions for opening and access to components for cleaning or repair, identify parts for maintenance, facilitate disassembly and substitution of critical parts)	Extending useful life of product

	Design modular structures	Facilitating disassembly of components for substitution or upgrading, consequently extending useful life
	Avoid designing products whose possible useful life is longer than that required	Reducing retirement of products still functionally efficient
	Conceive the product as being able to adapt to changes in required performance	Extending useful life of product
	Favor product reuse through design to avoid rapid obsolescence	Extending useful life of product, thus reducing manufacture of new products
Retirement and Disposal	Favor the reuse of product or components through the use of modular structures, standardized, easily uncoupled junctions, and give instructions for nondestructive disassembly	Extending life of product or components, consequently reducing manufacture of new components
	Favor the recycling of materials through the use of recyclable materials and those compatible with each other for recycling purposes, and reduce the number of materials used	Extending life of materials and decreasing consumption of virgin materials
	Provide for safe incineration, concentrating toxic materials and facilitating their rapid removal	Reducing polluting emissions in incineration processes

structure, materials and processes, and Life Cycle Assessment is the tool that visualizes the environmental and resource consequences of these choices" (Alting and Legarth, 1995). LCA will be considered in detail in Chapter 4.

The second category includes tools developed to meet a precise need, that of providing a rapid (though not detailed) evaluation of a product's impact, in the first phases of the design process. A complete LCA is not suited to this goal, since it requires a complex and expensive (in both time and money) process of analysis and evaluation, and also requires the acquisition of a considerable quantity of data that are not available during the early phases of design development. Numerous tools already exist in this category. Some, based on methods and metrics to evaluate environmental performance, were developed independently of LCA (Coogan, 1993; Horvath et al., 1995; Brezet and van Hemel, 1997; Hui et al., 2002; Zhou and Schoenung, 2004). Others were directly derived from LCA procedures, appropriately simplified, and can be either of a quantitative or qualitative type. These will be considered in detail in Chapter 4 where tools for Simplified LCA will be introduced. In all cases, the tools in this category can be used to evaluate the weak points of design proposals and quantify the fulfillment of environmental require-ments. They are also based on different metrics, which can be developed and integrated to allow more structured and complete evaluations (Muller et al., 2000; Takata et al., 2003).

Finally, the third category includes tools able to evaluate only specific prod-uct characteristics, using metrics that quantify only some aspects of perfor-mance in relation to diverse requirements considered singly, such as the recyclable fraction of a product, its energy content, the productivity of the resources used, and the flows of materials in the life cycle (Navin-Chandra, 1991; Watanabe et al., 1995; Giudice et al., 1999; Bailey et al., 2001). In their most evolved forms, these tools can include a wide range of evaluation crite-ria, providing information on the eco-efficiency not only of the product but also of the entire production system (Mosovsky et al., 2000; Hur et al., 2004). These tools can also integrate different metrics which, while not exclusively or specifically environmental, also take into account economic factors and product quality (Lye et al., 2001; Liu and Chen, 2003).

Finally, some metrics are specific to the orientation of the design interven-tion, such as the product's properties of recyclability and remanufacturabil-ity (Simon and Dowie, 1993; Bras and Hammond, 1996; Hiroshige et al., 2001), or its property of serviceability (Gershenson and Ishii, 1993; Dewhurst and Abbatiello, 1996); these metrics are included in the category of evalua-tion tools for the methodologies of Design for X, which will be treated in Chapter 7.

The advisability of implementing environmental evaluation metrics and methodologies in computer-aided tools is clear. It is, in fact, the best way to disseminate environmental protection and integrate it within the context of the process of product design and development, where the use of computer programs is increasingly common. This is also necessary in order to confront the growing emphasis on the reduction of development times and costs and the continual extension of functional requisites, leaving the designer with few resources to tackle environmental problems effectively within the wider context of an LCD intervention (Mani et al., 2001). The literature contains several useful reviews of software tools developed in relation to some issues of DFE, to be used within the more general context of LCD (de Caluwe, 1997; Vezzoli, 1999; Wanyama et al., 2003).

3.2.4 Life Cycle Simulation

As described in Section 3.2.1, the reference scheme for the LCD process shown in Figure 3.3 envisions an evolution of the product based on iterations of interventions revising the design solutions, which are in turn guided by a continuous process of evaluating the results. The alternative to this evolutionary approach (based on the feedback of information flows used to improve the design intervention) consists of predicting the consequences of design choices on the product's life cycle during the solution synthesis phase itself. This is equivalent to simulating the life cycle of the product, varying the design choices and taking account of various facets of the environment in which the life cycle unfolds.

The crucial question is how, during the earliest phases of the design process, to evaluate the consequences that design choices will have on the product's entire life cycle (life cycle consequences). In this regard, the need to resort to a "life cycle knowledge-based" approach has been emphasized (Xue et al., 1999), where the authors describe an approach that incorporates and elaborates information on a wide range of product characteristics describing it in relation to its life cycle, to add to those conventional characteristics describing it in physical–geometric terms. In this sense, it is interesting to note the proposal to aid the life cycle design of components through an extension of the now commonly used "product feature-based" approach, considering a new, analogous "life-phase feature-based" aspect (Borg and Giannini, 2003).

With regard to specifically environmental issues, life cycle simulation can draw on some of the premises described above. As was observed in Chapter 2, modeling the product–system in order to evaluate its environmental impact is based on breaking down the system into activities requiring flows of resources in input and output. The model of flows of material resources shown in Figure 2.9 interprets this approach, evidencing the potential of different recovery levels within the life cycle. To evaluate the environmental behavior of this system of elementary processes and resource flows, and therefore of the product–system it represents, it is necessary to describe the circulation of resources including the recovery flows. This can be realized through a simulation of the life cycle, understood as a simulation of the behavior of the product, its parts, and its constituent materials in relation to the distribution and circulation of the flows of material resources. Relating the product's behavior to its constructional characteristics, life cycle simulation becomes the most effective instrument for LCD because it already allows, in the design phase, a prediction of the product's performance as a function of the main design choices. To this end, a simulation system must be developed according to the following requirements:

- It must be able to represent in detail the entire system of circulating flows of resources.

- It must be based on a modeling of the product that allows the complete characterization of the design choices and of the consequent properties of the product itself.

- The product and life cycle models must interface appropriately, allowing direct correlation between design variables and product behavior in the different environments where it will be marketed and used.

- The performance behavior of the whole system and its parts must be modeled, translating the constructional characteristics (dependent on design choices) into potential behavior in the phases of product use and recovery.

In recent years, several significant studies have already confirmed the validity of life cycle simulation approaches in LCD (Tomiyama et al., 1997; Kimura et al., 1998; Hata et al., 2000; Sakai et al., 2001; Takata et al., 2003), and another by the authors (Giudice et al., 2004) will be proposed in this book (Chapter 16).

3.3   Life Cycle Management

Managing the complexity of a process of product design and development, and of the consequent wide range of performances required (functionality, producibility, reliability, maintainability, quality, recyclability, cost), constitutes the principal obstacle to the diffusion of the life cycle approach in the context of manufacturing reality. Compensating for this problem, recent years have seen an increasing acceptance of the concept of Life Cycle Management (LCM). Although this term assumes different connotations according to the context in which it is used, it is appropriate to consider it here as an approach to the management of the production activity which, taking into consideration the product's entire life cycle in the sense fully detailed previously, has the objective of optimizing the interaction between product design and the activities correlated with the various phases of the cycle (Westkamper et al., 2001).

Similar to LCD, LCM can be extended to consider the whole range of product requirements, or can highlight the environmental and socioeconomic aspect of product management. In the first case, it can be understood as a broader and more flexible life cycle–oriented structure for planning, design, and decision making, explicitly including costs and other basic company metrics, together with environmental factors, quality, safety, and other technological aspects (Fiksel, 1996). In the second case, it is understood in the sense of systematic, life cycle–based, sustainable product management, and can be considered "a framework for improving organizations and their respective goods and services in a sustainable direction," with the aim of "minimizing the environmental and socio-economic burdens associated with a product over its entire life cycle and value chain" (Astrup Jensen, 2003). LCM is intended, therefore, to be an additional structure to complement a preexisting management organization, and has the objective of regulating the interface between different aspects (financial, technological, and environmental) of production activity, all closely tied to the goals of competitiveness and sustainability (Hunkeler and Rebitzer, 2003).

In order to translate the principles of sustainable development in company practice, making environmental sustainability accessible, quantifiable, and operative, LCM can make use of the diverse tools developed for the life cycle approach. As well as LCD, these include the environmental and economic life cycle analysis tools of Life Cycle Assessment and Life Cycle Costing, which will be considered in Chapters 4 and 5. In particular, the most recent developments regarding LCM are considered in Chapter 6.

3.4 Summary

Life Cycle Design is the application of life cycle concepts to the design phases of the product development process. It denotes a design intervention that takes into consideration all the phases of a product's life cycle (development, production, distribution, use, maintenance, disposal, and recovery) in the context of the entire design process, from concept definition to detailed design development. Life Cycle Management, on the other hand, operates at the level of production activity management, with the objective of optimizing the interaction between product design and the activities correlated with the various phases of the cycle.

Emphasizing the environmental aspect, Life Cycle Design can be understood as an approach to design based on a systematic view of the life cycle which, integrating the environmental requirements in every phase of the design process, seeks the reduction of a product's overall impact to make it as ecologically sustainable as possible. This means that the use of resources must be planned, taking into account the environmental efficiency of the distribution of all the resource flows in the entire life cycle; the production system must be organized to minimize the discards and waste and optimize the energy efficiency of processes; the product must be designed for optimum environmental performance in the phases of use and at end-of-life; and the costs of the life cycle must include all the environmental costs associable with the phases of production, use, and disposal.

While intensifying the effort for environmental protection, Life Cycle Design and Management must always maintain their primary objective of enabling integrated design and development that takes into account a vast range of design requisites, not exclusively environmental. This is an ineluctable premise for the development of a product whose life cycle can be considered sustainable in the widest sense, maintaining adequate performance standards, and guaranteeing the product's economic sustainability.

3.5   References

Alting, L., Life-cycle design of products: A new opportunity for manufacturing enterprises, in *Concurrent Engineering: Automation, Tools and Techniques,* Kusiak, A., Ed., John Wiley & Sons, New York, 1993, 1–17.

Alting, L. and Legarth, J.B., Life-cycle engineering and design, *Annals of the CIRP,* 44(2), 569–580, 1995.

Asiedu, Y. and Gu, P., Product life cycle cost analysis: State of the art review, *International Journal of Production Research,* 36(4), 883–908, 1998.

Astrup Jensen, A., Life cycle management: A bridge to more sustainable products, in *Proceedings of 8th International Conference on Environmental Science and Technology,* Lemnos Island, Greece, 2003, 39–43.

Bailey, R., Bras, B., and Allen, J., Measuring material cycling in industrial systems, in *Proceedings of IEEE International Symposium on Electronics and the Environment,* Denver, CO, 2001, 4–9.

Billatos, S.B. and Basaly, N.A., *Green Technology and Design for the Environment,* Taylor & Francis, Washington, DC, 1997.

Borg, J.C. and Giannini, F., Exploiting integrated "product" and "life-phase" features, in *Feature Based Product Life-Cycle Modelling,* Soenen, R. and Olling, G.J., Eds., Kluwer Academic Publisher, Dordrecht, The Netherlands, 2003, 1–18.

Borg, J.C., Yan, X.-T., and Juster, N.P., Exploring decisions' influence on life-cycle performance to aid "design for multi-X," *Artificial Intelligence for Engineering Design, Analysis and Manufacturing,* 14(2), 91–113, 2000.

Bras, B. and Hammond, R., Design for remanufacturing metrics, in *Proceedings of 1st International Workshop on Reuse,* Eindhoven, The Netherlands, 1996, 5–22.

Brezet, H. and van Hemel, C., *Ecodesign: A Promising Approach to Sustainable Production and Consumption,* UNEP United Nations Environment Programme, Paris, 1997.

Coogan, C.O., Front end environmental analysis, in *Proceedings of IEEE International Symposium on Electronics and the Environment,* Arlington, VA, 1993, 132–137.

Cooper, J.S. and Vigon, B., Life Cycle Engineering Guidelines, EPA/600/R-01/101, U.S. Environmental Protection Agency, National Risk Management Research Laboratory, Office of Research and Development, Cincinnati, OH, 2001.

Coulter, S. and Bras, B., Reducing environmental impact through systematic product evolution, *International Journal of Environmentally Conscious Design and Manufacturing*, 6(2), 1–10, 1997.

Coulter, S., Bras, B., and Foley, C., A lexicon of green engineering terms, in *Proceedings of ICED 95 10th International Conference on Engineering Design*, Prague, Czech Republic, 1995, 1–7.

de Caluwe, N., Ecotools Manual: A Comprehensive Review of Design for Environment Tools, Report DFE/TR33, Manchester Metropolitan University, Manchester, UK, 1997.

Dewhurst, P. and Abbatiello N., Design for serviceability, in *Design for X: Concurrent Engineering Imperatives*, Huang, G.Q., Ed., Chapman & Hall, London, 1996, 298–317.

Fabrycky, W.J., Designing for the life cycle, *Mechanical Engineering*, 109(1), 72–74, 1987.

Fiksel, J., *Design for the Environment: Creating Eco-Efficient Products and Processes*, McGraw Hill, New York, 1996.

Gershenson, J. and Ishii, K., Life-cycle serviceability design, in *Concurrent Engineering: Automation, Tools and Techniques*, Kusiak, A., Ed., John Wiley & Sons, New York, 1993, 363–384.

Giudice, F., La Rosa, G., and Risitano, A., Indicators for environmentally conscious product design, in *Proceedings of EcoDesign '99: 1st International Symposium on Environmentally Conscious Design and Inverse Manufacturing*, Tokyo, 1999, 71–76.

Giudice, F., La Rosa, G., and Risitano, A., Simulation of product life cycle: Methodological basis and analysis models, in *Proceeding of Design 2004—8th International Design Conference*, Dubrovnik, Croatia, 2004, 1527–1538.

Glazebrook, B., Coulon, R., and Abrassart, C., Towards a product life cycle design tool, in *Proceedings of IEEE International Symposium on Electronics and the Environment*, San Francisco, 2000, 81–85.

Graedel, T.E. and Allenby, B.R., *Industrial Ecology*, Prentice Hall, Englewood Cliffs, NJ, 1995.

Hata, T. et al., Product life cycle simulation with quality model, in *Proceedings of 7th CIRP International Seminar in Life Cycle Engineering*, Tokyo, 2000, 60–67.

Hauschild, M., Wenzel, H., and Alting, L., Life cycle design: A route to the sustainable industrial culture?, *Annals of the CIRP*, 48(1), 393–396, 1999.

Hiroshige, Y., Nishi, T., and Ohashi, T., Recyclability evaluation method (REM) and its applications, in *Proceedings of EcoDesign 2001: 2nd International Symposium on Environmentally Conscious Design and Inverse Manufacturing*, Tokyo, 2001, 315–320.

Horvath et al., Performance measurement for environmentally-conscious manufacturing, in *Proceedings of ASME Manufacturing Science and Engineering Symposium*, San Francisco, 1995, MED2–2, 855–860.

Hui, I.K. et al., An environmental impact scoring system for manufactured products, *International Journal of Advanced Manufacturing Technology*, 19, 302–312, 2002.

Hundal, M.S., Introduction to design for the environment and life cycle engineering, in *Mechanical Life Cycle Handbook*, Hundal, M.S., Ed., Marcel Dekker, New York, 2002, 1–26.

Hunkeler, D. and Rebitzer, G., Life cycle costing: Paving the road to sustainable development?, *International Journal of Life Cycle Assessment*, 8(2), 109–110, 2003.

Hur, T., Kim, I., and Yamamoto, R., Measurement of green productivity and its improvement, *Journal of Cleaner Production*, 12, 673–683, 2004.

Ishii, K., Life-cycle engineering design, *Journal of Mechanical Design*, 117, 42–47, 1995.

Keoleian, G.A. and Menerey, D., Life Cycle Design Guidance Manual, EPA/600/R-92/226, U.S. Environmental Protection Agency, Office of Research and Development, Cincinnati, OH, 1993.

Keys, L.K., System life cycle engineering and DF"X," *IEEE Transactions on Components, Hybrids, and Manufacturing Technology*, 13(1), 83–93, 1990.

Kimura, F., Hata, T., and Suzuki, H., Product quality evaluation based on behavior simulation of used product, *Annals of the CIRP*, 47(1), 119–122, 1998.

Kusiak, A., Preface, in *Concurrent Engineering: Automation, Tools and Techniques*, Kusiak, A., Ed., John Wiley & Sons, New York, 1993, ix-xvi.

Liu, C.C. and Chen, J.L., Inventive principles and evaluation method for eco-innovative design of products, in *Proceedings of EcoDesign 2003: 3rd International Symposium on Environmentally Conscious Design and Inverse Manufacturing*, Tokyo, 2003, 369–376.

Luttropp, C. and Karlsson, R., The conflict of contradictory environmental targets, in *Proceedings of EcoDesign 2001: 2nd International Symposium on Environmentally Conscious Design and Inverse Manufacturing*, Tokyo, 2001, 43–48.

Lye, S.W., Lee, S.G., and Khoo, M.K., A design methodology for the strategic assessment of a product's eco-efficiency, *International Journal of Production Research*, 39(11), 2453–2474, 2001.

Mani, V., Das, S., and Caudill, R., Disassembly complexity and recyclability analysis of new designs from CAD file data, in *Proceedings of IEEE International Symposium on Electronics and the Environment*, Denver, CO, 2001, 10–15.

Mosovsky, J., Dickinson, D., and Morabito, J., Creating competitive advantage through resource productivity, eco-efficiency, and sustainability in the supply chain, in *Proceedings of IEEE International Symposium on Electronics and the Environment*, San Francisco, 2000, 230–237.

Muller, A., Lave, L.B., and Hendrickson, C.T., Product-based environmental performance metrics: A methodology for selecting and supporting environmental metrics, in *Proceedings of IEEE International Symposium on Electronics and the Environment*, San Francisco, 2000, 93–98.

Navin-Chandra, D., Design for environmentability, in *Proceedings of ASME Conference on Design Theory and Methodology*, Miami, FL, 1991, DE-31, 119–125.

Office of the Technology Assessment (OTA), Green Products by Design: Choices for a Cleaner Environment, Report OTA-E-541, Office of the Technology Assessment, Congress of the United States, Washington, DC, 1992.

Prasad, B., Survey of life-cycle measures and metrics for concurrent product and process design, *Artificial Intelligence for Engineering Design, Analysis and Manufacturing*, 14(2), 163–176, 2000.

Rivera-Becerra, A. and Lin, L., Measuring environmental consciousness in product design and manufacturing, *Concurrent Engineering: Research and Applications,* 7(2), 123–138, 1999.

Sakai, N., Tomiyama, T., and Umeda, Y., Life cycle simulation with prediction of incoming flow of discarded products, in *Proceedings of 8th CIRP International Seminar in Life Cycle Engineering,* Varna, Bulgaria, 2001, 65–73.

Simon, M. and Dowie, T., Quantitative Assessment of Design Recyclability, Report DDR/TR8, Manchester Metropolitan University, Manchester, UK, 1993.

Stuart, J.A., Materials selection for life cycle design, in *Proceedings of IEEE International Symposium on Electronics and the Environment,* Oak Brook, IL, 1998, 151–157.

Takata, S. et al., Framework for systematic evaluation of life cycle strategy by means of life cycle simulation, in *Proceedings of EcoDesign 2003: 3rd International Symposium on Environmentally Conscious Design and Inverse Manufacturing,* Tokyo, 2003, 198–205.

Tipnis, V.A., Evolving issues in product life cycle design: Design for sustainability, in *Handbook of Life Cycle Engineering: Concepts, Models and Technologies,* Molina, A., Sánchez, J.M., and Kusiak, A., Eds., Kluwer Academic Publisher, Dordrecht, The Netherlands, 1998, 413–459.

Tomiyama, T., Umeda, Y., and Wallace, D.R., A holistic approach to life cycle design, in *Proceedings of 4th CIRP International Seminar on Life-Cycle Engineering,* Berlin, 1997, 92–103.

Veroutis, A.D. and Fava, J.A., Framework for the development of metrics for design for environment assessment of products, in *Proceedings of IEEE International Symposium on Electronics and the Environment,* Dallas, TX, 1996, 13–18.

Vezzoli, C., An overview of life cycle design and information technology tools, *Journal of Sustainable Product Design,* 9, 27–35, 1999.

Wanyama, W. et al., Life-cycle engineering: Issues, tools and research, *International Journal of Computer Integrated Manufacturing,* 16(4–5), 307–316, 2003.

Watanabe, S., Noguchi, T., and Nagata, S., Resource productivity: New measure for environmental soundness of industry, *International Journal of Environmentally Conscious Design and Manufacturing,* 4(1), 99–105, 1995.

Westkamper, E., Niemann, J., and Dauensteiner, A., Economic and ecological aspects in product life cycle evaluation, in *Proceedings of the I MECH E Part B—Journal of Engineering Manufacture,* 215(5), 673–681, 2001.

Xue, D., Yadav, S., and Norrie, D.H., Knowledge base and database representation for intelligent concurrent design, *Computer-Aided Design,* 31, 131–145, 1999.

Zhang, D., Kuo, T.C., and Zhang, H.C., Life-cycle engineering: Concepts and researches, *Journal of Manufacturing Science and Engineering,* MED2(2), 827–846, 1995.

Zhou, X. and Schoenung, J.M., Development of a hybrid environmental impact assessment model: A case study on computer displays, in *Proceedings of IEEE International Symposium on Electronics and the Environment,* Scottsdale, AZ, 2004, 91–96.

Zust, R. and Caduff, G., Life-cycle modeling as an instrument for life-cycle engineering, *Annals of the CIRP,* 46(1), 351–354, 1997.

Chapter 4

Life Cycle Assessment

The identification of design alternatives that best satisfy environmental demands requires the use of instruments able to quantify the environmental performance of the product under development and guide the ameliorative measures. Moreover, only a systematic vision of the product over its life cycle can ensure that these measures reduce the environmental criticalities and so avoid simply transferring impacts from one phase of the life cycle to another. Life Cycle Assessment (LCA) is an objective procedure used to evaluate the environmental impacts associated with a product's entire life cycle, through the quantitative determination of all the exchange flows between the product–system and the ecosphere in all the transformation processes involved, from the extraction of raw materials to their return into the ecosphere in the form of waste.

This chapter provides an overview of Life Cycle Assessment, tracing a general picture from its origins to the latest developments. Having described the premises, principal characteristics, and reference methodological structures, it will be possible to specify its fields of application and limitations and to indicate the tools most commonly used in practice.

4.1 Environmental Analysis and Evaluation of the Life Cycle

Environmental impact can be defined as "any change to the environment, whether adverse or beneficial, wholly or partially resulting from an organization's activities, products or services" (ISO 14001, 1996). One of the greatest difficulties in an attempt to reduce the negative impact that a generic activity has on the environment is that of evaluating this impact qualitatively or quantitatively, so as to be then able to undertake appropriate initiatives to contain it. Although the methodologies developed for the evaluation of environmental impact (Environmental Impact Assessment—EIA) are numerous (Jain et al., 1993) and differ in their identification, measurement and interpretation of the impact, they generally have some significant limitations in common:

- They were developed in relation to specific cases.
- They do not include the possibility of assessing the reliability and stability of the results.
- They do not start from the premise of the life cycle approach in line with the ideas expressed in previous chapters.

Only the methodology known as Life Cycle Assessment differs with regard to precisely these limitations, in that it overcomes them by an approach to the evaluation of environmental impact based on the premises already noted in Chapter 2. In fact, LCA is an objective procedure to evaluate the consumption of resources and the emission of waste relative to a generic industrial activity (see Section 2.4, Figures 2.4 and 2.6). It addresses the entire life cycle of the product of this activity through the quantitative determination of all the flows of exchange between the product–system and the ecosphere necessitated by all the transformation processes involved, from the extraction of raw materials to their return into the ecosphere in the form of waste.

4.1.1   Origins and Evolution

The first studies regarding the quantification of the consumption of virgin materials and the energy efficiency of production processes date back to the 1960s. They were greatly influenced by the results of some provisional models of development that described a scenario characterized by a rapid increase in the world population and limited natural resources, with the consequent impoverishment of fossil fuels and mineral resources, accompanied by climate change caused by excessive thermal emissions from the processes of energy transformation (NAS, 1969; Goldsmith et al., 1972). As a direct consequence, these studies were mainly oriented toward the evaluation of the costs, and environmental implications associated with conventional energy production and alternative sources of energy, and the efficiency with which the energy produced was used in industrial processes. While these types of evaluations were based on energy analyses, they nevertheless required the compilation of balance sheets on the flows describing the processes under examination. Since the latter could not be separated from the quantification of the consumption of raw materials and the generation of solid waste, these early energy studies were also important analyses of the flows of all the resources in play. One of the most significant of these first concrete testimonies to the new approach to environmental issues was conducted in the field of industrial chemistry (Smith, 1969).

The same period saw the first significant investigation regarding products of mass consumption, consisting of a study conducted by the Midwest Research Institute, and later by Franklin Associates, on behalf of the Coca-Cola Company on different containers for soft drinks (Vigon et al., 1993; Curran, 1996; Hunt and Franklin, 1996). This study is frequently cited as the first example of an inventory analysis of resources and waste undertaken in the United States, since it had the goal of quantifying the raw materials, fuels, and environmental charges correlated with the production processes of each type of container under consideration.

Stimulated by the oil crisis of the early 1970s, the process of quantifying the consumption of resources and of the flows returning to the environment due to the manufacturing of industrial products began to acquire a definite methodological structure, becoming known as "Resources and Environmental Profile Analysis" (REPA) in the United States of America (Hunt et al., 1992). A noted REPA study on drink containers conducted by the U.S. Environmental Protection Agency (Hunt et al., 1974) can be considered a typical forerunner of modern LCA. At the same time, analogous studies in Europe introduced the practice known as "Ecobalance" (Vigon et al., 1993; White and Wagner, 1996). These focused on the consumption of material and energy resources and on the generation of waste, in accord with the issues characterizing the environmental debate at a time when information on the emissions of processes, the polluting properties of substances and the potential effects on the environment was still limited.

From these first studies, the idea began to take shape that the only effective route for a complete study of industrial production systems from the viewpoint of their environmental consequences was to examine their operation quantitatively, following the path of the raw materials, beginning with their extraction, through all the transformation processes until their return to the ecosphere under the form of waste ("from cradle to grave").

The main stimulus leading to this approach was the newly acquired understanding that the traditional approach of concentrating on rendering single production processes more efficient, without considering the entire system of interrelated activities, was totally inadequate. This result is even more true precisely in relation to the environmental question, since an industrial activity, considered singly, may be made to seem cleaner and more efficient by simply transferring the pollution elsewhere: the environmental benefits derived are counteracted by new environmental criticalities generated at some other part of the same industrial system, without obtaining any real overall improvement.

Because the oil crisis was no longer acting as a stimulus during the late 1970s and early 1980s, attention passed from the saving of resources to the management of polluting waste. Nevertheless research continued into the environmental profiles of products and processes, above all regarding two themes: the energy efficiency of industrial processes (Boustead and Hancock, 1979; Brown et al., 1980); the production and use of packaging in general

(Bridgwater and Lidgren, 1983; Boustead and Lidgren, 1984) and, in particular, drink containers (Lundholm and Sundstrom, 1985). This last problem was particularly felt in Europe, to such an extent that as soon as a specific body of the European Commission was created to address environmental questions (Environment Directorate), attention was immediately centered on the question of the environmental consequences of the production and diffusion of containers. A specific directive promoted monitoring of the consumption of material and energy resources and the generation of solid wastes is associated with this product typology (EC Directive 85/339, 1985).

4.1.2   Introduction of Life Cycle Assessment and Concept Development

The definitive development of a common methodology for the environmental evaluation of products appeared at the end of the 1980s, and was characterized by two successive phases. Initially, the environmental issue of the flows of solid waste was emphasized, resulting in the field of interest being extended to recycling and waste disposal, completely fulfilling the life cycle approach in inventory analysis (Life Cycle Inventory—LCI). Subsequently, there was heightened interest in a transition from inventory analysis alone to a more complete evaluation of the consequences to the environment (Life Cycle Impact Assessment—LCIA).

These new stimuli led to the widespread understanding that there was a clear and immediate need for a common structure in the methods of evaluation and analysis developed until then. The conference organized by the Society for Environmental Toxicology and Chemistry (SETAC) in Vermont in August 1990 was a direct response to this need. It was here that the term "Life Cycle Assessment" was coined and defined as "an objective process to evaluate the environmental burdens associated with a product or activity by identifying and quantifying energy and materials used and wastes released to the environment, and to evaluate and implement opportunities to affect environmental improvements" (Fava et al., 1991). The life cycle approach was expressly highlighted: "The assessment includes the entire life cycle of the product, process, or activity, encompassing extracting and processing raw materials; manufacturing, transportation and distribution; use, reuse, maintenance; recycling and final disposal."

From then on, as shown in Figure 4.1, the conferences and workshops organized by SETAC became an international forum for the discussion of the methodological foundations and more specific issues of LCA. One of the subsequent meetings, the workshop held in Portugal in April 1993, is considered particularly important because it saw the first definition of common guidelines for conducting an LCA (Consoli et al., 1993).

Recognition of the validity and utility of this methodology led, finally, to international standardization through the publication, from 1997 on, of the

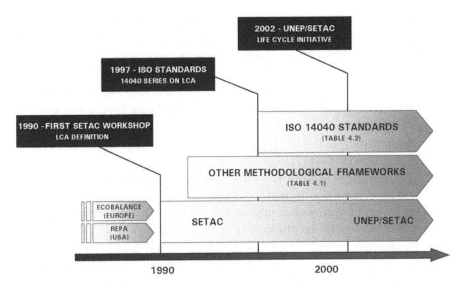

FIGURE 4.1 Evolution of LCA.

ISO 14040 series of norms (which are integrated into the greater corpus of ISO 14000 standards for Environmental Management Systems). The detailed definition of LCA evidences the implicit intention of the standardization to delineate a clear reference methodology considering it a technique for assessing the environmental impacts associated with a product by "compiling an inventory of relevant inputs and outputs of a product system; evaluating the potential environmental impacts associated with those inputs and outputs; interpreting the results of the inventory analysis and impact assessment phases in relation to the objectives of the study" (ISO 14040, 1997). As will be illustrated below, the group of ISO 14040 standards describes in detail the general criteria and underlying methodological framework for the main phases making up a complete LCA. Finally, attention should be drawn to one of the most recent important initiatives directed at the dissemination of LCA (Figure 4.1), that of the cooperation between SETAC and the United Nations Environmental Programme (UNEP), set up in 2002 under the name of UNEP/SETAC Life Cycle Initiative (Topfer, 2002).

4.2 Premises, Properties, and Framework of Life Cycle Assessment

The ISO 14040 series of standards are based on a preliminary structure that can be ascribed to that defined by SETAC in the early 1990s. The most basic definition of LCA is a concise summing up of the SETAC proposal, understood as the "compilation and evaluation of the inputs, outputs and the potential environ-mental impacts of a product system throughout its life cycle" (ISO 14040, 1997). The methodological structure is based on several premises:

- The analysis of the environmental interactions between the elemen-tary activities of the product–system starts from the "cradle to grave" perspective.
- The approach to the life cycle is holistic in the sense fully discussed in previous chapters.
- The analysis of the effects on the environment are based on a multi-criteria perspective, in that it evaluates the whole panorama of cate-gories of environmental impact and damage that may result from interactions with the product–system.
- The evaluation and comparison of activities are related to a functional unit and, therefore, are in line with the principle of equivalence for equal functionality produced (i.e., requiring the preliminary defini-tion of a reference functional unit, quantifying the unit of product–system performance).

4.2.1 Definition of Life Cycle and Product–System

The ISO standard definition of life cycle also fully reflects the SETAC outline and the concepts discussed in previous chapters. In fact, it defines the life cycle of a product as "consecutive and interlinked stages of a product system, from raw material acquisition or generation of natural resources to the final disposal." The product–system is understood to be the "collection of materi-ally and energetically connected unit processes which perform one or more defined functions" (ISO 14040, 1997).

Again, the subdivision of the main phases of the life cycle according to the conventional structuring of LCA (Fava et al., 1991) is substantially analogous to that described in Chapter 2:

- Raw Material Acquisition—Includes all the activities and processes required to obtain material and energy resources from the environ-ment, starting from the extraction of raw materials.
- Processing and Manufacturing—Includes all the activities and processes required to transform resources into the desired product.
- Distribution—Includes all the activities of transport, storage, and distribution that allow the product to arrive at the customer.
- Use, Maintenance, Repair—Includes the entire phase of product use, including all typologies of servicing operations.

- Recycle—Follows the phase of product use and includes all the recycling options, both internal (closed loop) and external (open loop) to the life cycle of origin.
- Waste Management—Concerns the nonrecyclable fraction of the product and consists of the management of final waste disposal.

Together, these phases constitute the complete system "from cradle to grave." However, it is possible to conduct partial LCA, defining one or more levels (gates) at which the complete system is interrupted so that LCA can be broken down into additional variants (Todd, 1996):

- Cradle to Gate—Analysis of the portion of life cycle upstream from the gate
- Gate to Grave—Analysis of the portion of life cycle downstream from the gate
- Gate to Gate—Analysis of the portion of life cycle between two gates

4.2.2 Methodological Framework of LCA

As was seen in Section 4.1, the first examples of LCA fundamentally consisted of the quantification of the material and energy resources in play and of the resulting solid waste. With respect to these first studies, the environmental requirements have changed, as has the level of completeness and refinement of the environmental analysis of the life cycle. This level has been raised considerably, starting from the first methodological structuring (Fava et al., 1991; Consoli et al., 1993) and subsequent developments (Vigon et al., 1993; Fava et al., 1993; EPA, 1995) to arrive at the final form defined by the ISO standards.

The main methodological frameworks predating the ISO standardization are summarized in Table 4.1 (see also Figure 4.1). A complete panorama of the present state of development of the LCA framework (according to the ISO standards, in particular with regard to the phases of Life Cycle Inventory and Life Cycle Impact Assessment) has recently been proposed (Rebitzer et al., 2004; Pennington et al., 2004) in studies evidencing the key issues still under debate and the main fields of practical application of LCA in its current state.

At the present time, a complete LCA is structured in four main stages:

- Goal and Scope Definition—The objectives of the analysis and the set of preliminary assumptions according to which it will be conducted are defined in this first phase. This requires the definition of the evaluation typology (aimed at system improvement or comparing alternative systems); the boundaries of the system under examination; the reference functional unit, assumptions, and parameters

TABLE 4.1   LCA methodological frameworks prior to ISO standards

METHODOLOGICAL FRAMEWORK	YEAR	REFERENCES
SETAC Society of Environmental Toxicology and Chemistry	1990	(Fava et al., 1991) (Consoli et al., 1993)
CML University of Leiden The Netherlands	1992	(Heijungs et al., 1992) (Guinée, 2002) (updated after ISO 14040)
EPS IVL Environmental Research Institute Sweden	1992	(Steen, 1996) (Steen, 1999) (updated after ISO 14040)
US EPA Environmental Protection Agency United States	1993	(Vigon et al., 1993) (EPA, 1995)
EIO-LCA Carnegie Mellon University United States	1995	(Cobas et al., 1995) (Hendrickson et al., 1998)
NORDIC GUIDELINES Nordic Council of Ministers Denmark	1995	(Lindfors et al., 1995)
UNEP United Nations Environment Programme	1996	(UNEP, 1996)
EDIP Technical University of Denmark and Danish Environmental Protection Agency Denmark	1997	(Wenzel et al., 1997)

for inventory and allocation operations; and the categories of impact to be considered.

- Life Cycle Inventory (LCI)—Includes the compilation and quantification of the inputs and outputs of the entire life cycle. The data can be obtained from various sources such as direct measurement as well as information from databases and the literature.

- Life Cycle Impact Assessment (LCIA)—Constitutes the phase of LCA where the inventory data are translated into potential environmental impacts, evaluating their size and significance. The conventional procedure consists of classifying the inventory flows (Classification) and characterizing them quantitatively in relation to different impact categories (Characterization). Impact categories include ozone depletion, acidification, eutrophication, climate change, depletion of resources, etc. In Classification, the inputs and outputs identified in LCI are assigned to impact categories. In Characterization, the impact potentials of each consumption or emission are calculated by multiplying the quantity consumed or emitted by the respective impact assessment factor (or characterization factor) relative to each impact category. The resulting impact data are then normalized (Normalization) and weighted (Weighting), to obtain single indices of environmental impact.

- Interpretation (according to ISO) or Improvement Analysis (according to SETAC)—In this final phase, the results of the LCA or LCIA are evaluated in relation to the planned objectives in order to formulate final considerations and directives for improvement.

For an overview of the parameters to be quantified in an inventory analysis, reference should be made to the considerations regarding the environmental impact of the product-system discussed in Section 2.3 (and, in particular, to Figure 2.3), where some introductory indications regarding the problem of evaluating environmental effects are also presented.

The organization into four main stages is common to both the first methodological framework of LCA proposed by SETAC and the standardization defined in the international ISO norms (Figure 4.2). The only significant differences can be found in the phase of evaluating the impacts (LCIA) and in the final phase (Interpretation or Improvement). Regarding the LCIA phase, the SETAC structure differs from the ISO standard in that:

- In the SETAC method, impact assessment is structured more rigidly, providing for not only Classification and Characterization, but also a final action of Valuation, where weighting procedures are applied to

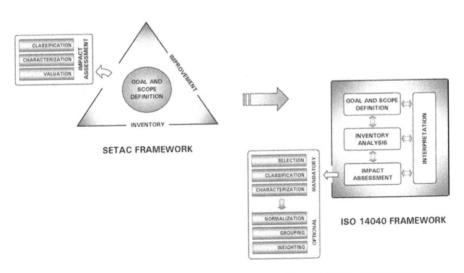

FIGURE 4.2   LCA framework and impact assessment according to SETAC and ISO 14040.

the different impact typologies with the aim of obtaining data for comparison or aggregation. These data are used as the basis for deciding on the actions to be taken (Fava et al., 1993).

- In the ISO standard, there is a sharp distinction between obligatory and optional procedures. The choice of the environmental effects to be taken into consideration (Selection), Classification, and Characterization are all obligatory. The optional procedures (Normalization, Grouping, and Weighting) concern the elaboration of the results of the Characterization phase to obtain concise indices that can be used to obtain an overall evaluation (ISO 14043, 2000).

With regard to the final phase, the SETAC structure differs from the ISO standard in that:

- In the SETAC method, this is indicated as Improvement Analysis and focuses attention on the possibility of formulating interventions to improve environmental performance (Fava et al., 1991).
- In the ISO standard, it is indicated as Interpretation and involves a more extensive intervention, including sensitivity analysis and assessment of the uncertainty of the results, and the formulation of final recommendations (ISO 14043, 2000).

4.2.3   Phases of LCA in ISO Standards

The ISO 14040 series of standards includes all the main phases comprising LCA. Table 4.2 shows the main headings of these standards and a summary of their specific contents. The lower section of the table also reports information on technical reports and preliminary documents belonging to the same group, but not yet made normative. This section presents a more detailed discussion of the standards published to date and held to be of particular importance in providing a more complete picture of the current scope and features of the methodological approach to LCA.

4.2.3.1   *ISO 14040:1997—Principles and Framework*

This standard focuses on some important preliminary questions:

- Goal and scope—The objectives and range of an LCA must be clearly defined and coherent with the intended application. An explicit statement of the purposes of the study is, therefore, essential.
- Functional unit—This is a reference unit of measurement used to treat and present the data and information of an LCA. According to the norm, a functional unit constitutes "a measure of the performance of

TABLE 4.2 ISO international standards and technical reports for LCA

DESIGNATION DOCUMENT TYPE YEAR	TITLE	CONTENTS
ISO 14040:1997 International Standard 1997	Environmental management: Life Cycle assessment— Principles and framework	General framework, principles, and requirements for conducting and reporting LCA studies
ISO 14041:1998 International Standard 1998	Environmental management: Life Cycle assessment— Goal and scope definition and inventory analysis	Requirements and procedures necessary for the compilation and preparation of the definition of goal and scope for LCA, and for performing, interpreting, and reporting a Life Cycle Inventory analysis (LCI)
ISO 14042:2000 International Standard 2000	Environmental management: Life Cycle assessment—Life cycle impact assessment	General framework for the Life Cycle Impact Assessment (LCIA) phase of LCA Key features and inherent limitations of LCIA Requirements for conducting the LCIA phase Relationship to the other LCA phases
ISO 14043:2000 International Standard 2000	Environmental management: Life Cycle assessment—Life cycle interpretation	Requirements and recommendations for conducting the Life Cycle Interpretation phase in LCA or LCI studies
ISO/TR 14047 Technical Report 2003	Environmental management: Life Cycle assessment— Examples of application of ISO 14042	Examples to illustrate practice in carrying out a LCIA according to ISO 14042
ISO/TS 14048: 2002 Technical Specification 2002	Environmental management: Life Cycle assessment— Data documentation format	Requirements and a structure for a data documentation format, to be used for transparent and unambiguous documentation and exchange of LCA and LCI data
ISO/TR 14049: 2000 Technical Report 2000	Environmental management: Life Cycle assessment— Examples of application of ISO 14041 to goal and scope definition and inventory analysis	Examples of practices in carrying out a LCI as a means of satisfying certain provisions of ISO 14041

the functional outputs of the product system." The main aim of the definition of the functional unit is to provide a reference unit to which input and output flows can be correlated.

- System boundaries—The definition of the confines is a delicate operation because these "determine which unit processes shall be included within the LCA." A first delimitation is obtained by taking into consideration the physical environments and production processes. It is then possible to exclude those components found to be largely irrelevant or difficult to represent in detail. It is clear, therefore, that the domain of application of an LCA is highly subjective and essentially depends on the intended depth of analysis. In all cases, "the criteria used in establishing the system boundaries shall be identified and justified in the scope of the study" (ISO 14040, 1997).

4.2.3.2   ISO 14041: 1998—Life Cycle Inventory (LCI)

The aim of the inventory is to provide objective data which only later can be elaborated and interpreted to obtain evaluations useful at the decision-making stage. The main steps of inventory analysis are:

- Modeling product system—Consists of developing a schematic model in sufficient detail to represent the system of operations performing the process in question. The degree of detail is determined in each case on the basis of the aims of the LCA, and the complexity and difficulty of the measurements that can be made on the system components.

- Data collection—The collection of inventory data "involves the quantitative and qualitative description of the inputs and outputs needed to determine where the process starts and ends, and the function of the unit process." This is achieved using appositely prepared records. The data in question are classified as "primary data," acquired through direct measurement, and "secondary data," derived data generally obtained from the literature.

- Allocation procedure—Real industrial processes rarely produce just one output; therefore, "the materials and energy flows as well as associated environmental releases shall be allocated to the different products according to clearly stated procedures." The procedures of allocation allow energetic and environmental charges to be associated with the various coproducts and byproducts of individual processes. Through these associations it is possible to undertake the study of complex systems using energy and environmental indices summarizing their behavior (ISO 14041, 1998).

4.2.3.3 *ISO 14042: 2000—Life Cycle Impact Assessment (LCIA)*

LCIA is directed at the evaluation of environmental impact, revealing the magnitude of the effects produced as a consequence of the consumption of resources and the emissions that result from the entire life cycle. The difficulty of assessment lies in identifying the correlation existing between the operations performed during the life cycle and the effects of these operations on the environment. The environmental effects (greenhouse effect, reduction of the ozone layer, acid rain, eutrophication, photochemical smog, toxicity, impoverishment of resources, etc.) are subdivided into local, regional, or global effects and are understood in terms of impact categories (i.e., classes of impact "representing environmental issues of concern to which LCI results may be assigned"). LCIA is structured in different stages, some obligatory and others optional. The obligatory stages are:

- Selection—Consists of choosing the environmental effects to be taken into consideration and the corresponding environmental indicators representing these effects.
- Classification—Consists of cataloging the inventory data, correlating it with different environmental effects, and thus associating it with the various impact categories.
- Characterization—Aimed at quantitatively determining the value of environmental indicators (category indicator) relating to the various impact categories. The calculation consists of converting the results of the LCI into common units and aggregating the converted results according to the impact category.

Additional optional stages can complete the LCIA. Their application depends on the objectives of the LCA. Known as Normalization, Grouping, and Weighting, these phases consist of elaborating the results of the characterization phase to obtain indices used to perform an overall evaluation of the process under examination. The methods of normalization and weighting are varied and not standardized—each refers to different parameters, often linked to artificial and debatable considerations (ISO 14042, 2000).

4.2.3.4 *ISO 14043: 2000—Life Cycle Interpretation*

The last phase of LCA provides for the interpretation of the data obtained in the preceding phases and, on the basis of this data, identifying the actions to be undertaken with the aim of lessening the environmental impact of the system. The approach is iterative in the sense that, after having modified the process, it is necessary to repeat the LCA to verify that the modification has improved, not worsened, the situation. This phase is usually understood as a step where it is possible to create a valid correlation between the results of inventory analysis and those of impact analysis in order to propose useful recommendations conforming to the aims and objectives of the study. This procedure is highly subjective and the ISO 14043 standard itself advises that the data obtained in the previous phases must be organized in a clear and understandable manner in order to provide useful indications for the unambiguous planning of any intervention of improvement (ISO 14043, 2000).

4.3 Fields of Application and Limitations of Life Cycle Assessment

The ISO reference standards themselves indicate the breadth and variety of the fields of LCA application. LCA can constitute a valid aid in diverse contexts, from product development to the planning of environmental policies.

In product development, LCA is used to identify opportunities for improving a product's environmental performance throughout its life cycle. In product development, depending on whether the objective is the analysis and improvement of a product–system or the comparison between different systems, LCA can aid in determining the environmental criticalities of the solution under examination. It can also help define the best design strategies and criteria, choose the most efficient (from the environmental viewpoint) alternatives; and identify the cleanest technologies. With regard to the planning of environmental policies, at both the level of single industrial activities and the government level, LCA is an effective instrument in decision making; it can be used to identify the environmental criticalities of complex systems, characterize environmentally critical groups of products, and define the most appropriate strategies to adopt in relation to each specific problem.

In contrast to this broad potential, some significant limitations of LCA should be noted. Although it can be considered the most effective instrument currently available for the evaluation of a product's environmental performance, most of the problems attributable to its underlying methodological approach (already apparent during the last decade) must be considered still unresolved:

- The preliminary assumptions at the base of every LCA (i.e., defining the system boundaries, selecting the sources of information, and deciding the impact categories to be examined) are all subjective and have a decisive effect on the final results of the assessment.

- The uncertainty over the quality and reliability of the wide range of data at the base of the inventory phase can condition the reliability of the whole analysis technique (Ayres, 1995; Weidema and Wesnaes, 1996; von Bahr and Steen, 2004).

- The very models and procedures underlying the phases of LCI and LCIA are limited by further assumptions and lacunae (Baumann and

Rydberg, 1994; Guinée and Heijungs, 1995; Ekvall and Finnveden, 2001), whose effects on the final result are difficult to control.

- The absence of spatial and temporal dimensions in the LCI phase leads to uncertainties in the results of impact assessment (Udo de Haes, 1996).

Because of these limitations, much effort has been directed at extending the potential of LCA (Udo de Haes et al., 2004). Some of the more important results will be considered in the following section. Nevertheless, the clear-cut suggestion made in the ISO standards themselves remains valid: "The information developed in an LCA study should be used as part of a much more comprehensive decision process or used to understand the broad or general trade-offs" (ISO 14040, 1997).

4.4   Overview of Practical Approaches and Tools for Life Cycle Assessment

The description of the methodological framework of LCA discloses its complexity and the vast number of problems involved. It is evident that implementing an analysis technique with such a complex structure requires the development of methodological and analytical instruments supporting its most important phases. Thus, beginning with the first studies codifying LCA in the 1990s, there has been a wide proliferation of practical approaches and tools for this purpose. The following overview distinguishes between the two main typologies of such tools:

- Tools for a complete LCA (Full LCA), evolving from the need to overcome the application difficulties in the phases of LCA, such as those described in Section 4.2, in particular those relative to the procedures of transforming inventory data into indicators of environmental impact typical of LCIA

- Tools for a simplified LCA (Simplified or Streamlined LCA), evolving from the need to operate environmental assessments in the first phases of product design and development, when it is not yet possible to marshal the mass of data required by a full LCA (which in any case would be prohibitive in terms of time and cost at the early stages of development)

As will become clearer below, while the first type of tools are generally quantitative, the second type can be further differentiated into quantitative and qualitative instruments. The quantitative tools use a greater amount of data (also empiric), which are elaborated to calculate functions measuring environmental performance. The evaluations are, therefore, more objective and verifiable, but require more detailed information that can only be obtained after a certain point of the design process has been reached. Conversely, the qualitative instruments are based on mostly subjective observations and evaluations. They are, therefore, easy to implement because they require only a limited mass of data, but their results are not unequivocal or easily corroborated.

4.4.1   Full LCA

As was noted in Section 4.2.2, the methodological framework of LCA is common to both the first SETAC structure and the ISO standards. An important difference can be found in the phase of Impact Assessment, where the SETAC structure differs from the ISO standards in that the evaluation of the impacts provides for Classification and Characterization as well as a conclusive step of Valuation, where weighting procedures are applied to the different impact typologies. These procedures are intended to obtain data to use as the basis for deciding on the actions to be taken (Fava et al., 1993). In the ISO standards, this final step consists of optional, rather than obligatory, procedures (ISO 14042, 2000).

In all cases, it is evident that an effective use of LCA as a decisional and selective method requires an elaboration of the results from the inventory phase, which provides concise indicators allowing even designers (for whom simple inventory data would be largely incomprehensible) to make a clear overall evaluation. The various methodological and analytical tools (of the quantitative type) proposed for this purpose differ not only in the typologies of environmental effects they take into consideration, but also in the procedures of Characterization, Normalization, and Weighting. Some of the best-known are reported in Table 4.3, together with the main references. A broader panorama and detailed description of the characteristics differentiating them can be found in several studies comparing different methods (Braunschweig et al., 1996; Finnveden, 1996; Hertwich et al., 1997; Powell et al., 1997; Dreyer et al., 2003).

4.4.2   Streamlined LCA

Although LCA is at present the most effective technique for evaluating the environmental performance of a product or system, its complexity limits its use in the design phase. Currently, it is used only at the end of the design phase, when the data available for the analysis are more complete and reliable, but when there is no longer the possibility of LCA guiding the process of product development (Bhander et al., 2003). The concept of a simplified LCA was developed to respond to the precise need to provide designers with

TABLE 4.3   Quantitative valuation methods

METHOD	ORIGIN	REFERENCES
CML 1992	University of Leiden	(Heijungs et al., 1992)
CML 2001	The Netherlands	(Guinée et al., 1993a/b)
	1992	(Guinée, 2002)
Ecopoint	BUWAL	(Ahbe et al., 1990)
Ecopoint 1997	Switzerland	(Braunschweig et al.,
	1990	1998)
Eco-indicator 95	Pré Consultants BV	(Goedkoop, 1998)
Eco-indicator 99	The Netherlands	(Goedkoop and
	1995	Spriensma, 2000)
EDIP	Technical University of	(Wenzel et al., 1997)
	Denmark	
	Denmark	
	1997	
EPS (Environmental	IVL Swedish Environmental	(Steen and Ryding, 1992)
Priority Strategies)	Research Institute	(Steen, 1996)
EPS 2000	Sweden	(Steen, 1999)
	1992	
MET-Points	TNO Industrial Technology/	(Kalisvaart and
(Material use, Energy use,	TU Delft	Remmerswaal, 1994)
Toxicity effects)	The Netherlands	
	1994	
MIPS	Wuppertal Institut	(Schmidt-Bleek, 1994)
(Material Intensity Per	Germany	(Ritthoff et al., 2002)
Service unit)	1994	

a rapid (though not detailed) evaluation of the product's impact in the early phases of design. This approach is intended to overcome the main drawbacks of a Full LCA, which are:

- The complexity and length of the analysis and evaluation process, contrasting with the ever-more-pressing need to reduce the product's time-to-market
- The necessity of using a considerable mass of data, usually not available in the first phases of design development, requiring the intervention of environmental experts for its acquisition
- The difficulty in interpreting the results, usually consisting of a wide range of heterogeneous data that are largely incomprehensible to designers and product development team members

With the aim of overcoming these drawbacks, various studies have examined the possibility of developing new methods for conducting a simplified

LCA, usually known as a Streamlined LCA (Todd, 1996; Graedel, 1998). This can be defined as the "identification of elements of an LCA that can be omitted or where surrogate or generic data can be used without significantly affecting the accuracy of the results" (Todd and Curran, 1999).

Of the different phases of LCA, the primary object of simplification is usually the inventory phase (i.e., the most costly and time-consuming phase, and that presenting the greatest potential for streamlining) (Christiansen et al., 1997). From this perspective, simplification encompasses all the stratagems directed at limiting the analysis boundaries, eliminating life cycle phases (as in the case of "cradle-to-gate" LCA, noted in Section 4.2.1), or reducing the parameters required in the description of the activities. This can be achieved by limiting the aims and scope of the analysis and by simplifying the models and procedures used in both inventory analysis and impact evaluation. Alternatively, again in order to facilitate the information-gathering phase, it is possible to use surrogate data as well as both quantitative and qualitative data. In practice, numerous stratagems are applied in order to "streamline" LCA, and descriptions of the most widely used are reported in the literature (Graedel, 1998; Todd and Curran, 1999).

At present, simplified LCA includes numerous tools, both quantitative and qualitative (Tolle et al., 1994; Graedel et al., 1995; Wenzel et al., 1997; Biswas et al., 1998; Bey et al., 1999; Sousa et al., 1999; Fleischer et al., 2001; Chen and Chow, 2003; Ryu et al., 2003). In general, they are directly derived from LCA procedures and often elaborate environmental data derived from previous full LCAs. Some of the methods for evaluating impact reported in Table 4.3 can themselves be used as tools for simplified LCA (particularly Eco-indicator, EPS, and MET-Points methods).

In all cases, these tools have the objective of evaluating the weak points of the developed product, quantifying the fulfillment of environmental requisites, and assessing the effectiveness of the design intervention. They differ in terms of typology (systems of guidelines, checklists, evaluation tables and matrices, mathematical models); procedure of implementation; phases of the life cycle subjected to analysis; type of environmental impact measured; and the form of the final result. The proliferation of these tools over recent years and the diversity of their statements and characteristics is confirmed by the publication of studies comparing different types (Hunt et al., 1998; Luo et al., 2001; Ernzer and Wimmer, 2002; Hochschorner and Finnveden, 2003; Lee et al., 2003).

However, it should be noted that despite the growing interest in and use of LCA simplification techniques, it is not always fully appreciated to what extent these simplifications can affect the validity of the results obtained, as shown by studies comparing the results with those of a Full LCA (Hunt et al., 1998). Thus, these techniques can sometimes undermine the very life cycle–oriented properties of this important instrument.

4.4.3 Alternative Approaches

Section 4.3 described how, in their present state, the reference methodological structure of LCA and the underlying approach itself manifest various critical aspects that can limit the validity of the results. Prominent among these principal weaknesses are the absence of spatial and temporal dimensions in LCI (leading, as noted above, to uncertainties in the results of the successive LCIA phase) and the lack of reference to economic and social aspects (constituting, together with the environmental aspect, the three principal factors of sustainability). One method of intervening to counteract this problem is an elaboration of LCA in the context of its traditional structure, in order to extend its potentials and goals (Udo de Heas et al., 2004).

Alternatively, it is possible to have recourse to radically different methodological approaches, such as the integrated approaches intended to combine the two analyses, environmental and economic. The most important of these integrated approaches are Activity-based LCA and Input–Output LCA, which are described below. A more detailed discussion of the integration between environmental and economic life cycle analyses is presented in Chapter 6.

4.4.3.1 *Activity-Based LCA*

Activity-based LCA is derived from the application in LCA of a methodology for cost estimation, already widely used since the mid-1980s and known as Activity-based Costing (Cooper, 1990). Differing from traditional cost analyses, where each unit of product consumes resources (the concept underlying process-based analyses such as the traditional LCA itself), this approach starts from the supposition that a product or service does not use resources directly but, rather, consumes activities. Activity-based LCA consists of an extension of Activity-based Costing expressly conceived as an aid to decision making in design and management operations, able to evaluate economic costs, energy consumptions and waste generation under conditions of uncertainty (Emblemsvag and Bras, 1997).

4.4.3.2 *Input–Output LCA*

Input–Output LCA (IO–LCA) is based on the methodological structure of input–output economic analysis (IOA), with which it is possible to reconstruct all the input and output flows of an entire economic system. This analysis methodology evidences the interdependence between the different economic sectors of a system and is, in turn, based on an economic model of equilibrium wherein the inputs that each sector requires of the other sectors in order to produce a unit of product are specified (Leontief, 1986). This involves positing the simplified hypothesis that an increase in production of a commodity by one sector leads to a proportional increase in the inputs this sector receives from all the other sectors fueling it.

IOA was originally used to model complex economic systems such as the economies of industrialized nations. In the early 1990s it came to be considered an instrument for use in quantifying not only the economic and cost flows of an industrial system, but also the resource flows (materials and energy) throughout their entire cycle of use, from extraction to disposal (Duchin, 1992). The same period saw the first studies of integrating IOA into LCA (Moriguchi et al., 1993), since when this approach has continued to arouse interest, as attested by more recent papers (Joshi, 2000). This interest arose from a recognition of IOA's potential for overcoming some limitations of traditional LCA, which, being a process-based type of analysis, circumscribes the assessments to a system closed within preset boundaries (i.e., it treats the product–system as isolated rather than in the context of a broader system). By contrast, in the IOA approach the analysis boundaries include the entire economic system in which the product is conceived and its life cycle develops. Precisely because of this characteristic, IOA can be used to compensate for the incompleteness of analysis techniques that reduce the product–system to a process-based system, like LCA. On the other hand, process-based analysis maintains its primacy in terms of the level of detail and specificity of the processes, a characteristic lacking in the IOA approach.

A well-known example of the use of input–output analysis as a methodological structure for the environmental analysis of a product's life cycle is represented by Economic Input–Output LCA (EIO–LCA) (Cobas et al., 1995; Hendrickson et al., 1998). In this technique, once the monetary flows characterizing the entire economic system of the product have been quantified (i.e., the economic system in which the phases of the life cycle of the product under examination develop), these flows can be translated into environmental impact terms (resource consumption, polluting factors). Although it was based on a model for economic analysis, EIO–LCA evolved as a technique for environmental analysis and, for this reason, is also included among the methodological structures for LCA listed in Table 4.1.

Finally, it is worth noting the recent use of approaches that integrate input–output and process-based analyses, called hybrid techniques (Hybrid LCA) (Joshi, 2000; Udo de Heas et al., 2004). Both the hybrid techniques and the input–output approach itself are considered tools for Simplified LCA (Rebitzer et al., 2004).

4.4.4 Software Tools

A clear and complete picture of the continuous production and diffusion of tools for LCA can be provided by an overview of the software tools developed and used in both academic and industrial sectors. In the majority of cases, they refer to the methodological structures and instruments supporting LCA discussed above, and are directed at facilitating the implementation of LCA techniques in design, production, and managerial contexts. Such software is designed to improve, wherever possible, the environmental performance of products, processes, or systems in their entirety. The numerous tools commercially available chiefly differ in terms of:

- The methodology used to translate inventory data into environmental effects, and these into environmental damage
- The control boundaries hypothesized in the life cycle analysis (e.g., some instruments go as far as considering the energy employed in the extraction of raw materials)
- The properties of the databases, their extension, and the quality of the information they contain

With regard to this last aspect, it should be noted that in order to calculate the sum of the wide variety of resources and emissions involved in a product's entire life cycle, it is necessary to acquire detailed information on a vast range of production processes, materials, and energy flows, and to make predictions regarding the product's use and the processes it will undergo at end-of-life. In general, it is impossible to compile a database with all the information necessary, and this has induced software developers to provide interactive databases that can be enhanced and customized according to the precise needs of the user. Furthermore, the quality and significance of the collected data are strictly dependent on the methodology used and on the economic and productive reality in which the user operates.

In conclusion, it must be said that it is not easy to evaluate the differences between the numerous tools available other than through their direct use and experience. It can, however, be difficult for potential users to select the most appropriate software tool for their needs. In this respect, it may be helpful to refer to the descriptions and comparisons reported in the literature (Menke et al., 1996; de Caluwe, 1997; Rice et al., 1997; Vezzoli, 1999; Gibson et al., 2001).

4.5 Summary

Life Cycle Assessment arose from the need to investigate the potential environmental effects of the entire system of activities and processes associated with the production and provision of goods and services. It has evolved into a complete methodological structure, able to estimate the environmental impacts attributable to the whole life cycle of a product, such as depletion of resources, toxicological stresses on human health and the ecosystem, ozone depletion, acidification, eutrophication, and climate change. The quantification and allocation of resources and emissions, together with the phase of translating inventory data into environmental impact, constitute the most complete approach to the evaluation of the overall impact of a product based on a holistic vision of the problem—precisely what is required for an orientation aimed at truly sustainable development.

Life Cycle Assessment can be a valid support in diverse contexts, from product development to the planning of environmental policies. Particularly in product development, it can help in determining the environmental criticalities of the solution under examination, defining the best design strategies and criteria, and choosing the most efficient alternatives. Nevertheless, as explicitly suggested by the ISO standards, the information developed in a Life Cycle Assessment study should be used as part of a much more comprehensive decision-making process or used to understand the broad or general trade-offs.

4.6   References

Ahbe, S., Braunschweig, A., and Müller-Wenk, R., Methodik für Oekobilanzen auf der Basis ökologischer Optimierung, Schriftenreihe Umwelt Nr. 133, BUWAL, Bern, Switzerland, 1990.

Ayres, R.U., Life cycle analysis: A critique, *Resources, Conservation and Recycling*, 14, 199–223, 1995.

Baumann, H. and Rydberg, T., A comparison of three methods for impact analysis and evaluation, *Journal of Cleaner Production*, 2(1), 13–20, 1994.

Bey, N., Lenau, T., and Larsen, M.H., Oil Point: Life cycle evaluation without the data problem, in *Proceedings of ICED 99 International Conference on Engineering Design*, Munich, 1999, 469–472.

Bhander, G.S., Hauschild, M., and McAloone, T., Implementing life cycle assessment in product development, *Environmental Progress*, 22(4), 255–267, 2003.

Biswas, G. et al., An environmentally conscious decision support system based on a streamlined life cycle assessment and a cost residual risk based evaluation, *Journal of Industrial Ecology*, 2(1), 127–142, 1998.

Boustead, I. and Hancock, G.F., *Handbook of Industrial Energy Analysis*, Ellis Horwood, Chichester, UK, 1979.

Boustead, I. and Lidgren, K., *Problems in Packaging: The Environmental Issue*, Ellis Horwood, Chichester, UK, 1984.

Braunschweig, A. et al., Developments in LCA Valuation, IWÖ-HSG series No. 32, University of St. Gallen, Switzerland, 1996.

Braunschweig, A. et al., Bewertung in Ökobilanzen mit der Methode der Ökologischen Knappheit: Ökofaktoren 1997, Methodik für Oekobilanzen, Schriftenreihe Umwelt Nr. 297, BUWAL, Bern, Switzerland, 1998.

Bridgwater, A.V. and Lidgren, K., *Energy in Packaging and Waste*, Van Nostrand Reinhold, Wokingham, UK, 1983.

Brown, H.L. et al., Energy Analysis of 108 Industrial Processes, U.S. Department of Energy, Washington, DC, 1980.

Chen, J.L. and Chow, W.K., Matrix-type and pattern-based simple LCA for eco-innovative design of products, in *Proceedings of EcoDesign 2003: 3rd International Symposium on Environmentally Conscious Design and Inverse Manufacturing*, Tokyo, 2003, 467–472.

Christiansen, F. et al., Simplifying LCA: Just a Cut?, Final report, SETAC EUROPE—LCA Screening and Streamlining Working Group, SETAC Society of Environmental Toxicology and Chemistry, Brussels, 1997.

Cobas, E. et al., Economic input–output analysis to aid life cycle assessment of electronics products, in *Proceedings of IEEE International Symposium on Electronics and the Environment*, Orlando, FL, 1995, 273–277.

Consoli, F. et al., Guidelines for Life-Cycle Assessment: A Code of Practice, SETAC Society of Environmental Toxicology and Chemistry, Brussels, 1993.

Cooper, R., Implementing an activity-based cost system, *Journal of Cost Management*, 3, 33–42, 1990.

Curran, M.A., The history of LCA, in *Environmental Life-Cycle Assessment*, Curran, M.A., Ed., John Wiley & Sons, New York, 1996, chap. 1.

de Caluwe, N., Ecotools Manual: A Comprehensive Review of Design for Environment Tools, Report DFE/TR33, Manchester Metropolitan University, Manchester, UK, 1997.

Dreyer, L.C., Niemann, A.L., and Hauschild, M., Comparison of three different LCIA methods: EDIP97, CML2001 and Eco-indicator 99—Does it matter which one you choose?, *International Journal of Life Cycle Assessment*, 8(4), 191–200, 2003.

Duchin, F., Industrial input–output analysis: Implications for industrial ecology, in *Proceedings of National Academy of Sciences, Colloquium on Industrial Ecology*, Washington, DC, 89(3), 1992, 851–855.

EC Directive 85/339, Containers of Liquids for Human Consumption, European Commission, 1985.

Ekvall, T. and Finnveden, G., Allocation in ISO 14041: A critical review, *Journal of Cleaner Production*, 9, 197–208, 2001.

Emblemsvag, J. and Bras, B., An activity-based life cycle assessment method, in *Proceedings of ASME Design Engineering Technical Conference*, Sacramento, CA, 1997, DFM-97-119.

EPA, Life-Cycle Impact Assessment: A Conceptual Framework, Key Issues, and Summary of Existing Methods, EPA-452/R-95-002, U.S. Environmental Protection Agency, Office of Air Quality Planning and Standards, Research Triangle Park, NC, 1995.

Ernzer, M. and Wimmer, W., From environmental assessment results to design for environment product changes: An evaluation of quantitative and qualitative methods, *Journal of Engineering Design*, 13(3), 233–242, 2002.

Fava, J. et al., A Technical Framework for Life-Cycle Assessment, SETAC Society of Environmental Toxicology and Chemistry, Washington, DC, 1991.

Fava, J. et al., A Conceptual Framework for Life-Cycle Impact Assessment, SETAC Society of Environmental Toxicology and Chemistry, Pensacola, FL, 1993.

Finnveden, G., Valuation Methods within the Framework of Life Cycle Assessment, Report B1231, IVL Swedish Environmental Research Institute, Stockholm, 1996.

Fleischer, G. et al., A semi-quantitative method for the impact assessment of emissions within a simplified life cycle assessment, *International Journal of Life Cycle Assessment*, 6(3), 149–156, 2001.

Gibson, T.L., Kumar, S., and Wheeler, C.S., Evaluation of life cycle assessment software for automotive applications, in *Proceedings of SAE Environmental Sustainability Conference*, Grat, Austria, 2001, 169–174.

Goedkoop, M., Eco-indicator 95: Final Report, Pré Consultants BV, Amersfoort, The Netherlands, 1998.

Goedkoop, M. and Spriensma, R., The Eco-indicator 99: Methodology Report, Pré Consultants BV, Amersfoort, The Netherlands, 2000.

Goldsmith, E. et al., A blueprint for survival, *The Ecologist*, 2(1), 1972.

Graedel, T.E., *Streamlined Life-Cycle Assessment*, Prentice Hall, Englewood Cliffs, NJ, 1998.

Graedel, T.E., Allenby, B.R., and Comrie, P.R., Matrix approaches to abridged life cycle assessment, *Environmental Science & Technology*, 29, 134–139, 1995.

Guinée, J.B., *Handbook on Life Cycle Assessment: Operational Guide to the ISO Standards*, Kluwer Academic Publisher, Dordrecht, The Netherlands, 2002.

Guinée, J.B. and Heijungs, R., A proposal for the definition of resource equivalency factors for use in product life-cycle assessment, *Environmental Toxicology and Chemistry*, 14(5), 917–925, 1995.

Guinée, J.B. et al., Quantitative life cycle assessment of products: Classification, valuation and improvement analysis, *Journal of Cleaner Production*, 1(2), 81–91, 1993b.

Guinée, J.B., Udo de Haes, H.A., and Huppes, G., Quantitative life cycle assessment of products: Goal definition and inventory, *Journal of Cleaner Production*, 1(1), 3–13, 1993a.

Heijungs, R. et al., Environmental Life Cycle Assessment of Products, CML Center for Environmental Science, University of Leiden, The Netherlands, 1992.

Hendrickson, C.T. et al., Economic input–output models for environmental life cycle assessment, *Environmental Science and Technology*, 32(7), 184A–191A, 1998.

Hertwich, E., Pease, W., and Koshland, C., Evaluating the environmental impact of products and production processes: A comparison of six methods, *The Science of the Total Environment*, 196, 13–29, 1997.

Hochschorner, E. and Finnveden, G., Evaluation of two simplified life cycle assessment methods, *International Journal of Life Cycle Assessment*, 8(3), 119–128, 2003.

Hunt, R. et al., Resource and Environmental Profile Analysis of Nine Beverage Container Alternatives, EPA/530/SW-91c, U.S. Environmental Protection Agency, Office of Solid Waste Management Programs, Washington, DC, 1974.

Hunt, R., Sellers, J., and Franklin, W., Resource and environmental profile analysis: A life cycle environmental assessment for products and procedures, *Environmental Impact Assessment Review*, 12, 245–269, 1992.

Hunt, R.G. and Franklin, W.E., Personal reflections on the origin and development of LCA in the USA, *International Journal of Life Cycle Assessment*, 1(1), 4–7, 1996.

Hunt, R.G. et al., Case studies examining LCA streamlining techniques, *International Journal of Life Cycle Assessment*, 3(1), 36–42, 1998.

ISO 14001, Environmental Management Systems—Specifications with Guidance for Use, ISO 14001:1996, International Organization for Standardization, Geneva, 1996.

ISO 14040, Environmental Management—Life Cycle Assessment—Principles and Framework, ISO 14040:1997(E), International Organization for Standardization, Geneva, 1997.

ISO 14041, Environmental Management—Life Cycle Assessment—Goal and Scope Definition and Inventory Analysis, ISO 14041: 1998(E), International Organization for Standardization, Geneva, 1998.

ISO 14042, Environmental Management—Life Cycle Assessment—Life Cycle Impact Assessment, ISO 14042:2000(E), International Organization for Standardization, Geneva, 2000.

ISO 14043, Environmental Management—Life Cycle Assessment—Life Cycle Interpretation, ISO 14043:2000(E), International Organization for Standardization, Geneva, 2000.

Jain, R.K. et al., *Environmental Assessment*, McGraw Hill, New York, 1993.

Joshi, S., Product environmental life cycle assessment using input–output techniques, *Journal of Industrial Ecology*, 3(2–3), 95–120, 2000.

Kalisvaart, S. and Remmerswaal, J., The MET-points method: A new single figure environmental performance indicator, in *Proceedings of SETAC Conference—Integrating Impact Assessment into LCA*, Brussels, 1994.

Lee, J. et al., Comparison of simplified LCA and matrix methods in identifying the environmental aspects of products, in *Proceedings of EcoDesign 2003: 3rd International Symposium on Environmentally Conscious Design and Inverse Manufacturing*, Tokyo, 2003, 682–686.

Leontief, W., *Input–Output Economics*, 2nd ed., Oxford University Press, New York, 1986.

Lindfors, L.G. et al., Nordic Guidelines on Life-Cycle Assessment, Nordic Council of Ministers, Copenhagen, 1995.

Lundholm, M.P. and Sundstrom, G., Tetra Brick Aseptic Environmental Profile, AB Tetra Pack, Malmo, Sweden, 1985.

Luo, Y., Wirojanagud, P., and Caudill, R.J., Comparison of major environmental performance metrics and their application to typical electronic products, in *Proceedings of IEEE International Symposium on Electronics and the Environment*, Denver, CO, 2001, 94–99.

Menke, D.M., Davis, G.A., and Vigon, B.W., Evaluation of Life-Cycle Assessment Tools, Environment Canada, Ottawa, Ontario, 1996.

Moriguchi, Y., Kondo, Y., and Shimuzu, H., Analysing the life cycle impact of cars: The case of CO_2, *Industry and Environment*, 16(1–2), 42–45, 1993.

National Academy of Science (NAS), *Resources and Man*, W.H. Freeman, San Francisco, 1969.

Pennington, D.W. et al., Life cycle assessment—Part 2: Current impact assessment practice, *Environment International*, 30(5), 721–739, 2004.

Powell, J., Pearce, D., and Craighill, A., Approaches to valuation in LCA impact assessment, *International Journal of Life Cycle Assessment*, 2(1), 11–15, 1997.

Rebitzer, G. et al., Life cycle assessment—Part 1: Framework, goal and scope definition, inventory analysis, and applications, *Environment International*, 30(5), 701–720, 2004.

Rice, G., Clift, R., and Burns, R., Comparison of currently available European LCA software, *International Journal of Life Cycle Assessment*, 2(1), 53–59, 1997.

Ritthoff, M., Rohn, H., and Liedtke, C., Calculating MIPS: Resource Productivity for Products and Services, Wuppertal Spezial 27e, Wuppertal Institute for Climate, Environment and Energy, Science Centre North Rhine-Westphalia, Germany, 2002.

Ryu, J. et al., Simplified life cycle assessment for eco-design, in *Proceedings of EcoDesign 2003: 3rd International Symposium on Environmentally Conscious Design and Inverse Manufacturing*, Tokyo, 2003, 459–463.

Schmidt-Bleek, F., *Wieviel Umwelt Braucht der Mensch? MIPS*, Birkhaüser Verlag, Basel, Switzerland, 1994.

Smith, H., The cumulative energy requirements of some final products of the chemical industry, *Transactions of the World Energy Conference*, 18(E), 1969.

Sousa, I. et al., A learning surrogate LCA model for integrated product design, in *Proceedings of 6th International CIRP Seminar on Life Cycle Engineering*, Kingston, Ontario, 1999, 209–219.

Steen, B., EPS: Default Valuation of Environmental Impacts from Emission and Use of Resources, Report 111, IVL Swedish Environmental Research Institute, Goteborg, Sweden, 1996.

Steen, B., A Systematic Approach to Environmental Priority Strategies in Products Development (EPS): General System Characteristics—Models and Data, CPM Reports 1999: 4 and 1999: 5, Chalmers University of Technology, Goteborg, Sweden, 1999.

Steen, B. and Ryding, S.O., The EPS Enviro-Accounting Method: An Application of Environmental Accounting Principles for Evaluation and Valuation in Product Design, Report B1080, IVL Swedish Environmental Research Institute, Göteborg, Sweden, 1992.

Todd, J.A., Streamlining, in *Environmental Life-Cycle Assessment*, Curran, M.A., Ed., John Wiley & Sons, New York, 1996, chap. 4.

Todd, J.A. and Curran, M.A, Streamlined Life-Cycle Assessment: A Final Report from the SETAC North America Streamlined LCA Workgroup, SETAC Society of Environmental Toxicology and Chemistry, Pensacola, FL, 1999.

Tolle, D. et al., Development and assessment of a pre-LCA tool, in *Proceedings of IEEE International Symposium on Electronics and the Environment*, San Francisco, 1994, 201–206.

Topfer, K., The launch of the UNEP-SETAC Life Cycle Initiative, *International Journal of Life Cycle Assessment*, 7(4), 191, 2002.

Udo de Haes, H.A., Discussion of general principles and guidelines for practical use, in *Towards a Methodology for Life Cycle Impact Assessment*, Udo de Haes, H.A., Ed., SETAC Society of Environmental Toxicology and Chemistry, Brussels, 1996, 7–30.

Udo de Haes, H.A. et al., Three strategies to overcome the limitations of life-cycle assessment, *Journal of Industrial Ecology*, 8(3), 19–32, 2004.

United Nations Environment Programme (UNEP), Life Cycle Assessment: What It Is and How to Do It, United Nations Environment Programme, Paris, 1996.

Vezzoli, C., An overview of life cycle design and information technology tools, *Journal of Sustainable Product Design*, 9, 27–35, 1999.

Vigon, B.W. et al., Life-Cycle Assessment: Inventory Guidelines and Principles, EPA/600/R-92/245, U.S. Environmental Protection Agency, Office of Research and Development, Cincinnati, OH, 1993.

von Bahr, B. and Steen, B., Reducing epistemological uncertainty in life cycle inventory, *Journal of Cleaner Production*, 12, 369–388, 2004.

Weidema, B.P. and Wesnaes, M.S., Data quality management for life cycle inventories: An example of using data quality indicators, *Journal of Cleaner Production*, 4(3–4), 167–174, 1996.

Wenzel, H., Hauschild, M., and Alting, L., *Environmental Assessment of Products*, Vol. 1–2, Chapman & Hall, London, 1997.

White, M.A. and Wagner, B., The ecobalance as a tool for environmental financial management, *Pollution Prevention Review*, 6(2), 31–46, 1996.

Chapter 5

Life Cycle Cost Analysis

A manufacturing company's level of competitiveness in the global market depends on its capacity to bring products to market at the right time, guaranteeing their functionality and quality, and limiting the cost. Independent of product typology, controlling only the costs of resource acquisition, production, and distribution can be considered an incomplete and obsolete approach in today's world. Only by including the costs of the entire life cycle among the parameters of the decision-making process is it possible to achieve a design that is effective in terms of economic feasibility. Life Cycle Cost Analysis, or Life Cycle Costing, is a methodology directed at the evaluation of all the costs associated with an activity or a product over its entire life cycle, thus assuming the dual role of a Life Cycle Assessment in economic terms. By virtue of their shared life cycle approach, Life Cycle Cost Analysis is the most appropriate economic analysis instrument in a Life Cycle Design intervention.

Tracing the evolution of cost analysis in the ambit of product development, this chapter will provide a full description of Life Cycle Cost Analysis, considering the more significant premises, outlining a methodological structure of reference, and introducing both traditional and alternative models and tools.

5.1 Cost Analysis and the Life Cycle Approach

Cost analysis and assessment are two of the principal factors guiding the process of product development, since they strictly condition the main decisional choices in a clear-cut manner. From the earliest theorizing on design intervention, it has been emphasized how the economic worthwhileness of a proposal (i.e., the property of making the final product acquire sufficient value to repay the expenditure incurred in the production phase) is one of the most rigid selection criteria (Asimow, 1962). On entering the market, a product manufactured through processes of transforming the resources employed must have increased in value such that it can be produced and commercialized.

From the earliest initial phases of needs analysis and their translation into product concept, the design team must assess at least two different typologies of economic validity, according to whether the viewpoint is that of the manufacturer or of the consumer. In the first typology, the economic efficiency of a proposal can be evaluated quantitatively. The producer directly acquires the necessary raw materials, energy resources, and workforce on the open market, transforms them into product, and introduces the product onto the market. By evaluating the costs of development, production, and distribution, and comparing them with the market value of the product, it is possible to accurately quantify its economic validity.

The evaluation of economic efficiency from the consumer's viewpoint is much more complex and subjective. In fact, it depends not only on the cost of the product on the market but also on the level of efficacy with which the product satisfies the needs that generated it. Clearly, this kind of value is subjective in that it cannot be measured by the market, but depends on the perceptions of the customer.

The most common economic models used in product design and development originated in relation to the first necessity, that of assessing the economic validity of a commodity during its definition and development; their primary aim is, therefore, to evaluate the production costs corresponding to different design alternatives (Dieter, 2000; Ulrich and Eppinger, 2000). These models are part of that approach to product analysis which, developed in relation to the interests of the manufacturer, generally stop at distribution without taking into consideration successive phases of the life cycle. In this case, the life cycle is understood as the set of phases consisting of development, production, and distribution, at most going so far as to consider product support services.

An assessment of product value as perceived by the consumer requires different models that are able to relate the functionality of a product with its cost, in a way that quantifies its capacity to meet the performance required per unit cost. This is the concept at the base of Value Engineering, a customer-oriented approach to the whole design process, formulated according to a view of the life cycle extended, of necessity, to include the phase of product use (Ullman, 2003).

Even more complex is the case where environmental performance becomes one of the factors in play, principally because of the aspects noted previously—the environmental performance of a product must be evaluated over its entire life cycle and is influenced by the interaction between the actors involved; an effective approach, therefore, requires the adoption of the concept of a product-system, wherein the product is understood as integral with its life cycle and within the environmental, social, and technological context in which this life cycle develops (Chapter 2). As will be discussed more fully in the following chapter, these aspects entail some inescapable prerequisites for an economic analysis of products that also takes account of environmental performance:

- The cost analysis, like environmental impact analysis, has to be extended to cover the entire life cycle, from conception and design to retirement and disposal.

- In its most complete form, it also requires the difficult assessment of the environmental costs borne by all the subjects involved in the life cycle, including society as a whole.

5.1.1 From Assessment of Production Costs to Economic Analysis of Life Cycle

Evaluating the cost of production has always been considered one of the main tasks of the designer engaged in the development of a new product. It is undertaken from the very first phases of the design intervention; initial estimates must be made at the beginning of the concept development phase and then elaborated ever more accurately as the design process evolves, until they lead to the final cost of the product. Conducting a cost analysis so early in the design intervention is appropriate given that up to 75% of the production cost has already been defined by the end of the concept design phase (Ullman, 2003). Besides identifying the most economic way of producing a new product, the evaluation of costs during product design is commonly used to determine the selling price and to guide programs aimed at reducing the costs of preexisting production processes.

The first extension of cost analysis beyond the production phase dates back to the mid-1960s, when the term "Life Cycle Costing" was first coined (LMI, 1965). In its original form, the analysis of life cycle costs was heavily conditioned by the context in which it was developed, that of defense procurement (i.e., the planning and acquisition of large pieces of military equipment and matériel characterized by their great expense and particularly long useful life) (Kinch, 1992). This area is particularly susceptible to the problem of establishing the right balance between the cost of acquisition and the cost of utilization, considering that the latter, consisting of operating and maintenance costs, is usually much greater than the former. Under this stimulus, Life Cycle Costing (LCC) became widely used to evaluate the advantage of developing and purchasing this particular type of matériel, which is expensive and must be kept at maximum efficiency for a long period. It was, therefore, understood as a technique for evaluating the comprehensive cost of a commodity—i.e., the sum of the costs of purchase) (Procurement Cost) and operation (Ownership Cost) (Dhillon, 1989), where the latter includes all the costs incurred during the useful life of the commodity itself.

Starting from these premises, by the mid-1970s the technique known as LCC, by then well-accepted in the field of military procurement, began to spread into the more general arena of industrial activity (Harvey, 1976; Blanchard, 1979). Thus the concept of "product life cycle" began to take form also in the context of economic analysis, and it was immediately extended; the category of procurement cost was enlarged to include the phases of research and development, evaluation and choice of solutions, and product support. The category of ownership costs, in some cases, went so far as to also include the cost of disposal.

In this context, therefore, there was also a maturation of a "life cycle thinking" approach, understood as "a decision-making framework that encompasses the identification of all the revenues and costs associated with a product or service as it moves time-wise through predictable stages and phases of evolution" (Shewchuk, 1992).

While maintaining the original concept, that of reducing the costs over the entire life of a complex project (such as the development of new military hardware or the construction of large civilian facilities), the spread of the "life cycle thinking" approach led to an understanding of its potential in the planning and control of resources regarding any product typology (Czyzewski and Hull, 1991). In this way the original framework merged with the life cycle approach presented here in relation to the needs of environmental analysis, strictly linked to the quantification, allocation, and management of the various forms of resources in play.

5.2   Product Life Cycle Cost Analysis

As noted above, LCC (also known as Life Cycle Cost Analysis—LCCA) originated as an instrument of cost assessment, control, and reduction, to be used in much wider contexts than that of product development. First of all, the subject of the analysis can be a generic commodity in the broadest sense (product, appliance, equipment, facility, system, etc.). Moreover, starting from the optimal balance between the costs of production and those of use and maintenance of the commodity, LCCA extends to cover a wider level—that related to the organization and management of the life cycle of the commodity itself and of the reality that produces and/or uses it. LCCA becomes an instrument aiding the management of a manufacturing company's activities and, more generally, of all the types of organizations that handle large resources. In this broad context, the typologies of use of LCCA can be various:

- Definition of guidelines and strategies for decision making
- Identification of the main guiding factors in the development of costs
- Prediction of economic flows and of future financial requirements
- Evaluation of the costs of acquiring different resources, selection between the alternatives, and optimization of resource management
- Assessment of the efficiency of new technologies
- Optimization of logistic support
- Evaluation and forecast of the commodity's useful economic life, and planning its retirement

While product design is, therefore, a relatively restricted field of application, the importance of including economic modeling in the design intervention was already being emphasized in the early 1970s (Pugh, 1974). Since then, the assessment of the economic consequences of design choices and compromises on the entire life cycle of the product has been repeatedly shown to be one of the key factors in the design and development phase (Wierda, 1988; Fabrycky and Blanchard, 1991; Alting, 1993). LCCA is recognized as a valuable instrument in:

- Facilitating the identification and improving the comprehensibility of the determinant parameters in the whole design process
- Managing the conflicts and identifying the most efficient interventions of trade-offs

The importance of estimating and controlling costs during the design process, with the aim of limiting the cost of producing a product, is now considered an ineluctable factor in the development of an efficient product. Such products are able to respond to a market demanding high standards of quality and ever-shorter development times combined with contained costs (Weustink et al., 2000). LCCA plays a primary role in this specific context due to the fact that not only production costs, but also those costs incurred during use and disposal are greatly conditioned by the initial design choices. By some assessments, more than half of the total cost of a product's life cycle is determined by the concept design phase alone (Fabrycky and Blanchard, 1991), and up to 85% can be considered fixed by the end of the completed design phase (Dowlatshahi, 1992), although only a limited fraction of this cost will have actually been spent on these phases of the development process. This is clearly shown by the curves in Figure 5.1, where it should be noted that the trend of the expenditure curve (incurred costs) varies appreciably according to the product typology (Fixson, 2004). In fact, comparing large-scale products (very high total cost, limited production volume, and long life measurable in decades) with medium-scale products (medium total cost and production volume, and life measurable in years), it can be seen how in the first case a large part of the total cost incurred corresponds to the phase of use (operation and maintenance), while in the second case it corresponds to the production phase.

Figure 5.1 also highlights an important paradox—the effectiveness of design choices in controlling the costs of the life cycle is greatest in the preliminary phases of product development, and decreases as the design level evolves. On the other hand, the possibility of establishing a relation between design choices and costs is lower in the preliminary phases of product development, and increases as the design level evolves. This is a direct consequence of how adequate knowledge and information about the design problem and the product under development is obtained only at the end of

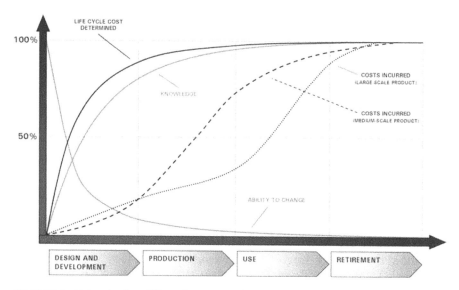

FIGURE 5.1   Determination of life cycle cost, costs incurred, information acquisition, and possibility of change as the life cycle develops.

the design process (as shown in Figure 5.1 by the curve of knowledge acquisition). In parallel, the possibility of effecting corrective interventions decreases (as shown by the curve of the possibility of making changes to the design) (Ullman, 2003).

   This dual aspect precisely characterizes the link between design choices and their consequences on life cycle costs (i.e., how important it is to succeed in defining them as soon as possible and, at the same time, how difficult it is to establish this link in the initial phases of design). This has led to the adoption of different tools used in product development with the specific purpose of reducing costs. These tools are ascribable to the approach known as Design for Cost, which can be defined as "the orientation of the engineering process to reduce life cycle cost while satisfying, and hopefully exceeding, customer demands" (Dean and Unal, 1992).

5.2.1   Premises and Definition of LCCA

A manufacturing company's level of competitiveness in the global market depends on its capacity to bring out its products with the right timing, guarantee their functionality and quality, and limit their cost. Independent of the product typology, the control and reduction of only the costs of resource acquisition, production, and distribution can nowadays be considered an incomplete and obsolete approach.

As discussed above, the "life cycle thinking" approach was introduced into economic analysis precisely to meet this incompleteness and inadequacy. Economic competitiveness can only be achieved through a life cycle approach, and only by including the costs of the entire life cycle among the parameters of the design process is it possible to achieve an effective Design for Economic Feasibility (Fabrycky and Blanchard, 1991). In this context, life cycle costs are defined as the sum of the economic resources expended, directly or indirectly attributable to a product, beginning from its conception and including the phases of production, use, and retirement. Life Cycle Cost Analysis (LCCA) or Life Cycle Costing (LCC) are the methodologies used to evaluate all the costs associated with a product over its entire life cycle (Fabrycky and Blanchard, 1991). In the design and development phases, LCCA "provides a framework for specifying the estimated total incremental cost of developing, producing, using, and retiring a particular item" (Asiedu and Gu, 1998).

An important consideration delineates the complexity of the problem of cost assessment and analysis that this approach entails for the designer. As stated at the beginning of Section 5.1, a truly effective design must, in fact, assess the validity of the choices and solutions proposed from the dual perspectives of the manufacturer and of the consumer. This is even more true where the life cycle approach is concerned, since the two actors in question (the producer and the buyer) have different perceptions of the life cycle itself, as shown in Figure 5.2. For both actors, the life cycle can be considered to consist of two principal phases: acquisition and use. This division, which mirrors the original structure of cost categories (Procurement/Ownership) typical of early studies in LCCA, can still be considered valid within the limits of traditional economic analysis.

For the producer, the acquisition phase begins with the identification and analysis of consumer needs and the definition of the design objectives, followed by the different design levels (conceptual, system or embodiment,

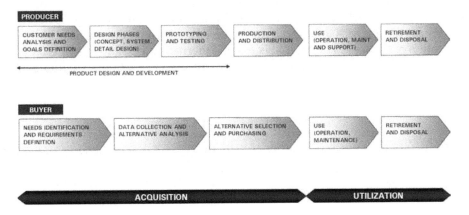

FIGURE 5.2   Perception of life cycle: Producer versus buyer.

and detailed design), prototyping, and testing. This first part of the acquisition phase constitutes what is commonly considered to be the product design and development process (Ulrich and Eppinger, 2000). It is followed by production, with distribution concluding the acquisition phase. The utilization phase is broken down into two stages: operation, maintenance, and any support services, and product retirement and disposal.

For the buyer, the utilization phase remains substantially the same, while the acquisition phase is completely different. The buyer's acquisition phase begins with the perception, analysis, and evaluation of needs, and the consequent identification of the requirements demanded of the product. Subsequently, the buyer gathers information on what the market has to offer and evaluates the alternatives. The phase ends with the selection of the best alternative and purchase of the product.

5.2.2   General Framework for LCCA

The methodological framework of LCCA has evolved from simpler forms (such as that developed specifically for military systems) to more general forms. The advantage of the former is that they are relatively inexpensive and rapid to use, but they are not adequate for the development of radically new systems (Dhillon, 1989). The methodological framework shown in Figure 5.3 can be considered a reference procedure. It is characterized by four main phases and is inspired by proposals made in more general terms by other authors (Fabrycky and Blanchard, 1991).

5.2.2.1   *Preliminary Definitions*

The preliminary definitions includes the definition of the problem necessitating the application of LCCA, identification of the possible alternatives to be analyzed, and the development of the structure for allocating the costs (Cost Breakdown Structure—CBS). The detailed definition of the problem is necessary for the analysis to be structured correctly, which requires a clear identification of the subject of the analysis itself (alternative technologies, operation and utilization scenarios, servicing and maintenance strategies, different

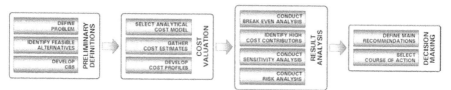

FIGURE 5.3   LCCA framework.

production approaches, etc.). This is followed by the identification of the possible alternatives, which must be guided by the definition of the requisites relative to the main activities comprising the life cycle and which must include forecasts of the consequences that possible alternatives will have on the entire life cycle. Each alternative, in fact, has different implications for the development of the various activities of the cycle. Finally, after the definition of alternative configurations and of the activities associated with them, a structure of cost allocation and collection is developed (Cost Breakdown Structure), which must allow the classification of the different cost typologies, relating them to the main life cycle activities. From the analytical viewpoint, this is achieved through the definition of relations estimating the costs (Cost Estimating Relationships—CERs) consisting of mathematical expressions of varying complexity that express the costs as functions of one or more significant variables (cost-driving variables).

5.2.2.2   Cost Valuation

Cost estimating provides for the choice of the cost model best suited to the case under examination, estimation of the costs according to the CBS, and the development of cost profiles (these bring together the future cost projections in relation to each alternative under consideration). The choice of calculation method and models is one of the key steps in the whole procedure, since models that are not adequate for the purposes of the investigation may be insensitive to the problem set as the objective. As noted below, different models are proposed in the literature. It is necessary to examine their effectiveness with respect to the case under examination, as well as to the preliminary definitions of the problem. The cost estimation must be made in strict relation to the CBS and CERs, making use of a combination of different evaluation methods of varying typology (e.g., engineering procedures, assessments by analogy, parametric methods). Finally, the development of cost profiles is determinant in the comparison of the various alternatives under consideration, since they quantify the influence of the alternatives over the entire life cycle through future cost projections.

5.2.2.3   Results Analysis

This phase gathers together the different procedures for analyzing the results (breakeven, sensitivity, risk analysis) and identifies the most influential cost factors (high cost contributors). The results of the cost estimating phase must be evaluated in different ways. With a breakeven analysis it is possible to compare the performance of different alternatives over time and to determine the moment when one alternative is better than another on the basis of the cost projections. By identifying the main cost factors it is possible to reveal the criticalities of each alternative, indicating which factors may be modified to improve the overall economic performance. This can also help identify the most influential input data of the cost analysis; particular attention must be paid to the validity and reliability of this data, and their effective influence on the final results of the evaluation must be quantified through sensitivity analysis. Finally, risk analysis allows the identification and management of potential areas of risk. Although these can usually be traced back to the critical factors at the source, they are not always completely eliminated by corrective actions. Their potential emergence can be characterized through probabilistic evaluations and must, therefore, be managed appropriately.

5.2.2.4 *Decision Making*

The LCCA process concludes with the decision-making phase—choosing the alternatives considered best, and defining the principal recommendations and actions for improvement.

5.2.3 Decomposition of Costs and Cost Breakdown Structure

A key step in the preliminary definition phase of LCCA, the decomposition of costs, must make it possible to classify the different typologies of economic flows and to relate these flows to the main life cycle activities. As noted above, this is achieved through the development of a structure to collect and allocate costs and revenues, known as Cost Breakdown Structure (CBS). Because it must include all the phases of the product's life cycle, CBS has a tree structure with the branches usually corresponding to the organization of the elementary activities of the life cycle itself. For the purposes of economic analysis, these elementary activities comprise the four phases of research and development, production and distribution, operation and support, and retirement and disposal. In fact, the main categories of a product's life cycle costs correspond to these four phases:

- Research and development costs—Include the costs related to the initial planning of the project (market analysis, feasibility studies, product research, and innovation), development of the design document, evaluation and improvement of the results, and all the associated management operations.
- Production and distribution costs—Include all the costs of product engineering, production planning, manufacturing, testing and quality control, and the distribution costs (marketing, shipping, customer sales).
- Operation and support costs—Include the costs of utilization and operation on the part of the buyer, and after-sales support (customer service, maintenance, updating).

- Retirement and disposal costs—Include the costs of retiring the product at end-of-life, of any recovery of parts and recycling of materials, and of disposal and waste management.

The depth of decomposition of the CBS depends on the purposes of the analysis to be performed. Figure 5.4 shows an example of breakdown structure development which, beginning with an initial subdivision following the four cost categories defined above, branches according to the elementary life cycle activities defined in Figure 2.9 (Chapter 2), the schematic representation of a product's life cycle on the basis of the flows of material resources. This demonstrates how in LCCA, in a manner completely analogous to that of the approach to environmental analysis of the life cycle, the cycle can be schematized in terms of its main phases and elementary activities. This schematization is mirrored in the branching structure of CBS, where all the costs and revenues of a product over its entire life are quantified. As shown in Figure 5.4, the roots of the tree structure correspond to the cost functions, expressed by CERs. The various design alternatives can then be characterized within the same CBS through appropriate local economic parameters.

5.2.4 Life Cycle Cost Models

Clearly, the choice of model for the calculation of costs is fundamental to the entire LCCA procedure. The model consists of a set of assumptions, rules, equations, constants, and variables defining the mechanisms of the system of monetary flows to be examined. From the first steps in the study and development of life cycle cost analysis, a wide variety of models have been proposed, both of a specific type (i.e., developed in relation to the need to evaluate the costs of specific systems) and those of a more general nature (Dhillon, 1989). None of these has ever evolved to become or been accepted as a standard reference model, for diverse reasons: substantial differences in the nature of the

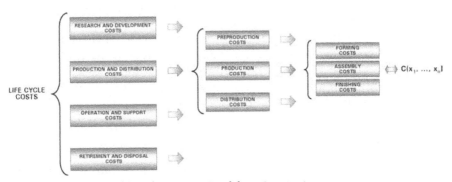

FIGURE 5.4 Decomposition of costs (Cost Breakdown Structure).

problems motivating the analysis, different typologies of products and systems under analysis, and the existence of different systems of data collection. Despite the variety, all the models must unavoidably share some aspects:

- They must represent the system under examination in the most effective manner possible, with respect to the objectives of the analysis.
- They must include predictive properties, usually obtained through stochastic processes involving some important parameters (e.g., interest rates, system reliability) and using simulation techniques that allow the application of different types of analysis (break-even, sensitivity analysis).

Not even the classification of the models is unequivocal. One of the first distinctions, still made today, is between conceptual, analytical, and heuristic models (Kolarik, 1980; Gupta, 1983):

- Conceptual models are based on a set of hypothetical relationships between the economic parameters in play, defined qualitatively. These models are particularly flexible but manifest limitations when a more detailed analysis is required.
- Analytical models are based on mathematical relations developed with the aim of representing particular aspects of the product or system under examination, and in relation to specific preliminary assumptions. These assumptions constitute the main limitation of this model typology, since they usually reduce the accuracy with which the behavior of the real system is represented.
- Heuristic models are commonly defined as ill-structured versions of analytical models, based on an approach that allows the identification of a solution to a specific problem while not guaranteeing that this is the optimal solution.

A large collection of early cost models developed in the 1970s and 1980s, divided into specific and general models, is available in the literature (Dhillon, 1989). Obviously, they are strictly dependent on the definition of the cost categories and on their breakdown structure. In subsequent years this development has continued in the context of an evolution of the models of LCCA itself and, as will be seen below, has led to the definition of new approaches and new methodological frameworks. These have required the adaptation of preexisting cost models and the development of new models.

5.2.5   Cost Estimating

A production activity does no more than combine and transform, through the use of various technologies, different types of limited resources (not only natural resources, but also labor, equipment, and capital) to obtain goods and services. The objective of cost estimation in general is, therefore, "to describe adequately the cost of using the appropriate combination of the limited resources to achieve the required level of goods or services for the society" (Stewart, 1991). Thus, an effective cost estimate, in itself, makes possible an effective use of resources.

Undertaking a complete cost analysis requires a collection of such detailed data on the economic flows that it can be considered practicable only at an advanced stage of the product development process. However, as noted previously, interventions of correction and improvement are more effective if they can be made in the first phases of design. In order to be able to introduce LCCA in these first stages, it is possible to make use of specific methods of cost estimating, conceived to fill the lacunae in an inevitably incomplete collection of data. In this case, cost estimation is intended as a general evaluation, based on the analysis of previous experience and on cost previsions. Given that these are characterized by subjectivity, the estimation must be considered a tentative valuation that can only approximate the real behavior of the monetary flows. Cost estimation can make use of the different methods and instruments reviewed in the literature (Dhillon, 1989; Greer and Nussbaum, 1990), developed to satisfy the need for cost evaluation and prevision in the context of various company operations (manufacturing, engineering, marketing, sales). When conducted in relation to the specific design context, cost estimating is usually assimilated into the discipline known as Cost Engineering (Roy et al., 2001). Three main typologies of methods for cost estimating are used in the field of product development (Fabrycky and Blanchard, 1991; Asiedu at Gu, 1998). On the basis of their particular characteristics, they are applicable at different levels of the development process (Farineau et al., 2001) and can be summarized as:

- Estimating by detailed models (or engineering procedures)—Based on the definition of the elementary operations making up the activity under examination; the definition of the material (including also equipment and instrumentation) and human (working times) resources in play; and the definition of the subsequent allocation of the corresponding costs. These are quantified at the highest level of detail possible and are then elaborated to determine the total cost of the activity. This approach is the most detailed but requires the greatest expenditure of time and the collection of the most detailed information (Wierda, 1988). It is, therefore, the method least suited to the preliminary phases of product development and should only be used at the stage of detailed design phase and beyond. Furthermore, although it is the most accurate, this method is not without imprecision. In particular, the evaluation of the cost of a complex activity obtained by combining a large number of estimates of this type, even though detailed, is usually subject to a greater error than the sum of the errors in the single estimations (Stewart, 1991).

- Parametric estimating—Based on determining the functional relationships between cost variations and the factors on which these depend (product characteristics). These relationships are expressed using mathematical functions primarily obtained through the statistical evaluation of previous design experiences; they allow the evaluation of the costs of a product or activity associated with various important parameters expressing measurable attributes (system performance, physical characteristics of the product, efficiency, etc.). This approach is more appropriate for an LCCA at the stage of product concept development and makes it possible to directly relate technical and economic parameters (DoD, 1999). While it can be considered a valid method for the entire design process, it cannot be used when new technologies are employed.

- Estimating by analogy—Consists of the identification of existing products or components similar to those under examination, and of adapting the costs of the former (already acquired in previous experience) in relation to the differences displayed by the latter. The main disadvantage of estimating by analogy relates to the subjectivity implicit in evaluating the differences between similar activities and in the consequent correction of the costs. This evaluation is undoubtedly the characteristic aspect of the approach and is essentially based on experience (Greves and Schreiber, 1995). The method is, however, suitable for application in the preliminary phases of product development.

The constant need to shorten product development times has significantly increased the importance of cost estimating, favoring the use of ever-more-refined techniques: feature-based cost estimation (in the ambit of parametric estimations); case-based reasoning (in the ambit of estimating by analogy); and neural network-based cost estimating. A complete overview of the most recent techniques, compared with more traditional ones and placed in the context of the product development phase where they can be used, is available in the literature (Rush and Roy, 2000).

Independent of the method used, cost estimations are examples of decision making under conditions of uncertainty, such that each action (evaluation) can lead to more than one final state (result), and each of them can occur with unknown probability (Jelen and Black, 1983). This is because such estimations are previsions of future costs and, as such, the certainty of the evaluation decreases with the increasing time span between the estimate and the actual incurring of the costs. Thus, the problem cannot be ignored in cost estimations of advanced phases of the life cycle (such as that of use) or for products characterized by a long useful life. In fact, these cases present diverse sources of uncertainty (Woodward, 1997)—the differences between the predicted and the real performance of components and subsystems can condition future operating and maintenance costs; the modes of utilization on the part of the consumer can change over time and conflict with expected modes of use; and the arrival of new technologies can accelerate the process of product obsolescence and shorten product life. Additional sources of uncertainty can be identified with errors in the definition of the CERs, in the previsions of the cost of resources, in estimating interest rates, and so on.

On the other hand, the accuracy of cost estimating is a critical factor in product competitiveness, given that an inexact evaluation (either too high or too low) can have serious consequences for the final cost (Freiman, 1983). For this reason, purely deterministic cost models (i.e., those treating economic parameters as terms with stable values that can be evaluated with certainty) are of limited use. They can be applied only as instruments for evaluations of first approximation and can provide reliable information only in cases where a product's life cycle is not prolonged over time. In more general cases, accurate cost estimating requires instead the use of probabilistic models.

5.3 Evolution of Models for Product Life Cycle Cost Analysis

Since the link between design choices and their economic implications was first identified as one of the key factors in the design and development process, considerable effort has been directed at providing the designer with appropriate instruments to evaluate the life cycle costs of a product during the design process. Several important studies date back to the period when LCCA techniques were first used and developed (Gupta and Chow, 1985). Complete and general procedures for LCCA began to be introduced from the early 1990s (Greene and Shaw, 1990; Fabrycky and Blanchard, 1991); these served as inspiration for the reference scheme presented in Section 5.2.2.

Over the past 10 years the development of LCCA models has continued, providing a wide range of new approaches developed from different initial premises and with specific objectives and characteristics. Some refer to particular fields of application—e.g., the design of production systems (Dahlén and Bolmsjo, 1996; Westkamper and von der Osten-Sacken, 1998). Others were developed to aid cost analysis expressly in the design phase, but taking into consideration specific activities of a product's life cycle, such as manufacturing (Boothroyd, 1994), servicing (Gershenson and Ishii, 1993), or retirement (Navin-Chandra, 1993; Ishii et al., 1994).

Alongside all these detailed cost analysis models intended for use at advanced stages of the design process, approximate LCCA methods have also been proposed. These are not presumed to substitute more detailed models but, rather, to use the results obtained in previous analyses to provide the designer with general indications in the early phases of a new project (Seo et al., 2002).

Given the proliferation of models for LCCA developed with the objective of fully integrating cost analysis into product design, the literature contains complete studies that provide an overview of the state of the art as well as comparative information about the characteristics and limitations of the various approaches (Asiedu and Gu, 1998; Kumaran et al., 2001). Among those having a common objective of guiding the development of a generic product (not limited to the examination of a specific activity but potentially applicable to the entire life cycle), it is worth detailing some of the more interesting approaches that are considered alternatives to those underpinning the more traditional models discussed above.

5.3.1 Function Costing

Function Costing consists of breaking the product down into the functions that it must provide and then evaluating the total cost as the sum of the cost of all the functions characterizing the system. It is based on the presupposition that a large number of elementary functions can be quantified and that their associated costs can be correlated to quantitative and qualitative aspects that can be easily valued, regardless of the specific product under examination (French, 1990). The particular features of Function Costing allow this approach to be used from the beginning of conceptual design phase, where it is precisely those functions that must characterize the product that are defined and where alternative ways of obtaining these functions are identified and developed.

5.3.2 Activity-Based Costing

Activity-based Costing (ABC) consists of identifying the activities that will comprise the product's life cycle, and then associating to each activity the main cost factors (cost drivers) and the degree of intensity with which they are consumed (consumption intensities). This approach differs from traditional cost analysis (wherein each unit of product consumes resources) by taking as its starting point the presupposition that a product or service does not use resources directly but, rather, consumes activities. The cost of a product's life cycle or of its life cycle phases can be evaluated as the sum of the costs of all the activities that must be performed.

Although this approach was first adopted as an instrument for cost estimating in the mid-1980s (Cooper, 1990), it can be considered to have evolved into a true and proper method for LCCA. In this respect, clear examples are given by more recent formulations (Greenwood and Reeve, 1992), some of which were expressly developed for use in the ambit of Life Cycle Design (Bras and Emblemsvag, 1996). They also allow the improvement of cost estimating by specifically taking account of the uncertainty factor (see also Chapter 4, Section 4.4.3.1).

5.3.3 Feature-Based Costing

Feature-based Costing (FBC) is a methodology for estimating the cost of a product based on the analysis of a series of its elementary characteristics, called product features. Stimulated by the widespread use of CAD programs in design, it refers to the well-known feature-based approach to product modeling (Wierda, 1991). FBC extends the concept of "feature" from the main geometric element characterizing a component to a characteristic of varying typology (geometric, physical, of process, of activity) (Taylor, 1997). With the FBC approach it is possible to evaluate the consequences that choosing to include or exclude the feature will have not only on the costs of a single component, but also on the system of costs of the entire life cycle of a product consisting of several components.

The advantages of this approach are evident: a clear link between design choices and their implications in terms of cost, and consequent increase in the potential capacity for correcting and optimizing the design; ease of application compared to other approaches, by virtue of a simplification of the collection of data required to calculate the cost of a product; and the transverse nature of the main feature typologies and the resulting possibility of applying the method even when no similar studies exist and no previous data are available.

FBC originated as one of the more recent approaches to parametric estimating (Section 5.2.5) and is considered a valid alternative to the ABC approach, in that it is simpler to apply. FBC has also evolved into a structured method for cost analysis (Brimson, 1998) and has shown potential for use in a "life cycle thinking" perspective (Starling, 1999).

5.4 Reference Standards and Codes of Practice

As observed in Section 5.2, a complete LCCA includes a wide range of different activities. A similarly large number of codes and standards have been developed in relation to specific aspects of the complex LCCA procedure outlined in Figure 5.3. However, none of these has evolved to become the reference standard. As with the cost models discussed in Section 5.2.4, there are two main reasons for this: substantial differences in the nature of the problems motivating the analysis, and different product and system typologies under analysis. Table 5.1 itemizes the more important LCCA reference standards. Although these standards developed in relation to specific industrial sectors (aerospace, petroleum, and natural gas), they (more than others) delineate a methodological

TABLE 5.1   LLC standards and relationship with main activities as illustrated in Figure 5.3

DESIGNATION DOCUMENT TYPE YEAR	TITLE	LCCA ACTIVITIES (FROM Figure 5.3)									
		(1)	(2)	(3)	(4)	(5)	(6)	(7)	(8)	(9)	(10)
SAE-ARP 4293 SAE-ARP 4294 American Standards 1992	Life cycle cost: Techniques and applications (4293) Data formats and practices for life cycle cost information (4294)	✓	✓	✓	✓	✓			✓	✓	✓
NORSOK O-CR-001 NORSOK O-CR-002 Norwegian Standards 1996	Life cycle cost for systems and equipment (001) Life cycle cost for production facility (002)	✓			✓	✓					
ISO 15663-1: 2000 ISO 15663-2: 2001 ISO 15663-3: 2001 International Standards 2000–2001	Petroleum and natural gas industries - Life cycle costing Part 1: Methodology (15663-1) Part 2: Guidance on application of methodology and calculation methods (15663-2) Part 3: Implementation guides (15663-3)	✓	✓	✓	✓	✓			✓	✓	
IEC 60300–3–3 International Standard 1996 and 2004	Dependability management - Part 3-3: Application guide - Life cycle costing	✓		✓	✓	✓			✓	✓	✓

(1) Problem definition; (2) alternatives identification; (3) CBS development; (4) cost model definition; (5) cost estimation; (6) cost profiles development; (7) breakeven analysis; (8) high cost contributors identification; (9) sensitivity analysis; (10) risk analysis.

structure and treat the greatest number of the multiple activities of a complete LCCA in a more or less detailed manner. The right-hand columns of the table show exactly which of the first 10 activities of the LCCA procedure summarized in Figure 5.3 are treated by each of the standards listed.

SAE-ARP 4293/4294 is an American standard developed by the Society of Automotive Engineers for the aerospace industry (ARP stands for Aerospace Recommended Practice) (SAE-ARP, 1992). It introduces LCCA techniques to be applied in the development of a complete program of research, development, testing, and evaluation; procurement; operation and support; and disposal of systems, equipment, and services. The second part of the standard treats the specific case of estimating the costs involved in the development of aerospace propulsion systems.

The NORSOK O-CR-001/002 standard was developed in Norway with the support of the Norwegian Oil Industry Association (OLF), expressly for the offshore petroleum and natural gas industry (NORSOK, 1996). The first part deals with systems and equipment in general, while the second concerns production plants. This standard has recently been substituted by the international standard ISO 15663, explicitly developed in relation to offshore extraction installations (ISO 15663, 2000). Divided into three parts, this standard introduces a methodological structure for LCCA, the main instruments, and guidelines for implementation.

Finally, IEC 60300–3-3 is an international standard introduced by the International Electrotechnical Commission, with a first edition in 1996 and a second in 2004 (IEC 60300, 2004). It is part of the IEC 60300 standards for dependability management and is not expressly oriented toward specific industrial typologies; instead, it presents a general methodological basis for LCCA. Because of its generality, some of its more important subject matter is summarized below.

In conclusion, it should be noted that precisely in order to provide a methodological basis and common application procedures appropriate to any production sector, an apposite Working Group SETAC–Europe was set up at the end of 2002, with the aim of developing a reference code of practice for LCC (Rebitzer and Seuring, 2003).

5.4.1 Standard IEC 60300–3-3: Life Cycle Costing

This standard provides a general introduction to the concept of LCCA, covering a vast field of applications and particularly emphasizing the aspects linked to product dependability (IEC 60300, 2004). After explaining the aims and the importance of LCCA, it outlines the methodological approach; identifies the main cost factors in the development of a project; gives a general guide to conducting an LCCA (including indications for the development of the cost model); and completes the overview with illustrative examples.

TABLE 5.2   Decomposition and definition of cost elements according to IEC 60300–3-3

MAIN CATEGORIES		COST ELEMENTS
Concept and Definition		Market research
		Project management
		System concept and design analysis
		Preparation of a requirement specification of the product
Design and Development		Project management
		System and design engineering (including reliability, maintainabity, and environmental protection activities)
		Design document
		Prototype fabrication
		Software development
		Testing and evaluation
		Productivity engineering and planning
		Vendor selection
		Demonstration and validation
		Quality management
Manufacturing and Installation	Nonrecurring costs	Industrial engineering and operations analysis
		Construction of facilities
		Production tooling and test equipment
		Special support and test equipment
		Initial spares and repair parts
		Initial training
		Documentation
		Software
		Qualification testing
	Recurring costs	Production management and engineering
		Facility maintenance
		Fabrication (labor, materials, etc.)
		Quality control and inspection
		Assembly, installation, and checkout
		Packing, storage, shipping, and transportation
		Ongoing training
Operation and Maintenance	Operation	Labor and training
		Materials and consumables
		Power
		Equipment and facilities
		Engineering modification
		New software release
	Maintenance	Labor and training
		Facilities
		Contractor services
		Software maintenance

TABLE 5.2 (continued)

MAIN CATEGORIES		COST ELEMENTS
	Supply	Labor and training
		Spare parts and repair materials
		Warehousing facilities
		Package, shipping, and transportation
Disposal		System shutdown
		Disassembly and removal
		Recycling or safe disposal

Source: IEC 60300, 2004.

According to this standard, the life cycle should be understood as the "time interval between a product's conception and its disposal," and the life cycle cost as the "cumulative cost of a product over its life cycle" (IEC 60300, 2004). LCCA then becomes a "process of economic analysis to assess the life cycle cost of a product over its life cycle or a portion thereof." The field of application is particularly wide and includes evaluation and comparison of alternative designs; assessment of economic viability of projects and products; identification of cost drivers and cost-effective improvements; evaluation and comparison of alternative strategies for product use, operation, and maintenance; evaluation and comparison of different approaches for replacement, rehabilitation, life extension, and disposal; optimal allocation of available funds to activities in a process of product development; and long-term financial planning.

Table 5.2 shows the definitions of the cost elements proposed by the standard. Given its generality and the level of detail, it can be considered a good reference example for the decomposition of costs in LCCA, irrespective of product typology. Costs are divided according to five principal categories but the decomposition essentially mirrors the division into four categories proposed in Section 5.2.3 and shown in Figure 5.4. The sole difference is the fact that the first cost category described in Section 5.2.3 (research and development costs) is further subdivided in the IEC 60300-3-3 standard (concept and definition costs, design and development costs).

5.5 Summary

Life Cycle Cost Analysis originated as an instrument for the assessment and reduction of costs in much broader contexts than that of product development. It has become a valid aid in the management of the activities of manufacturing companies and, more generally, of all the typologies of organizations that handle and transform resources. On the other hand, the importance of cost estimation and control during the design process with the aim of forecasting the costs involved in the various phases of a product's life cycle is today an inescapable factor that must be taken into account in the development of an efficient product able to succeed in a highly competitive market.

The fact that Life Cycle Cost Analysis is based on the life cycle approach means that this instrument has a primary role in precisely the design context; it is particularly appropriate in relation to Life Cycle Design, with which it has a common basis. Production costs as well as those incurred in the phases of use and disposal are, in fact, strongly conditioned by the first design choices, and this renders Life Cycle Cost Analysis a valuable instrument for managing conflicts and identifying the most effective trade-off strategies and interventions.

5.6   References

Alting, L., Life-cycle design of products: A new opportunity for manufacturing enterprises, in *Concurrent Engineering: Automation, Tools and Techniques*, Kusiak, A., Ed., John Wiley & Sons, New York, 1993, 1–17.

Asiedu, Y. and Gu, P., Product life cycle cost analysis: State of the art review, *International Journal of Production Research*, 36(4), 883–908, 1998.

Asimow, M., *Introduction to Design*, Prentice Hall, Englewood Cliffs, NJ, 1962.

Blanchard, B.S., Life cycle costing: A review, *Terotechnica*, 1(1), 9–15, 1979.

Boothroyd, G., Product design for manufacture and assembly, *Computer-Aided Design*, 26(7), 505–520, 1994.

Bras, B.A. and Emblemsvag, J., Activity-based costing and uncertainty in designing for the life cycle, in *Design for X: Concurrent Engineering Imperatives*, Huang, G.Q., Ed., Chapman & Hall, London, 1996, 398–423.

Brimson, J.A., Feature costing: Beyond ABC, *Journal of Cost Management*, 12(1), 6–12, 1998.

Cooper, R., Implementing an activity-based cost system, *Journal of Cost Management*, 3, 33–42, 1990.

Czyzewski, A.B. and Hull, R.P., Improving profitability with life cycle costing, *Journal of Cost Management*, 5(2), 20–27, 1991.

Dahlén, P. and Bolmsjo, G.S., Life-cycle cost analysis of the labor factor, *International Journal of Production Economics*, 46–47, 459–467, 1996.

Dean, E.B. and Unal, R., Elements of designing for cost, in *Proceedings of AIAA 1992 Aerospace Design Conference—American Institute of Aeronautics and Astronautics*, Irvine, CA, 1992, paper AIAA-92–1057.

Dhillon, B.S., *Life Cycle Costing: Techniques, Models and Applications*, Gordon and Breach Science Publishers, New York, 1989.

Dieter, G.E., *Engineering Design: A Materials and Processing Approach*, 3rd ed., McGraw-Hill, Singapore, 2000.

DoD, Parametric Estimating Handbook, U.S. Department of Defense, Joint Industry–Government, Parametric Cost Estimating Initiative, 1999.

Dowlatshahi, S., Product design in a concurrent engineering environment: An optimisation approach, *International Journal of Production Research*, 30(8), 1803–1818, 1992.

Fabrycky, W.J. and Blanchard, B.S., *Life Cycle Cost and Economic Analysis*, Prentice Hall, Englewood Cliffs, NJ, 1991.

Farineau, T. et al., Use of parametric models in an economic evaluation step during the design phase, *International Journal of Advanced Manufacturing Technology*, 17(2), 79–86, 2001.

Fixson, S.K., Assessing product architecture costing: Product life cycles, allocation rules, and cost models, in *Proceedings of ASME Design Engineering Technical Conference*, Salt Lake City, UT, 2004, DETC2004-57458.

Freiman, F.R., The FAST cost estimating models, *AACE Transactions*, G.5, 1983.

French, M.J., Function costing: Potential aid to designers, *Journal of Engineering Design*, 1(1), 47–53, 1990.

Gershenson, J. and Ishii, K., Life-cycle serviceability design, in *Concurrent Engineering: Automation, Tools and Techniques*, Kusiak, A., Ed., John Wiley & Sons, New York, 1993, 363–384.

Greene, L.E. and Shaw, B.L., The steps for successful life cycle cost analysis, in *Proceedings of IEEE NAECON 1990 National Aerospace and Electronics Conference*, Dayton, OH, 1990, 1209–1216.

Greenwood, T.G. and Reeve, J.M., Activity-based costing management for continuous improvement: A process design framework, *Journal of Cost Management*, 5, 22–40, 1992.

Greer, W.R. and Nussbaum, D.A., *Cost Analysis and Estimating: Tools and Techniques*, Springer-Verlag, Berlin, 1990.

Greves, D. and Schreiber, B., Engineering costing techniques in ESA, *European Space Agency Bulletin*, 81, 63–69, 1995.

Gupta, Y.P., Life cycle cost models and associated uncertainties, in *Electronics Systems Effectiveness and Life Cycle Costing*, Skwirzynski, J.K., Ed., Springer-Verlag, Berlin, 1983, 535–549.

Gupta, Y.P. and Chow, W.S., Twenty-five years of life cycle costing—Theory and applications: A survey, *International Journal of Quality and Reliability Management*, 2(3), 51–76, 1985.

Harvey, G., Life cycle costing: A review of the technique, *Management Accounting*, 1(1), 9–15, 1976.

IEC 60300, Dependability Management—Part 3–3: Application Guide—Life Cycle Costing, *IEC 60300-3-3*, 2nd ed., International Electrotechnical Commission, Geneva, 2004.

Ishii, K., Eubanks, C.F., and Di Marco, P., Design for product retirement and material life cycle, *Materials and Design*, 15(4), 225–233, 1994.

ISO 15663, Petroleum and Natural Gas Industries—Life Cycle Costing—Part 1: Methodology, ISO 15663–1:2000, International Organization for Standardization, Geneva, 2000.

Jelen, F.C. and Black, J.H., *Cost and Optimization Engineering*, 2nd ed., McGraw-Hill, New York, 1983.

Kinch, M.J., Life cycle costing in the defence industry, in *Life Cycle Costing for Construction*, Bull, J.W., Ed., Blackie Academic and Professional, London, 1992, chap. 5.

Kolarik, W.J., Life cycle costing and associated models, in *Proceedings of 1980 Spring Annual Conference—American Institute of Industrial Engineers*, Atlanta, GA, 1980, 58–64.

Kumaran, D.S. et al., Environmental life cycle cost analysis of products, *Environmental Management and Health*, 12(3), 260–276, 2001.

LMI, Life Cycle Costing in Equipment Procurement, LMI Task 4C-5, Logistics Management Institute, Washington, DC, 1965.

Navin-Chandra, D., Restar: A design tool for environmental recovery analysis, in *Proceedings of ICED '93—9th International Conference on Engineering Design*, The Hague, The Netherlands, 1993, 780–787.

NORSOK, Life Cycle Cost for Systems and Equipment, NORSOK O-CR-001, Standards Norway, Lysaker, Norway, 1996.

Pugh, S., Manufacturing cost information: The needs of the engineering designer, in *Proceedings of 2nd International Symposium on Information Systems for Designers*, Southampton, UK, 1974, 12.1–12.8.

Rebitzer, G. and Seuring, S., Methodology and application of life cycle costing, *International Journal of Life Cycle Assessment*, 8(2), 110–111, 2003.

Roy, R. et al., Quantitative and qualitative cost estimating for engineering design, *Journal of Engineering Design*, 12(2), 147–162, 2001.

Rush, C. and Roy, R., Analysis of cost estimating processes used within a concurrent engineering environment throughout a product life cycle, in *Proceedings of CE 2000—7th ISPE International Conference on Concurrent Engineering: Research and Applications*, Lyon, France, 2000, 58–67.

SAE-ARP, Life Cycle Cost: Techniques and Applications, SAE-ARP 4293, Society of Automotive Engineers, Warrendale, PA, 1992.

Seo, K.K. et al., Approximate estimation of the product life cycle cost using artificial neural networks in conceptual design, *International Journal of Advanced Manufacturing Technology*, 19, 461–471, 2002.

Shewchuk, J., Life cycle thinking, *CMA Certified Management Accountant*, 66(4), 34–36, 1992.

Starling, S.L., Feature-based proactive estimation of producer's life cycle costs, in *Proceedings of 28th Annual Meeting of Western Decision Sciences Institute*, Puerto Vallarta, Mexico, 1999, Paper 8.

Stewart, R.D., *Cost Estimating*, 2nd ed., John Wiley & Sons, New York, 1991.

Taylor, I.M., Cost engineering: A feature-based approach, in *Proceedings of 85th AGARD SMP Meeting*, Aalborg, Denmark, 1997, 14.1–14.9.

Ullman, D.G., *The Mechanical Design Process*, 3rd ed., McGraw-Hill, New York, 2003.

Ulrich, K.T. and Eppinger, S.D., *Product Design and Development*, 2nd ed., McGraw-Hill, New York, 2000.

Westkamper, E. and von der Osten-Sacken, D., Product life cycle costing applied to manufacturing systems, *Annals of the CIRP*, 47(1), 353–356, 1998.

Weustink, I.F. et al., A generic framework for cost estimation and cost control in product design, *Journal of Materials Processing Technology*, 103, 141–148, 2000.

Wierda, L.S., Product cost-estimation by the designer, *Engineering Costs and Production Economics*, 13, 189–198, 1988.

Wierda, L.S., Linking design, process planning and cost information by feature-based modelling, *Journal of Engineering Design*, 2(1), 3–19, 1991.

Woodward, D.G., Life cycle costing: Theory, information acquisition and application, *International Journal of Project Management*, 15(6), 335–344, 1997.

Chapter 6

Integrated Economic–Environmental Analysis of the Life Cycle

Life Cycle Cost Analysis (LCCA), the methodology for estimating all the costs associated with a product over its entire life cycle, is perfectly suited to the "cradle to grave" approach that must characterize environmental product design. Precisely because of this life cycle approach, with an appropriately oriented Life Cycle Cost Analysis it is possible to quantify all the environmental costs resulting from the product–system. Some essential differences between the economic and environmental analysis of the life cycle have, however, led to a traditional separation in the study of the two aspects. In the specific context of design, this separation can be held responsible for the absence of a clear vision of the relationships between economic and environmental performances resulting from different design alternatives. This lack of vision has resulted in the limited relevance so far given to Life Cycle Assessment (LCA) in decision making.

The subjects of this chapter are the suitability and possibility of extending the scope of Life Cycle Cost Analysis to cover environmental aspects in product development, and the most important issues arising in attempts to establish an integrated economic–environmental approach to life cycle analysis. Particular emphasis will be given to the possibility of a direct integration of Life Cycle Cost Analysis with Life Cycle Assessment, and to how this integration is incorporated in the domain of Life Cycle Management.

6.1 Life Cycle Cost Analysis and Environmental Aspects

Environmental issues are generally perceived as constraints, impeding the development and planning of industrial activities. Life Cycle Cost Analysis (LCCA) can assume a particularly important role in relation to this problem, since it acts as a primary link between environmental demands and the production strategies of manufacturing companies (Hunkeler and Rebitzer, 2003). In this role, LCCA becomes an operational instrument used to implement one of the basic strategies for achieving sustainable development, that of integrating economic and environmental considerations into the decision-making process (WCED, 1987).

One of the fundamental questions in this context is how to use the instruments of economic analysis more effectively so as to extend their scope to include the environmental aspects of activities. In this regard, in Chapter 5 (Section 5.1) it was noted that an effective approach to the integrated economic-environmental analysis of a product's life cycle requires the adoption of the product–system concept, wherein the product is considered integral with its life cycle and within the environmental, social, and technological context in which this life cycle develops (Chapter 2). The implications for an economic analysis of products that also takes environmental performance into account are evident:

- The analysis of costs, such as those of environmental impacts, must be extended to cover the entire life cycle, from conception and design to retirement and disposal.
- Such an analysis also requires the problematic evaluation of the environmental costs incurred by all the actors affected by the life cycle, including society as a whole.

Furthermore, it should be noted that not all the environmental aspects of an activity are as explicitly linked to factors of economic cost as they are, for example, in the case of an efficient management of resources, where both environmental and economic benefits accrue from a reduction in the consumption of energy and materials. A quite different treatment is required for other environmental aspects, such as the generation of waste and emission of pollutants. While the relevant costs can be associated with these examples, it is not unequivocally certain how these costs should be paid, or by whom. These environmental costs are usually "externalized" (i.e., transferred to various segments of the community under the form of costs of damage to human health and to the well-being of the ecosystem). Alternatively, they can be "internalized" (i.e., assimilated by manufacturers, who invest in measures to avoid damage to the environment or to redress damage caused, thus translating environmental costs into economic costs of product life cycle management). Only effective governmental policies can adequately administer the problem in an eco-efficient way, reducing external environmental costs through legal constrictions, taxation, and economic incentives, and by favoring internalization.

The full integration of environmental aspects into cost analysis, therefore, requires:

- Overcoming the distinction between the different actors involved in a product's life cycle (producer, consumer, society), which creates ambiguity in how environmental costs are perceived

- Introducing a system of eco-costs, understood as a set of costs deterring damage to the environment, and of direct and indirect costs of the environmental impacts caused by a product over its entire life cycle

With these premises, LCCA becomes the assessment of all costs associated with the life cycle of a product "that are directly covered by the any one or more of the actors in the product life cycle (supplier, producer, user/consumer, end-of-life actors), with complimentary inclusion of externalities that are anticipated to be internalized in the decision-relevant future" (Rebitzer and Hunkeler, 2003).

6.1.1 Scenario of LCCA Extended to Environmental Aspects

With the introduction of the environmental question, LCCA fully assimilates the "life cycle thinking" approach. LCCA thus represents a clear evolution from the first limited vision of the economic life of a product, wherein the cost of the product's life cycle is the sum of the economic resources expended on it from conception and manufacture until the end of its useful life (Woodward, 1997). The new, more complete, perspective extends the concept of life cycle and includes within it the potential, already noted, of cost estimating as an instrument for the efficient use of limited resources. LCCA then becomes "a systematic analytical process for evaluating various designs or alternative courses of actions with the objective of choosing the best to employ scarce resources" (Kumaran et al., 2001). The ultimate objective of the product's LCCA is "to provide a framework for finding the total cost of design/development, production, use and disposal of the product with an intention of reducing the total cost" (Kumaran et al., 2001).

This view also introduces a further evolution in the perception of life cycle: that it is no longer variable according to whether it is considered from the viewpoint of the producer or the consumer (as discussed in Chapter 5, Section 5.2.1 and Figure 5.2). Instead, it is considered from a single viewpoint, that of the product itself as it passes through the main phases of its complete life cycle (design and development, production and distribution, use, retirement and disposal). It should be emphasized, however, that although LCCA incorporates this perception of the life cycle (which lies outside the jurisdiction of the main actors involved) necessary for a full understanding of the economic flows associated with environmental phenomena, it still remains an instrument for decision making. In its practical application, therefore, LCCA always takes the viewpoint of one actor—the decision maker.

With these premises, it is possible to construct the reference scenario of LCCA proposed in Figure 6.1, showing the actors and systems involved and the different typologies of cost flows (Rebitzer and Hunkeler, 2003). Here,

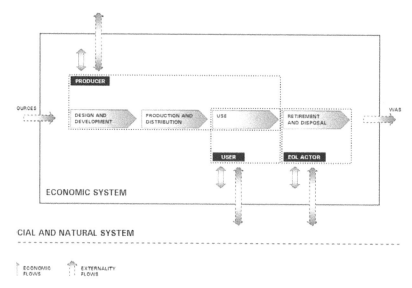

FIGURE 6.1   Scenario of environmental LCCA: Systems, actors, cost flows.

economic flows represent the costs and revenues internal to the economic system, and externality flows indicate environmental cost flows between the product's life cycle and the social and natural system.

Referring to the scenario in Figure 6.1, some aspects should be noted:

- Economic flows are understood as internal costs (i.e., attributable to a precise economic actor, and, therefore, quantifiable in monetary flows of cost or revenue). Externality flows are understood as external costs, quantifying the monetary effects of impacts on the environment and society, and are not directly ascribable to economic actors.

- Each phase of the life cycle traversed by the product involves one or more actors. In particular, the first two phases (design and development, production and distribution) involve the producer; use involves both producer and the user; and retirement involves an actor to be defined (end-of-life actor) who may be the producer and/or the user.

- Each actor interacts with the two systems (economic and natural) through the cost flows internal and external to the economic system.

- The field of application of LCCA is limited to the economic system; therefore, to take account of external costs they must be internalized within this system and translated into monetary flows.

Table 6.1 shows the main life cycle costs, subdivided into the four principal categories introduced in Section 5.2.3 (each corresponding to a phase of the life cycle shown in Figure 6.1) in relation to the three actors involved (producer, user, society), as suggested by various authors (Alting, 1993; Asiedu and Gu, 1998; Perera et al., 1999). The costs befalling society are represented by the externalities of the economic system. They are due to the impacts of activities on the environment and on human health, and they can be monetized using some evaluation methods included in the approach known as "damage cost approach," which consists of quantifying the loss suffered by the damaged actor (Gale and Stokoe, 2001).

6.2 Environmental Costs and Environmental Accounting

From the strictly environmental perspective, LCCA becomes a procedure for the inventory and analysis of the economic implications of the environmental impact of a product over its entire life cycle (Frankl and Rubik, 1999). In this sense it assumes the dual role of an LCA in economic terms, and differs from conventional cost analyses in estimating the entire spectrum of all environmental costs, including hidden and external costs, ascribable to a product's entire life cycle. As noted above, when applying cost analysis intended as a decision-making tool, the perspective adopted is always that of one actor, the decision maker. In the most common

TABLE 6.1   Product life cycle costs and life cycle actors

MAIN COST CATEGORIES/ LIFE CYCLE PHASES	INTERNAL COSTS		EXTERNAL COSTS
	PRODUCER	USER	SOCIETY
RESEARCH AND DEVELOPMENT	Market recognition Research Design and development		
PRODUCTION AND DISTRIBUTION	Materials, energy, labor Facilities Processing Packaging Transport, storage		Resource depletion Waste Pollution Health damages
OPERATION AND SUPPORT	Warranty Service	Materials, energy Maintenance Breakdown	Pollution Health damages
RETIREMENT AND DISPOSAL	Disassembly, reprocessing, recycling Disposal	Recycling or disposal dues	Waste Pollution Health damages

case, where the manufacturing company is considered the reference actor, the whole system of costs can be divided into two main categories (EPA, 1995):

- Private costs (internal costs)—Consist of all the costs incurred by the producer, or by those who can be considered financially responsible.
- Societal costs (external costs or externalities)—Indicate all the costs related to the impact of the production system on the environment and society, for which the producer actor cannot be considered financially responsible.

Both these categories include environmental costs. To the external environmental costs defined in the previous section, it is necessary to add those internal to the production system. These are the environmental costs belonging to the category of private costs, which in turn can be divided into (EPA, 1995):

- Potentially hidden costs—Hidden costs generally refer to all the terms that may be overlooked in the cost analysis of an activity because they are indirect costs; various environmental costs may fall into this category. Hidden costs can, in turn, be divided into upfront costs (acquisition and preproduction costs), back-end costs (costs incurred after the end of useful life), regulatory costs (costs required to comply with environmental legislation), and voluntary costs (costs that go beyond strict compliance with environmental laws). This category can also include costs known as conventional costs (raw materials, supplies, equipment, etc.), given that reducing their use results in obvious environmental benefits, although these are not easily monetized.
- Contingent costs (or liability costs)—Include environmental costs that cannot be planned for because they may be incurred as a result of unpredictable future events (accidental release of pollutants, infractions of future legislative restrictions, etc.). Because of their particular nature, these costs can be expressed in probabilistic terms.
- Image and relationship costs—Include costs related to actions directed at influencing the perceptions of various actors, such as employees, consumers, communities, and controlling bodies regarding the environmental performance of the production activity. Such costs might include preparation and distribution of environmental reports, public relations events in the community, advertising, and so forth. Although the expenses attributable to this type of investment are clearly definable, usually the benefits that may result

(improved company image) are not, and for this reason they are considered less-tangible costs.

6.2.1 Environmental Accounting

The study of the wider system of environmental costs described above is the subject of Environmental Accounting, a term understood as "the definition, assessment and allocation of environmental costs and expenditures for the purpose of cost and resource management, compliance reporting, and capital budgeting, planning, and operational decision making" (Gale and Stokoe, 2001). However, environmental accounting assumes different connotations according to the context in which it is used:

- In the context of financial accounting, the components of internal costs associated with environmental aspects are emphasized and analyzed. In this case environmental accounting is used, for example, in drawing up reports informing an external audience (investors, creditors, clients) of the company's condition and performance.

- In the context of management accounting, it is used as an aid to the company's internal decision-making processes (capital investment choices, performance evaluation, product and process design decisions), and takes into consideration a broader spectrum of environmental costs, also including externalities.

With regard to the latter context, which includes product design and development, reference may be made to the term Environmental Cost Accounting, understood as "the addition of environmental cost information into existing cost accounting procedures and/or recognizing embedded environmental costs and allocating them to appropriate products or processes" (EPA, 1995).

6.2.2 Typologies of Environmental Accounting

While sharing the same conceptual premises, different typologies of environmental accounting have recently been introduced into company practice. They differ principally in the depth of the cost analysis.

The term Total Cost Assessment (or Total Cost Accounting—TCA) describes a method of cost accounting oriented toward the long-term, including in the analysis the entire range of private costs (both conventional and environmental) and paying particular attention to the hidden, less-tangible, and liability costs associated with an activity or product (Gale and Stokoe, 2001). TCA is frequently used to make investment choices directed at improving the environmental performance of production activities (cleaner production and pollution prevention investments) that can be correctly evaluated only through an analysis characterized by a particularly extended time frame.

Full Cost Accounting (FCA) extends the analysis field of TCA to also include external costs (societal costs). It consists of the identification, evaluation, and allocation of the entire system of costs (conventional and environmental) of the life cycle of a product or activity, irrespective of the direct responsibilities of the actors involved. The related term Full Cost Environmental Accounting (FCEA) is sometimes used to specify that the evaluation of external costs focuses on the environmental aspect, ignoring any externalities consisting of other types of societal costs.

6.3   Integration between LCCA and LCA

The appropriateness of integrating cost analysis and environmental impact evaluations, by virtue of the shared life cycle approach, has been emphasized since the beginning of the methodological development of LCA (Henn and Fava, 1994). It has also been noted how this integration is necessary in order to achieve the real environmental effectiveness of systems. Only by investigating the relationships between the economic and environmental performances of a product's life cycle, and stressing the importance of reconciling these two, can environmental analysis be elevated to the level of a strategic decision-making tool for design and for all other company operations.

A first approach in this regard can consist of simply adding cost data to a traditional LCA, and this is sometimes what is intended by the term Life Cycle Cost Assessment (which originated with the precise aim of emphasizing economic aspects in an LCA). It has been defined as "a systematic process for evaluating the life cycle costs of a product, product line, process, system, or facility by identifying environmental consequences and assigning monetary value to those consequences" (EPA, 1995).

A much more complete approach is one that refers to the concept of LCCA as described in Section 6.1, involving the identification and evaluation of all the costs associated with the entire life cycle of a product, process, or activity, including costs deriving from its environmental impact. This concept, which retraces the SETAC definition of LCA, does not limit the economic analysis to the realm of environmental costs alone, as happens in the case of Life Cycle Cost Assessment. Instead, it extends the analysis to cover the entire system of costs associable with the life cycle.

The observation of some similarities between LCA and LCCA could be misleading, making their integration seem simple or even intrinsic to their very nature:

- Their "life cycle thinking" approach, reflected in the names of both the analysis instruments

- Their system-based approach, manifested in LCA by the distribution of physical flows (resources, polluting substances, waste) in the context of the system of elementary activities, and in LCCA by the breakdown of economic flows in a Cost Breakdown Structure (CBS), in some cases structured according to the product's physical life cycle, as shown in Figure 5.4 of the previous chapter
- The opportuneness of operating both analyses, environmental and economic, in the preliminary phases of product development, in that both the environmental impacts and costs of the life cycle are strongly conditioned by the first design choices

In contrast to these affinities, LCA and LCCA are divided not only by the different characteristics of the subjects under analysis (physical flows in the first, monetary flows in the second), but also by important conceptual premises that can hinder their integration (Norris, 2000):

- Aim of analysis—The aim of LCA is to evaluate the environmental performance of products and alternative choices guaranteeing the same functions, from the perspective of benefit for the whole of society. The aim of LCCA is to evaluate the economic effectiveness of investments and alternative choices from the perspective of the actor making the decisions.
- Extension and typology of life cycle model—LCA examines the entire system of processes and activities connected with the product's physical life cycle. LCCA considers the set of activities resulting in costs or revenues for the decision maker. This results in a difference not only in the extension of the life cycle (which in the case of LCA can also include activities and physical flows not associated with direct economic consequences for the decision maker) but also in the typology of the model describing the life cycle, because not all the monetary flows can be directly attributed to the physical flows considered in LCA.
- Influence of the time variable—The life cycle model for LCA does not usually take account of the temporal distribution of resources, wastes, and emissions. Instead, in the analyses for LCCA the time variable assumes a determining role in the evaluation of monetary flows; LCCA requires a clear definition of the duration of the time span, beyond which these flows are ignored.

The traditional separation between environmental and economic analyses of the life cycle is the direct consequence of these fundamental differences. In the specific context of design, this separation can be held responsible for the lack of a clear vision of the relationships between environmental and economic performances involved in different design alternatives, and for the consequent limited relevance given to LCA in decision making until today.

These considerations clearly emphasize the importance of an integrated approach that manages to combine the two analyses, environmental and economic. Such an approach, requiring the development of a model able to handle the different aspects of the two analysis techniques (e.g., manage variables not strictly dependent on physical flows, and introducing the temporal dimension), would allow evaluations of particular importance to decision making (Norris, 2000), such as:

- Identification of design variables whose modification has major effects on the combined environmental-economic performance
- Evaluation of the economic effects resulting from choices directed at improving the environmental performance, and vice versa

In recent years, various important studies have been conducted with exactly these objectives. One of the first proposes applying, in parallel, an LCCA and one of the quantitative methods for LCA (Eco-indicator 95 method, reported in Chapter 4, Table 4.3), with the aim of obtaining an economic/environmental profile of a product, and evaluating the correspondence between these two aspects (van Mier et al., 1996). The same period saw studies based on radically different arrangements, such as that regarding Economic Input–Output LCA (Cobas et al., 1995; Hendrickson et al., 1998), discussed in Chapter 4, Section 4.4.3, or regarding Activity-based Costing presented in Chapter 5, Section 5.3.2, and such as those regarding Activity-based LCA (Emblemsvag and Bras, 1997), introduced in Chapter 4, Section 4.4.3.

More recently, other approaches have been proposed. One of these begins with traditional models for LCA and seeks an evolution directed at also meeting the particular exigencies of LCCA (Norris, 2000). In some cases, this evolution involves broadening the field and the depth of analysis of the model, including the entire system of material and monetary flows ascribable to all the economic actors involved in the life cycle of the product under examination (Murayama et al., 2001). Another approach uses the methodology of Environmental Cost Accounting or of Total Cost Accounting (Section 6.2) as a basis for the integration of the two aspects, environmental and economic, in product development (Beaver, 2000; Norris, 2000; Shapiro, 2001; Tomita, 2001).

Also significant, from the perspective of Life Cycle Design, is the need expressed in recent years to further extend the capacity to analyze design alternatives. One clear example is the attempt to integrate instruments for the environmental and economic evaluation of design choices in the context of customer-oriented design tools, such as Quality Function Deployment (QFD) (Zhang et al., 1999). This is aimed at the evolution of new methodologies of product development or improvement, able to take into consideration environmental, economic, and quality requisites. Finally, as a confirmation of the importance now given to achieving an integrated LCCA–LCA approach, it should be noted that this issue has been identified as one of the priorities in the activity program of the previously cited Working Group on LCC, recently set up by SETAC-Europe ("Analysis of Methodological Challenges in LCC and LCA Integration," Rebitzer and Seuring, 2003).

6.3.1 Integrated Economic–Environmental Approach in Life Cycle Management

An integrated LCCA–LCA is fully qualified for inclusion within the scope of the approach to the integrated management of the life cycle known as Life Cycle Management (LCM), previously introduced in Chapter 3. Understood as a conceptual structure to place alongside preexisting management structures, LCM makes it possible to control the interface between diverse aspects of the production activity—financial, technological, and environmental aspects which, although differing, are all closely linked to the goals of competitiveness and sustainability (Hunkeler and Rebitzer, 2003). Only in the case where LCM acquires a full integration of LCCA and LCA, which make up its operating instruments, is it possible to reach one of its aims: that of interpreting the principles of sustainable development in company practice, making environmental sustainability accessible, quantifiable and operational (Rebitzer and Hunkeler, 2003).

It could indeed be concluded that the idea of appropriately integrating cost analysis and environmental impact evaluation following the life cycle approach, which was first debated in the early 1990s, reaches its full maturation in the context of LCM. Although it maintains more general connotations in some contexts (Chapter 3, Section 3.3), LCM seems increasingly to be associated with environmental protection, as confirmed by the orientation of an apposite Working Group on LCM set up by SETAC in 1998. This group was established with the objective of providing methodological support and guidelines for the implementation and dissemination of LCM in company management; it recently presented its first results, beginning with an eloquent definition of LCM: "A flexible integrated framework of concepts, techniques, and procedures to address environmental, economic, technological and social aspects of products and organizations to achieve continuous environmental improvement from a life cycle perspective" (Hunkeler et al., 2004).

6.4 Other Approaches to Economic–Environmental Analysis: Eco-Cost Models

Although it is clear that the direct integration of LCCA and LCA can be considered the most complete approach to providing a valid aid to the designer or, more generally, to the decision maker who wants to take into account the economic and environmental performance of a product's life cycle, there are some very interesting alternatives. As a conclusion to this chapter, it is worth considering two in particular which, while differing substantially in their premises, content, and aims, have in common the idea of defining functions of environmental cost (eco-cost) (i.e., functions translating different typologies of environmental impact, both direct and indirect, into economic terms).

The first alternative is a reference model called Virtual Eco-costs. It is based on the results of LCA and proposes the evaluation of a single indicator of environmental impact, expressed by the sum of five eco-cost components—three direct (material depletion, energy consumption, toxic emissions) and two indirect (labor, production assets) (Vogtlander et al., 2001). These eco-costs differ from the external costs introduced above. In fact, the latter express in monetary terms the damage caused to the environment, while eco-costs are to be understood as prevention costs (i.e., costs that must be spent preventively in order to take the product system to a condition of environmental sustainability). In this sense, therefore, they are thought of as virtual costs and are not part of the system of monetary flows examined by LCCA. Particularly interesting is the way in which the eco-costs of polluting emissions are evaluated, taking a specific model for the evaluation of pollution prevention costs and making reference to seven different classes of emission effects (these classes mirror the impact categories of Life Cycle Impact Assessment) (Vogtlander and Bijma, 2000). The Virtual Eco-costs model, combined with the analysis of the value chain applied to the life cycle, becomes an integral part of the Eco-costs/Value Ratio (EVR) model, which can be used as a useful metric to evaluate the eco-efficiency of a product, a service, or the service-product system.

The second example of an eco-cost system is represented by a model for LCCA explicitly formulated for the analysis of environmental costs (Kumaran et al., 2001). This model, called Life Cycle Environmental Cost Analysis (LCECA), is based on the traditional methodological framework of LCCA discussed in Chapter 5 (Section 5.2), and in particular (Fabrycky and Blanchard, 1991). Its Cost Breakdown Structure (CBS) is expanded to include a specific structure of eco-costs, which in this case are understood as the direct and indirect costs of the environmental impacts caused by the product over its life cycle. The model provides for different environmental cost terms: costs for controlling discharges; for the treatment of waste discharged into the ecosystem; costs of implementing environmental management systems (EMSs); costs due to environment-related taxes; costs for environmental cleanup operations; energy costs; and cost savings resulting from strategies of recycling and reuse. Each of these terms is added, together with the other monetary flows, to the four principal categories of life cycle costs: research and development, production and distribution, operation and support, retirement and disposal costs (Chapter 5, Section 5.2.3). Finally, using an appropriate mathematical model, it is possible to correlate the total cost of the product to the various eco-costs of its life cycle. Formulated in this way, the approach allows the identification of the best design alternatives, in both environmental and economic terms, and can help the designer reduce the total cost of a product through the adoption of environmental strategies applicable to its life cycle.

6.5   Summary

The life cycle approach upon which Life Cycle Cost Analysis is based invests this instrument with a primary role in the context of Life Cycle Design, where it is a valuable technique for the identification of the most efficient strategies and interventions. Of particular note is the role that Life Cycle Cost Analysis can have in relation to the environmental question, the latter being generally perceived as a constraint hampering the development and planning of industrial activities. In relation to this problem, Life Cycle Cost Analysis emerges as a primary link between environmental requirements and the other production strategies of a company. In this role, it becomes a functional instrument used to achieve one of the sine qua non strategies for sustainable development—to integrate economic and ecological considerations in decision-making processes. In particular, the integration of Life Cycle Cost Analysis and Life Cycle Assessment is fully entitled to be considered part of the integrated life cycle approach known as Life Cycle Management. Only when Life Cycle Management assimilates a complete integration of Life Cycle Cost Analysis and Life Cycle Assessment is it possible to achieve one of its main aims, that of translating the principles of sustainable development into company practice to make environmental sustainability operational.

6.6   References

Alting, L., Life-cycle design of products: A new opportunity for manufacturing enterprises, in *Concurrent Engineering: Automation, Tools and Techniques*, Kusiak, A., Ed., John Wiley & Sons, New York, 1993, 1–17.

Asiedu, Y. and Gu, P., Product life cycle cost analysis: State of the art review, *International Journal of Production Research*, 36(4), 883–908, 1998.

Beaver, E., LCA and total cost assessment, *Environmental Progress*, 19(2), 130–139, 2000.

Cobas, E. et al., Economic input–output analysis to aid life cycle assessment of electronics products, in *Proceedings of IEEE International Symposium on Electronics and the Environment*, Orlando, FL, 1995, 273–277.

Emblemsvag, J. and Bras, B., An activity-based life cycle assessment method, in *Proceedings of ASME Design Engineering Technical Conference*, Sacramento, CA, 1997, DFM-97-119.

EPA, An Introduction to Environmental Accounting as a Business Management Tool: Key Concepts and Terms, EPA 742/R-95–001, U.S. Environmental Protection Agency, Office of Pollution Prevention and Toxics, Washington, DC, 1995.

Fabrycky, W.J. and Blanchard, B.S., *Life Cycle Cost and Economic Analysis*, Prentice Hall, Englewood Cliffs, NJ, 1991.

Frankl, P. and Rubik, F., *Life Cycle Assessment in Industry and Business: Adoption Patterns, Applications and Implications*, Springer-Verlag, Berlin, 1999.

Gale, R.J.P. and Stokoe, P.K., Environmental cost accounting and business strategy, in *Handbook of Environmentally Conscious Manufacturing*, Madu, C.N., Ed., Kluwer Academic Publishers, Norwell, MA, 2001, 119–137.

Hendrickson, C.T. et al., Economic input–output models for environmental life cycle assessment, *Environmental Science and Technology*, 32(7), 184A-191A, 1998.

Henn, C.L. and Fava, J.A., Life cycle analysis and resource management, in *Environmental Strategies Handbook: A Guide to Effective Policies and Practices*, Kolluru, R.V., Ed., McGraw-Hill, New York, 1994, 541–642.

Hunkeler, D. and Rebitzer, G., Life cycle costing: Paving the road to sustainable development?, *International Journal of Life Cycle Assessment*, 8(2), 109–110, 2003.

Hunkeler, D. et al., *Life Cycle Management*, SETAC Press, Pensacola, FL, 2004.

Kumaran, D.S. et al., Environmental life cycle cost analysis of products, *Environmental Management and Health*, 12(3), 260–276, 2001.

Murayama, T. et al., Life cycle profitability analysis and LCA by simulating material and money flows, in *Proceedings of IEEE International Symposium on Electronics and the Environment*, Denver, CO, 2001, 139–144.

Norris, G.A., Integrating economic analysis into LCA, *Environmental Quality Management*, 10(3), 59–64, 2000.

Perera, H.S.C., Nagarur, N., and Tabucanon, M.T., Component part standardization: A way to reduce the life-cycle costs of products, *International Journal of Production Economics*, 60–61, 109–116, 1999.

Rebitzer, G. and Hunkeler, D., Life cycle costing in LCM: Ambitions, opportunities, and limitations, *International Journal of Life Cycle Assessment*, 8(5), 253–256, 2003.

Rebitzer, G. and Seuring, S., Methodology and application of life cycle costing, *International Journal of Life Cycle Assessment*, 8(2), 110–111, 2003.

Shapiro, K.G., Incorporating costs in LCA, *International Journal of Life Cycle Assessment*, 6(2), 121–123, 2001.

Tomita, H., Environmental accounting in product design, in *Proceedings of EcoDesign 2001: 2nd International Symposium on Environmentally Conscious Design and Inverse Manufacturing*, Tokyo, 2001, 658–661.

Van Mier, G.P., Sterke C.J., and Stevels, A.L., Life-cycle costs calculations and green design options: Computer monitor as example, in *Proceedings of IEEE International Symposium on Electronics and the Environment*, Dallas, TX, 1996, 191–196.

Vogtlander, J.G. and Bijma, A., The Virtual Pollution Prevention Costs '99: A single LCA-based indicator for emissions, *International Journal of Life Cycle Assessment*, 5(2), 113–124, 2000.

Vogtlander, J.G., Brezet, H.C., and Hendriks, C.F., The Virtual Eco-costs '99: A single LCA-based indicator for sustainability and the Eco-costs/Value Ratio (EVR) model for economic allocation, *International Journal of Life Cycle Assessment*, 6(3), 157–166, 2001.

WCED, *Our Common Future*, World Commission on Environment and Development, Oxford University Press, New York, 1987.

Woodward, D.G., Life cycle costing: Theory, information acquisition and application, *International Journal of Project Management*, 15(6), 335–344, 1997.

Zhang, Y., Wang, H.P., and Zhang, C., Green QFD-II: A life cycle approach for environmentally conscious manufacturing by integrating LCA and LCC into QFD matrices, *International Journal of Production Research*, 37(5), 1075–1091, 1999.

Part II

Methodological Statement

Chapter 7

Product Design and Development Process

In recent years, innovations in the design process and the management of production have been necessary to reduce the time required and the resources used in the design, production, and distribution of products having increasingly elevated and more diversified performance requirements. Methodological approaches have evolved to aid designers faced with the increasing complexity of design problems and of the system of factors influencing design problems in various ways. The new design challenges require a systematic, integrated, and simultaneous intervention on a product and its correlated processes, according to the new methods known as Concurrent Engineering and Design for X. These design approaches start from different premises, but both tend to embrace the life cycle approach.

This chapter offers a general overview of the product design and development process and the principal issues involved, and considers recent developments that have aspects linked to the life cycle approach. The intent here is to outline the context in which it is possible to introduce, in the most efficient manner possible, a design intervention oriented toward environmental protection.

7.1 Product Design and Development

The role played by technology in relation to the main factors of a process of sustainable development (sociocultural, economic, environmental) has already been discussed. Together with scientific research, technology can provide the instruments to achieve a condition of equilibrium between these factors (i.e. the condition of real sustainability) (Section 1.1.2). Technology can, in fact, be identified as one of the principal products of human activities, able to transform human society and the environment in which it exists. Design, in turn, can be understood as the keystone of technology: "It is how we solve our problems, fulfill our needs, shape our world, change the future, and create new problems. From extraction to disposal in the life cycle of a product, the design process is where we make the most important decisions" (Devon and van de Poel, 2004).

In entirely general terms, "design" can be understood as any activity directed at changing existing realities in such a way as to create the conditions one prefers (Simon, 1981). In relation to the technological dimension of human activity, design becomes a process of organizing and managing human resources and the information developed by them during the evolution of a product (Ullman, 2003). Particularly in cases involving the physical dimension of a concrete industrial product with engineering content, the design activity is a process of transforming resources (cognitive, human, economic, and material) with the aim of translating a set of functionality requirements into the description of a physical solution (product or system) satisfying these requirements.

Although the terms "product design" and "product development" are sometimes considered interchangeable, they are commonly used as complementary terms, giving rise to the expression "product design and development." This implies a possible distinction between the specific activity of design, in the sense expressed above, and a more extended activity which, while including design, encompasses a wider arena that begins with the identification of a need or market opportunity and concludes with the start-up of product manufacture (Ulrich and Eppinger, 2000). Sometimes "product development" is used to indicate an even broader arena, covering the entire process of transforming a market opportunity into a commercially available product, thus also including the production, distribution, and marketing of a product (Krishnan and Ulrich, 2001). Such a complete process thus involves all the main company operations (marketing, design, and production) and relates them to the demands of the consumer.

7.1.1 Contexts and Perspectives of Product Development: General Overview

Product development is, therefore, the entire process of translating needs into technical and commercial solutions (Whitney, 1990). The capacity for innovation in production is now obligatory in the context of a market ever-more-subjected to the pressures of globalization and technological evolution. This means that an effective process of product development has now become an ineluctable requisite in manufacturing (Cooper, 1993). The study and understanding of this process is the subject of considerable interest. Although these research activities have spread widely over recent decades, they began with the first studies defining models and methods for design dating back to the early 1960s (Jones, 1970), emphasizing the aspect associated with decision-making processes (Starr, 1963). This important aspect has remained the core of subsequent studies (Krishnan and Ulrich, 2001), which frequently drew on the ideas and preliminary statements of the early models.

Beginning with the first studies on methodological structuring, research into the process of product development has diversified in relation to the two disciplinary areas most concerned with the question (Smith and Morrow, 1999): engineering design and management science. Although the two different typologies of investigation originate in different areas, they are nevertheless complementary:

- In engineering design, attention is focused on the formal structures and procedures that can guide the designer in the decision-making process, for the purpose of realizing the product in terms of its physical dimension.
- In management science, attention is focused on the wide range of organizational issues related to product development and on the actors involved (design team, project leaders, senior managers, suppliers, customers), with particular regard to the rational planning of activities, the communication network between the different actors, and the function of problem solving.

Within these two different areas, the study of the design and development process has diversified according to perspectives that differ in some important aspects, such as the success factors of the development process, decision variables, performance metrics, and the very way in which the product is perceived (Krishnan and Ulrich, 2001). These perspectives, and the main aspects differentiating them, can be summarized in the following four points (indicating, for each perspective, some general overviews available in the literature and considered particularly significant):

- Marketing perspective (Green and Srinivasan, 1990; Mahajan and Wind, 1992)—According to this perspective, the product is understood as a set of attributes which, together with the final price, constitute the most significant decision variables. The performance metrics are market adherence, customer satisfaction, and profit.
- Organizational perspective (Brown and Eisenhardt, 1995)—In this perspective, the product is understood as the result of an organizational process. Here, the spectrum of decision variables is very broad, since it includes the set of organizational issues associated with the entire product development process whose principal performance metric is the success of the product itself.
- Engineering design perspective (Finger and Dixon, 1989a; Finger and Dixon, 1989b; Braha and Maimon, 1997)—In this case, the product assumes its physical dimension and is understood as a system of interacting components. The decision variables mirror this viewpoint and include function, configuration, shape, and dimensions. The

performance metrics become function and form, technical performance, innovative features, and cost.

- Operations management perspective (Smith and Morrow, 1999)—
This last perspective, which encompasses the competencies of both the engineering and managerial company functions, sees the product as the sequence of or, more generally, the set of activities necessary for its production and commercialization. The decision variables are those that can influence the planning and organization of the different design and development activities (the organization of the design process), and the main performance metric is the efficiency of planning the activities as expressed, for example, by development times and costs.

Finally, it should be noted that the authors proposing this subdivision into four perspectives also provide a complete and detailed review of recent research on product development, transversal to the different perspectives and centered on the decision-making aspects of the different phases of the development process (Krishnan and Ulrich, 2001).

7.1.2   Summary of the Product Development Process

The process of product development is the sequence of phases or activities that must be performed in order to ideate, design, and introduce a product into the market (Ulrich and Eppinger, 2000). The study of this process is directed at defining a schematic pathway common to the vast typology of possible applications. The aim is to delineate a reference model that describes the process through which the ideas about needs are transformed into ideas about things, which in turn are translated into technical prescriptions for the transformation of the most suitable resources into useful material products (Asimow, 1962).

Although there exists no single model that can include the great variety of possible product development processes (and each of these can be considered unique), it is possible to identify activities and elements they have in common. The identification and understanding of these shared factors, as well as allowing a descriptive summarization of the main activities involved and a reference modeling for the comprehension, management, and control of the entire process, is also considered the most effective guide for enabling the evolution of future product development processes.

To describe the product development process, recourse can be made to models available in the literature. According to the traditional viewpoint, product development is an essentially sequential process and can be summarized in the main stages shown in Figure 7.1, combining the

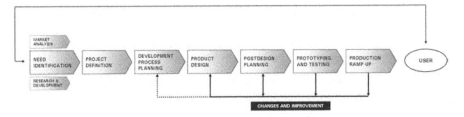

FIGURE 7.1 Product development process: Sequential model.

suggestions of several authors (Dieter, 2000; Ulrich and Eppinger, 2000; Ullman, 2003):

- Need identification—This phase consists of acquiring information on the needs of the customer, identifying the user typology and competing products on the market, and evaluating the most appropriate strategy (improvement of a preexisting product, development of new technologies). Precisely to highlight the close ties this phase has with knowledge of the market and the opportunities afforded by technological innovation, the parallel activities of market analysis and research and development are also included in Figure 7.1.
- Project definition—This is the phase where the project is approved, and constitutes the true and proper beginning of product development. It summarizes the company strategies, market reality, and the technological developments in what is called the project mission statement, which describes the market goal of the product, the company objectives, and the main restraints of the project.
- Development process planning—This phase involves planning the entire design and development process, through the decomposition, planning, and distribution of activities, the definition and distribution of resources (temporal, financial, and human), and the acquisition and distribution of information.
- Product design—This phase includes the specifically design-related activities, from the definition of product requirements and concept generation to the translation of the latter into a producible system. This phase is, therefore, in turn divided into a further subprocess, the design process, which will be discussed in the next section.
- Postdesign planning—This generic title is used to indicate the specific phase regarding the planning of the production–consumption cycle. For some authors, this planning is limited to solely production needs. In this case, it involves the definition and complete planning of product manufacture, from the sequence of machining the components,

to the preparation of tools and machinery, to planning the assembly. Together with production planning, some authors also add the necessities of distribution, use, and retirement of the product (Asimow, 1962; Dieter, 2000).

- Prototyping and testing—This phase requires the development of product prototypes that are then tested in order to evaluate how well the proposed solution satisfies the prescribed requisites, the performance levels offered by this solution, and its reliability. Clearly, this phase has a preeminent influence on interventions to improve the product and, therefore, on the evolution of the design solution.

- Production ramp-up—This phase of starting up production completes the product development process. It consists of manufacturing the product using the planned production system (this is not the case in the prototyping phase, where the product is created in another way). The main aim is to verify the suitability of the real production process, to resolve any problems that arise, and to identify any remaining defects in the final product. Ramp-up is then followed by a phase of transition toward actual production and the definitive launch of the product on the market.

As shown in Figure 7.1, the improvement of the final product is guided by feedback processes which, across the entire development process, transmit information from the postdesign planning and production phases to the product design phase (and, if necessary, also to the preceding phase of development process planning). This mechanism for improving the final result, based on feedback assessments and corrections, has its origins in the first studies theorizing about the design process (Asimow, 1962) and has always been considered its engine (Dieter, 2000)—the true and proper evolutionary mechanism leading to the final solution.

Once distributed and marketed, the product arrives at the final user. This actor is closely tied to the initial phase of the process, that of need identification, because the user interacts with the market and technological innovation, influencing and being influenced by them.

7.2   Product Design

As already noted, within the product development process the specific design phase is divided into a further subprocess (the product design process), which transforms the set of functional specifications and product requirements into the detailed description of the constructional system interpreting them. This transformation is achieved through a design route that begins with the definition of the problem and identification of the requirements, continues with the definition of the product concept, and concludes with the detailed specification of a producible design solution. The literature contains numerous important contributions regarding the conceptual premises and general methodological frameworks for design (Asimow, 1962; Starr, 1963; Alger and Hays, 1964; Jones, 1970; Glegg, 1973; Cross, 1984).

Referring to concrete products with an engineering content, the specifically design-related activities of the product development process essentially lie within a more restricted context, that of engineering design, while not excluding the important contributions that other approaches (in particular, industrial design) can offer in the development of product concept and in the aesthetic, ergonomic, and functional characterization of forms and materials.

7.2.1   Engineering Design

In the context of engineering science, design is the activity that enables the creation of new products, processes, systems, and organizational structures through which engineering contributes to society, satisfying its needs (Suh, 1990). Product design is understood as a process whereby an organizational structure defines a problem and translates it into a feasible solution, making a series of design choices that each depend on the preceding choices and on a set of variables that collectively define the product, how it is made, and how it functions (Simon, 1981; Steward, 1981; Pahl and Beitz, 1996; Smith and Eppinger, 1997).

As with the product development process, the engineering design process cannot easily be assigned a single common scheme due to the great variety of possible design experiences. To summarize this variety, some authors distinguish between product design processes according to the principal categories of design intervention (Sriram et al., 1989):

- Creative design—This typology includes design studies constrained by specific requirements (functionality, performance, producibility) but with no specifications regarding the transformation of the idea into product or the realm of possible solutions.

- Innovative design—In this case, the overall design problem and its possible decomposition into simpler subproblems is already known. Intervention then consists of synthesizing the possible alternatives for each constructional subunit, and can be reduced to a simple originative combination of preexisting components.

- Redesign—This category includes interventions altering and improving preexisting designs. This is necessary when a product does not fully meet the prescribed requirements or when changes in the environmental context for which the product was destined produce new requirements to which the product must be adapted if it is to remain on the market.

- Routine design—In this case, different characteristic design factors such as the form of the product, the method of design approach, and the production system are all known before the design process begins. Intervention is then reduced to the choice of the best alternative with respect to each subunit of the product.

Each of these different categories of design intervention has a corresponding different statement of the design problem (configuration design, selection design, parametric design) (Ullman, 2003).

While it is difficult to determine a single reference procedure, the process of engineering design is also characterized by certain aspects. Foremost among all of these is the evolutionary nature of the process, which is generally understood as a process of evolutionary transformation based on the iteration of successive steps (Simon, 1981). This evolutionary process is, in turn, characterized by:

- An underlying pattern or paradigm consisting of the three phases: analysis–synthesis–evaluation (Jones, 1963; Braha and Maimon, 1997). Analysis allows the definition and comprehension of the problem and its translation into design requirements. Synthesis operates in the selection of the best solutions from all the feasible alternatives. Finally, evaluation compares the best solutions with the specifications and requisites demanded in order to evaluate their validity.

- An evolutionary mechanism that improves the final result based on iterative feedback assessments and corrections (Steward, 1981; Smith and Eppinger, 1997; Dieter, 2000). Each iterative cycle of generating and verifying solutions fully realizes the analysis–synthesis–evaluation paradigm described above. This mechanism of verification and improvement is usually also extended to the product development process.

7.2.2 Organization and Decomposition in Product Design

The phase of development process planning consists of the decomposition, planning, and distribution of all the activities, resources, and information involved in the entire process under consideration. This phase plays a determinant role in relation to the ever-increasing complexity of design problems and sees the emergence of the viewpoint known as the operations management perspective, where the product is perceived as the set of activities necessary for its manufacture and marketing. This viewpoint is implemented through the application of the general principles of Organization Theory, wherein the particular function of product development consists of the planning and organization of design and development activities (i.e., organization of the design and development process).

Underpinning this specific function of design process organization (which by its very nature encompasses both engineering and managerial competences) is the now commonly used practice of modeling the product design process—breaking it down into single tasks and determining the structure and the interactions linking these together. This decomposition makes it possible to reduce the design problem to simpler subproblems, and thus becomes an approach to the management of the ever-greater complexity of design (Kusiak and Larson, 1995). Furthermore, the study of individual design tasks can be an effective approach to the analysis of alternative design strategies, and ultimately to an improvement of the overall design process (von Hippel, 1990; Eppinger et al., 1994).

The importance of organizing and managing the design process is thus clear. This process must then be supported by three different typologies of knowledge: knowledge to generate the ideas, to assess the ideas, and to structure the design process itself (Ullman, 2003). Within the study of the decomposition of the design process, there is a distinction between the two disciplines most involved in product development—engineering design and management science. In engineering design, attention is directed at the structure of the constructional system (study of the product architecture); management science focuses on the structure of the organization managing the project (study of the division and organization of the activities). In a complete perspective, the principle of decomposition can be extended to three different domains (Eppinger and Salminen, 2001): product, process, and organization. In the product domain, decomposition consists of splitting the complex system into subsystems, subassemblies, and components. In the process domain, it consists of dividing the design process into tasks, activities, and work units. Finally, in the organization domain the decomposition involves structuring the human resources into teams and workgroups and assigning them individual tasks.

Product and process domains are of particular relevance to this book, and for this reason it is worth considering two important issues in greater detail.

7.2.2.1 Integration and Decomposition of Product Architecture

The structure of the constructional system and its decomposition into subunits and components are considered key factors in effective product design (Suh, 1990; Pahl and Beitz, 1996; Ulrich and Eppinger, 2000). This structure is linked to the concept of product architecture, which is understood as the scheme through which the functions of a product are allocated to physical components (Ulrich, 1995). Product architecture defines the decomposition of a product in terms of subdivision into constructional units (the functional units or physical blocks comprising the product), the geometric arrangement of these constructional units, and the system of interactions linking them together. It is precisely the system of interactions that assumes a determinant role. The factor best characterizing an architecture is, in fact, the level of dependence between the constructional units; depending on this level, the product architecture can be modular or integral. The constructional units of a modular product contain only one, or a few, functional elements that are linked together by a limited number of well-defined interactions. Conversely, the constructional units of an integral product each contain various functional elements that are linked together by a complex system of interactions which are not always clearly defined.

The implications deriving from the characteristics of product architecture have been examined by various authors (Ulrich, 1995; Erixon, 1998; Kamrani and Salhieh, 2000; Ulrich and Eppinger, 2000) in relation to different aspects: product variety, standardization of the constructional units, product performance and quality, and management of the development process. A broad investigation into the benefits and limitations of varying the degree of modularity (Oosterman, 2001) has demonstrated how an effective design must succeed in finding the appropriate equilibrium between modular and integral architecture, in relation to a vast range of determining factors closely tied to the requisites demanded of the product and to the organizational and technological characteristics of the production reality.

7.2.2.2   Integration and Decomposition of Design Process

The approach to the decomposition of the design process appears to conflict with the principle of the integration of activities which, as will be discussed later, is the basis of simultaneous design and Concurrent Engineering. As noted by various authors (von Hippel, 1990; Smith and Eppinger, 1997; Smith and Morrow, 1999), the integration of activities does not in itself constitute the universal answer for the improvement of the design process. Rather, its effectiveness depends on the complexity of the problem. In the case of elevated complexity, excessive integration can result in weighing down the coordination mechanisms and in the overlapping of activities, disadvantages that can be managed only by the decomposition of the problem.

The decomposition–integration of the design process activities must, therefore, be appropriately balanced in relation to the objectives, typology, and complexity of the design problem, in a manner analogous to that required when defining the degree of modularity-integration of the product architecture.

7.2.3   Product Design Process

Design becomes the instrument linking functional requirements (which are part of the functional domain) to the physical solution (characterized by design parameters and belonging to the physical domain). The process of product design begins, therefore, with defining the functional requirements that satisfy a given set of needs and translating them into design parameters. Product design finishes with the creation of the physical object satisfying these requirements (Suh, 1990).

As noted in Section 7.2.1, the great variety of possible design experiences makes it difficult to arrive at a common model for the product design process. This explains the wide range of theoretical investigations in the field of engineering science regarding this aspect, and the proliferation of different methodological frameworks beginning in the 1960s (Buhl, 1960; Asimow, 1962; Woodson, 1966; Hill, 1970). Over the last 10 years, a large number of models for the engineering design process have been proposed with the aim of improving the understanding and practice of the design activity (Pugh, 1990; Hubka and Eder, 1992; Cross, 1994; Ertas and Jones, 1996; Pahl and Beitz, 1996; Ullman, 2003). Although critical analyses of these have underlined the difference between how the design process is theorized in the proposed models and how it is actually conducted in industrial practice (Maffin, 1998), many of these methodological frameworks have not only contributed to the evolution of the concept of "design" but have also become the structures of reference procedures.

7.2.3.1 Typologies of Design Process Models

In the final analysis, methodological frameworks for design are targeted at structuring the activities and orienting decision making during the entire design process in order to improve its effectiveness. Their variety is such that they, and the corresponding design process models they generate, can be divided into several main categories. A first distinction is made between design models focusing on the design process and those focusing on the artifact itself (Konda et al., 1992). Regarding models that focus on the design process, it is possible to make a further distinction between two different categories (Finger and Dixon, 1989a; Blessing, 1996):

- Describing methods—These are based on models that describe how the design process actually develops; thus, they summarize how it is (describing the process—how is). The use of this type of model is limited to cases where the problem requires an acceptable solution but not necessarily the optimal solution. This is because the design process represented in this way is of a heuristic type, based on previous studies and guidelines orienting the designer toward a solution without any guarantee that it is the most valid.

- Prescribing methods—These are a collection of methodological procedures that prescribe an operational modality structured in a rational manner, and thus suggest how the design process should be (prescribing the process—how should be). These models prescribe a sequence of activities to be undertaken following a systematic procedure, and are therefore better able than other types to summarize appropriate design methodologies.

With regard to the design models focusing on the artifact rather than on the design process, these are based on the premise that the design stems from a complete functional specification and that there are universal methods that can be used in all cases to achieve the product requirements. Various methods belong to this category, a typical and well-known example being that of Axiomatic Design (Suh, 1990).

7.2.3.2   Reference Model

Many of the methodological approaches referred to at the beginning of this section are considered prescriptive models. Despite belonging to the same category, they are not easily represented by a common scheme since they are generally characterized by different statements and by particular aspects. The different models do, however, have some elements in common:

- The design process is described as a sequence of activities consisting of several main stages, and is characterized by intermediate results. Usually three or four stages are considered: problem definition, conceptual design, embodiment design, and detail design (Konda et al., 1992; Blessing, 1996; Maffin, 1998).

- The sequence of activities is conceived in such a way that the product design proceeds from the abstract to the concrete. In this way, it is possible to initially operate in a solution space as vast as possible, and subsequently make the process converge toward a concrete, achievable solution.

- In the design process, the function of assessing the results assumes a determinant role, turning it into a process of evolutionary transformation based on the iteration of successive cycles of analysis–synthesis–evaluation (Section 7.2.1).

The reference scheme proposed here to describe the product design process is inspired by the most commonly used prescriptive models, and is based on the following premises:

- "Product" is understood as a concrete artifact that has a clear physical dimension and engineering content, and therefore belongs in the realm of technical systems whose functional behavior is based on physical principles and governed by the laws of physics.

- "Product design" is understood as an activity that applies scientific techniques and principles to a set of information (needs, requirements, constraints), with the aim of defining the constructional system comprising the product in a manner sufficiently detailed to allow its physical realization.

With these premises, the product design process then becomes a process of transforming information from the state where it describes and characterizes a product demand (needs, constraints, consumer requirements, market conditions, available technology) to a state where it fully describes the technical systems able to satisfy the initial demand (Hubka and Eder, 1996). This process of transformation is achieved through the use of various types of resources (cognitive, human, economic, material) that fuel the main phases of the design process. As summarized in Figure 7.2, these phases consist of:

- A preliminary phase of specifying the problem and defining the product requirements
- Phases of design at different levels (concept, system, detail) and of assessment, making up the iterative cycles of analysis–synthesis– evaluation (Pahl and Beitz, 1996; Dieter, 2000; Ulrich and Eppinger, 2000; Ullman, 2003)

In particular, it is worth considering the first four phases, typical of prescriptive models, in greater detail:

- Problem Specification—In this phase all the information relative to the project in question is elaborated to develop and define the requisites that must characterize the product. Information describing the

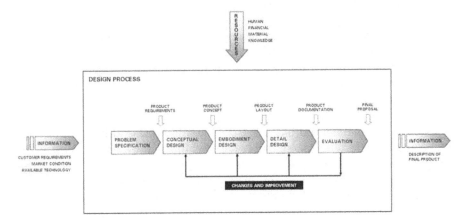

FIGURE 7.2 Product design process.

needs to be satisfied, the consumer requirements, the market conditions, and the company strategies must be clarified (and integrated, if necessary) and used to generate the specifications that will guide the subsequent design phases.

- Conceptual Design—Having defined the project specifications, it is necessary to develop ideas that will allow the creation of a product with the desired requisites. In this phase, "product" is understood in the abstract sense, as a set of attributes that must be embodied in the product concept. This is achieved through a first step of generating ideas (concept generation) and a second step of assessment and selection (concept evaluation) (Dieter, 2000; Ullman, 2003). In fact, a certain number of general concepts are proposed (description of the shape, functions, and main characteristics of the product) and then assessed in order to determine which of them best meets the intended target.

- Embodiment Design—Having identified the most appropriate concept, the next phase is the preliminary interpretation of the design idea in a physical system. The concepts formulated in the previous phases are developed, their feasibility is verified, and finally they are translated into a general product layout that defines subsystems and functional components. In this phase, the physical elements are combined to achieve the required functionality and the product architecture thus takes shape. This phase also includes a preliminary study of the shape of the components and a first selection of materials.

- Detail Design—The layout developed in the previous phases must be translated into geometric models and detailed designs. This requires the application of methods and tools aiding a correct definition of the design details. The choice of materials, study of the shapes, definition of the geometry of components and assemblies, and the development of the assembly sequences and definition of the junction systems must all be guided by the entire range of product requirements (performance, economic, environmental, etc.). To complete this phase, some authors provide for the comprehensive planning of the production process (Ulrich and Eppinger, 2000), while others suggest including instructions for production, assembly, shipping, and use in the final documentation (Pahl and Beitz, 1996).

Figure 7.2 shows the importance of the evaluation phase at the base of each iteration of the design process. This phase evaluates the degree to which the proposed solution corresponds to the design specifications defined in the problem specification phase, and guides modifications and improvements that can make the process evolve toward the definitive solution (i.e., the product that best satisfies the desired requisites). With this aim, it is necessary to analyze the critical aspects of the design in order to predict how the chosen solution will behave over time in relation to environmental factors (socioeconomic conditions, consumer tastes, competing offers, availability of raw materials). Technological factors (technological progress, deterioration in performance) and verification programs (modeling, initial prototyping) must also be incorporated. Among the verification techniques, modeling to evaluate product performance assumes a particularly important role. This usually consists of the simplest method (analytical, physical, graphical) to compare the detailed solution with the engineering targets, generally on the basis of numerical values.

The evaluation procedures must, however, take into account the effect of disturbance factors due, for example, to production or to changes in environmental conditions. A high-quality solution must be as robust as possible (i.e., such that its performance is not affected by disturbance factors).

A great variety of formal methods are frequently cited in the literature as tools supporting the various phases constituting the design process. They include Quality Function Deployment (QFD) matrices for the definition of design specifications and product requirements; the development of graphical or physical mockups for concept evaluation and selection; techniques of functional decomposition and the use of morphological charts for function–concept mapping; and various techniques of product generation for the development of the detailed design. For a complete review of the various formal methods prescribed in relation to the different phases of the design process, together with evaluations on their effective use in the design practice of manufacturing companies, refer to studies already proposed in the literature (Maffin, 1998).

7.2.4 Product Design in the Context of the Product Development Process

In the reference scheme proposed for the product development process, summarized in Figure 7.1, the phase of product design is contextualized as part of a sequential process that is separate from the development process planning phase (upstream) and the postdesign planning phase (downstream). Expanding the product design process following the scheme of Figure 7.2, it is possible to obtain the general overview of the product design and development process shown in Figure 7.3. While this simply duplicates the phases of the entire process presented in Figures 7.1 and 7.2, it does evidence some particularities regarding the relation between the product design process and the two phases of development process planning and postdesign planning.

7.2.4.1 *Relation with the Development Process Planning Phase*

As already noted in Sections 7.1.2 and 7.2.2, the development process planning phase consists of the decomposition, planning, and distribution of the

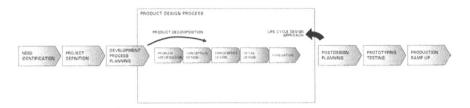

FIGURE 7.3   Product design and development process: Overview.

activities, resources, and information in play within the entire development process under consideration. It is appropriate, therefore, to emphasize that the part of this phase concerning the planning of strictly design-related activities is sometimes considered an integral part of the design process. In a vision extended to include the management of the human resources involved in the design process, this phase inevitably tends to assimilate typical models of project management, focusing on the decomposition, planning, and distribution of the design activities, and on the management of the information and interactions. For this reason, in Figure 7.3 it is shown as a phase that is not entirely outside the product design process.

This phase also provides for the organization of the design process through a modeling of the process itself and its decomposition into single tasks in order to reduce the design problem to a set of simpler subproblems. Section 7.2.2 noted how the principle of decomposition is extended to cover not only the process and organization domains but also the product domain. In the context of the latter, the decomposition consists of dividing the complex system into subsystems and components and, since it is the basis for the definition of the product architecture, this decomposition is one of the key factors of effective design. It is evident how this aspect places the development process planning phase in close relation with the phases of product design and, in particular, with that of embodiment design.

7.2.4.2   Relation with the Postdesign Planning Phase

In Section 7.1.2 it was noted that the phase of postdesign planning, which in the general scheme of the product development process follows the product design phase (Figure 7.1), concerns the planning of the production–consumption cycle. For some authors, this planning is limited to the necessities of production only and is considered to be part of the product design phase, in particular at the detail design level (Ulrich and Eppinger, 2000). In this case, it includes the definition and complete planning of product manufacture, from the sequence of machining components to the preparation of tools and machinery, including the design of any particular equipment that may be required. It is even more interesting to note that some authors also include in production planning the necessities of product distribution, use, and retirement (Asimow, 1962; Dieter,

2000). Although this planning is distinct from the primary activities of the product design process, it can be considered an integration of the specifically design-related phases and, therefore, of the product design process itself. From this perspective, production–consumption cycle planning includes the following aspects:

- Programming the production process—Constitutes the phase linking the design to the engineering requirements. It includes the detailed definition of the machining process, subassembly, final assembly, the definition and programming of the production plant system, the programming of production control and the quality control system, and financial programming.
- Programming for distribution—Assesses the effect of distribution problems on the design. It includes the design of packaging and programming of marketing operations.
- Programming for consumption—Incorporates into the design important characteristics of servicing and provides a rational basis for improvement and redesign. It includes design for maintenance, uniformity of operation, safety, ease of use, aesthetic characteristics, and economy of servicing.
- Programming for product retirement—Harmonizes obsolescence and the anticipated fully functional lifespan, and evaluates hypotheses of reuse and recovery. It includes the analysis of the factors determining the duration of the product, forecasts on the process of obsolescence, the programming of useful life, provision for different levels of use requiring different performances, the programming of recovery interventions, and analysis of products no longer in use in order to acquire information useful for redesign.

It is clear how the complete integration of production–consumption cycle planning, as conceived above, into the product design process would fully achieve the presuppositions of Life Cycle Design (Chapter 3) as highlighted in Figure 7.3. This aspect will be further considered in relation to the concepts of concurrent design and Design for X.

7.3 Methodological Evolution in Product Design

In general, the apparent complexity of the design process and of the system of factors influencing it in various ways is counterbalanced by a limited freedom of choice over the materials to be used and the ways in which these materials can be processed and assembled in the product in order to obtain the required functions. This condition complicates the intervention of the designer, whose objective is to balance the various specifications of function, cost, and reliability to achieve the appropriate compromise allowing the attainment of an ever-broader spectrum of required performances. As a direct consequence, the design process and production management have required considerable innovation in recent years, principally with the aim of reducing the time and resources employed in the design, production, and distribution of products.

In this context, one of the more important considerations is the need to extend the vision of the problems to be taken into account to cover the entire product design and development process. This involves moving from a conventional approach limited to considering the sale of the product as the final step of the analysis, to an innovative approach where the phase of product use is also considered, going so far as to include the product end-of-life and disposal. This is in complete harmony with the potential scope the postdesign planning phase can assume in cases where the planning of the entire production–consumption cycle is included (Section 7.2.4.2), and with the basic premises of Design for Environment and Life Cycle Design, and more generally with the product life cycle approach.

These new necessities, and the consequent increase in the level of complexity of the design problem, have revealed the inadequacy of the sequential structure of the design and development process as represented in Figure 7.1. It is, in fact, limited by two types of disadvantages:

- Prolonged development times due to the sequential nature of the different functions
- Limited capacity for product improvement because of the poor communication between the various functions and the consequently reduced and fragmentary information flows

The product design and development models described above, initially characterized by rigidly sequential structures, must therefore be inserted in new methodological contexts that provide for design actions of analysis and synthesis that are simultaneous and in close interaction, in relation to all the phases of product development. The sequential model, represented schematically in Figure 7.1, thus evolves into the simultaneous/integrated product development model shown in Figure 7.4. In this model, the phases of process development planning, product design, production–consumption cycle planning, and results evaluation are fused in a single, simultaneous intervention that draws information from a shared source and takes into account a wide variety of aspects (functionality, producibility, reliability, and cost).

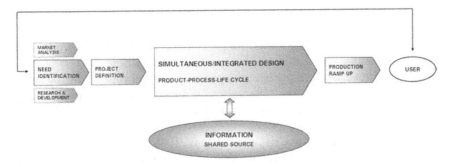

FIGURE 7.4 Product development process: Simultaneous/integrated model.

In the context of this new, simultaneous/integrated statement, three approaches are currently the subject of much of the research regarding design methodologies:

- Concurrent Engineering (CE)—Aims at a full harmonization between the increase in product quality and the reduction of development times and costs through a structuring of product development that involves a large design team conducting simultaneous and interconnected analysis and synthesis actions, in relation to all the phases of development.
- Design for X (DFX)—Involves a flexible system of design methodologies and tools, each directed at the attainment of a particular product requirement.
- Life Cycle Design (LCD)—Extends the field of design analysis to the entire life cycle of the product, from the production and use of materials to disposal.

While this latter approach has already been discussed in Chapter 3, the first two approaches deserve more detailed discussion, given their importance in the current state of engineering design. Their particular characteristics make them specially relevant to the issues considered in this book.

7.3.1 Concurrent Engineering

The ever-shorter useful life of products, a phenomenon characteristic of current market dynamics, demands a constant reduction in the time and costs required for the product development process. Conflicting with this necessity, in recent years development times have increased due to the growing complexity of design problems and the need to involve specialists from different disciplines. It is precisely this second aspect that requires the decomposition of design.

Efficient and competitive industrial production must produce products of high performance and quality, at low cost, in a short time (McGrath et al., 1992). To achieve these requirements, product development must be structured and managed as a simultaneous and multidisciplinary process, engaging a suitably structured design team able to cover a broad spectrum of competencies.

Concurrent Engineering (CE) (Winner et al., 1988; Nevins et al., 1989; Syan and Menon, 1994; Prasad, 1996), also called Simultaneous Engineering (Allen, 1990), developed in response to this need. It is directed at reconciling an increase in product quality with a reduction in development times and costs (Sohlenius, 1992). It can be defined as a systematic approach to the integrated and simultaneous design of products and processes that includes production problems and user support. The aim of this approach is to have, from the very beginning, the product development team simultaneously consider all the determinant factors operating on the product's life cycle, from concept development to retirement, including the requisites of quality, cost, and production planning, together with the exigencies and requests of the user (Winner et al., 1988; Kusiak and Wang, 1993).

The foundations of CE are frequently represented by several essential principles, which can be summarized in the following points (Jo et al., 1993):

- Highlighting the role of production process planning and its influence on the decisions of the product design process
- Emphasizing the multidisciplinary dimension of the design team engaged in the product development process
- Paying greater attention to customer demands and satisfaction
- Considering the reduction of development times and of time to market as factors of product success and competitiveness

Clearly, these principles are neither surprising nor radically innovative but, rather, are based on common sense. For this reason, CE can be considered an evolution of product development practice based on the criterion of efficiency; it can be seen as "a summary of best practice in product development, rather than the adoption of a radical new set of ideas" (Smith, 1997).

7.3.1.1 Characteristic Features of Concurrent Engineering

Reducing product development times is one of the primary objectives of CE. It is easy to imagine that one positive effect resulting from the simultaneous approach to design intervention is precisely that of shorter times. Figure 7.5 compares the structure of the traditional process of product development

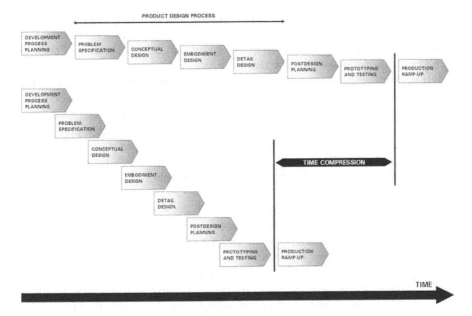

FIGURE 7.5 Comparison between product development processes typologies: Sequential Engineering—Concurrent Engineering. (Adapted from Yazdani, B. and Holmes, C., Four models of design definition: Sequential, design centered, concurrent and dynamic, *Journal of Engineering Design*, 10(1), 26, 1999.)

(sequential in nature) with that of a simultaneous or concurrent development process (Yazdani and Holmes, 1999). It should be noted, however, that the integrated approach leads to an increase in the complexity of the design problem, which may itself prolong development times. The primary objective of CE is, therefore, to reconcile this increased complexity, due to the level of integration between design and other company functions (principally that of manufacturing) with the control and reduction of the times and costs of development and production (Kusiak and Wang, 1993).

Furthermore, the possibility of overlapping the activities of the design and development process (as shown in Figure 7.5), in order to make it as simultaneous and integrated as possible, depends on how each of these activities is correlated with the others (Eppinger, 1991). Some activities may be completely independent of the others and can thus be performed simultaneously. Others will require information produced by activities that clearly must be performed first, thereby setting up a sequential structure. Other activities can be intimately linked, with one requiring information produced by the other and vice versa, so that they must be performed iteratively. Thus, the possibility of rendering the design process simultaneously is strictly dependent on the decomposition and planning of the design activities (Section 7.2.2).

As well as through simultaneous design and development activities, shorter development times can also be obtained through a more efficient design intervention, with a consequent reduction in the number of corrective iterations necessary for the design process to converge on the optimal solution (Bras and Mistree, 1991). The cyclical iteration of evaluation and improvement is, in fact, a basic paradigm of engineering design processes. An integrated design, following the scheme shown in Figure 7.4, can reduce the number of these iteration cycles since its very nature makes it possible to increase the effectiveness of each design choice. Integrated design considers several aspects simultaneously and takes into account a wider spectrum of determinant factors, ultimately reducing the number of necessary corrective interventions.

A fully integrated design must also achieve the integration of all the main phases of the product development process, beyond those of planning and design, to ultimately become a single, simultaneous intervention taking account of a wide range of aspects (functionality, producibility, reliability, costs, market conditions). This requires eliminating the traditional separation between product ideation and design (resulting from the design function) and its physical manufacture (resulting from the production function). Overcoming this divide, hoped for since the first critical analyses of design practice (Smith, 1997), is fully achieved precisely with CE, which removes the barrier separating design not only from production but also from other important company functions (marketing, research, financial management, etc.). In practice, this is achieved through the creation of multidisciplinary teams whose members are experts in different company functions; for this reason they are called cross-functional teams. This criterion of integrating company functions into product development, typical of CE, is now a common presupposition in the most recent methodological frameworks and arrangements of the development process (Ulrich and Eppinger, 2000; Ullman, 2003).

7.3.1.2   *Concurrent Engineering and Life Cycle Approach*

In light of what has been said regarding the need to eliminate the barriers separating the main company functions, the principle of integrating design and production is one of the fundamental precepts of CE. This is confirmed by the fact that this principle is the basis of the first definitions of CE itself (Whitney, 1988; Smith, 1997), and frequently leads CE to be compared with an integrated product and process design intervention known as Integrated Product and Process Development (IPPD) (Usher et al., 1998). Its importance is implicit in the observation that the limitations and defects of the design solution only come to light in the contexts of production and use (i.e., at a stage when corrective interventions are extremely expensive). This observation suggests that a decisive increase in the effectiveness of the design intervention requires a design approach that takes account of production necessities, and goes even further.

The continually broadening spectrum of design requirements and the increasing importance of producing customer-oriented products, together with new demands associated with the phases of product retirement and disposal, have led to extending the principle to address questions related to production issues as well as the requisites that must condition design choices. This extension is achieved through the life cycle approach, which promotes the consideration of not only the necessities of production but also those of the subsequent phases of the product's life cycle. The life cycle approach significantly broadens the range of factors to be taken into account: functionality, manufacturing, assembly, testing, maintenance, reliability, cost, and quality (Abdalla, 1999).

This aspect clearly brings CE closer to Life Cycle Design (LCD). As was noted in Chapter 3, LCD is a design intervention that considers all the phases of a product's life cycle (development, production, distribution, use, maintenance, disposal, and recovery) within the context of the entire design process (from the phase of concept definition to that of detail design development). Clearly, LCD has a close affinity with the CE approach; the two have some important features in common:

- An integrated design statement, not limited to considering only the primary product requirements but also taking into account various other factors (mainly producibility in CE)
- The presupposition that the most effective interventions are those conducted in the first phases of the design process
- The simultaneity of the analysis and synthesis of the various aspects of the design problem

For this reason, CE and LCD are sometimes considered to resemble one another (Yan et al., 1999); however, they can be separated on the basis of a substantial difference in the aims of the design intervention:

- The aims of CE are, as noted above, more markedly oriented toward the compression of the product development process (Tipnis, 1998).
- The aims of LCD are predominantly oriented toward the optimization of the product's performance over its entire life cycle (Alting, 1993).

As a direct consequence, it is also possible to point out differences in their premises and methodological statements:

- CE, by its very name, emphasizes the simultaneity of analysis and synthesis of design features, and the multidisciplinary nature of the design and development team.

- LCD emphasizes the extension of the analysis of design requirements and interventions to optimize performance over the product's
  entire life cycle.

Given that the full realization of the conceptual premises of LCD implies
simultaneity in the design intervention, it can be considered an extension of
CE, as some authors have suggested (Keoleian and Menerey, 1993).

7.3.2   Design for X and Design-Centered Development Model

The costs of modification and improvement in a rigidly sequential development process are often heavy, to the detriment of the competitiveness of
the manufacturing company. This is due to the possibility that that main
defects in the design solution are discovered too late (i.e., in the testing
and verification phases downstream of the design intervention, or even in
the production phase). The separation of different company functions and
the consequent limitation of design specifications to the primary requisites
demanded of the product make the design intervention even more
inefficient.

This aspect, combined with the need to reduce development times, has
resulted in an evolution of the structure of the development process from
a sequential to a concurrent model. One structure tending toward the
concurrent model while partly maintaining the sequential dimension of
some phases, and giving particular emphasis to the vast range of requisites demanded of the product in relation to the various phases of the life
cycle, is called "design-centered" (Yazdani and Holmes, 1999) and is
presented in Figure 7.6. In this model, the principle of simultaneity is
applied to the specifically design-related phases, thus emphasizing the
importance of a high level of information-sharing in the design analysis.
That being said, the development process does not require the direct
involvement of experts in various company functions, as does CE. The
increased effectiveness of design choices and the extension beyond the
primary product requirements are obtained through reinforcing the primary
design phases themselves (conceptual, embodiment, detail design). This
is realized introducing a series of conceptual and analytical tools and techniques, differing according to product requirement, and applied at different
design levels: Performance Analysis (PA); Design for Manufacturing and
Assembly (DFM, DFA); and Life Cycle Cost Analysis (LCCA).With the
design-centered model, therefore, it is possible to introduce and prioritize in
product development a flexible system of integrated design methodologies
and tools, each targeted at the attainment of a particular product requisite.
This is known as Design for X.

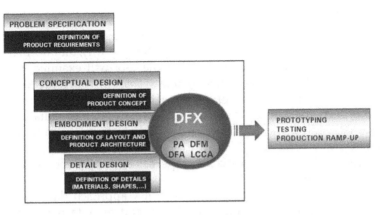

FIGURE 7.6 Product design and development process: Design-centered model. (Adapted from Yazdani, B. and Holmes, C., Four models of design definition: Sequential, design centered, concurrent and dynamic, *Journal of Engineering Design,* 10(1), 29, 1999.)

7.3.2.1 The Design for X System

In cases where a generic industrial product must pass through all the phases comprising its life and where design strongly influences the product's performance in each phase, there has recently been growing interest in a new methodological approach called Design for X (DFX). X stands for different properties of the product, characterizing it in relation to one or more phases of its life cycle (Gatenby and Foo, 1990). Generally, DFX is an approach to product design expressly directed at maximizing the requisites demanded of the product (functionality and performance, manufacturability, quality, reliability, serviceability, safety, user friendliness, environmental friendliness, short time to market) and, at the same time, at minimizing costs (Bralla, 1996). Since these requisites characterize the behavior of the product over its entire life cycle, these premises have led to the identification and classification of criteria for design oriented toward the production, assembly, use, and recovery of the product, with the objective of developing an aid to design as complete and flexible as possible that takes into account the exigencies deriving from all the phases of the product's life cycle.

Ultimately, DFX becomes a design system made up of tools and techniques aiding decision making, conducting different types of analysis and, through appropriate metrics, quantifying the performance of the design choices. These tools can assume different forms, from a set of guidelines to a detailed procedure using analytical models, sometimes implemented in software programs (Herrmann et al., 2004).

7.3.2.2   Objective Properties and Design for X Tools

Each tool of the DFX system must provide designers with a structured method allowing them to use, in the most effective way, an information set specific to the product requirement that the tool will examine. DFX can, therefore, be understood as "a knowledge system in which the knowledge about ways to reach individual properties of technical systems during designing have been (or can be) collected and organized into suitable forms" (Hubka, 2000). The characteristics of the technical system to be obtained during the design process, which are expressions of the requisites demanded of the product, become the objective properties of the different tools of DFX, taking their names.

Although some studies resembling the DFX approach date back to the 1950s and concern the integration of design and manufacturing, the technique that is generally recognized as the first belonging to this design system is Design for Assembly (DFA), used from the early 1970s (Huang, 1996; Kuo et al., 2001) and now well-known. Stimulated by a growing awareness of the product life cycle approach and by the increasing diversification of product requirements, a vast range of DFX tools and techniques have been developed, each characterized by a specific objective property to be attained (Kuo et al., 2001; Herrmann et al., 2004).

Some of the main objective properties (and, therefore, also the corresponding tools and techniques of the DFX system) can be grouped in relation to the phase of the product's life cycle where they are most precisely applied. Regarding the production phase, the most common DFX techniques are:

- Design for Producibility/Manufacturability—Design of components and constructional systems as a function of machining processes (Corbett et al., 1991; Bralla, 1998; Boothroyd et al., 2001; Poli, 2001)

- Design for Assembly—Design for facilitating assembly, reducing assembly times and the possibility of error (Andreasen et al., 1988; Miyakawa et al., 1990; Boothroyd et al., 2001)

- Design for Variety—Design oriented toward product variety, achieved through definition and optimization of basic model typologies, to provide a flexible product architecture guaranteeing variety in relation to market demand (Robertson and Ulrich, 1998; Martin and Ishii, 2002)

- Design for Robustness/Quality—Design of components directed at reducing the sensitivity of product performance to uncontrollable disturbance phenomena (machining errors and unforeseen operating conditions), and guaranteeing product quality (Phadke, 1989; Sanchez et al., 1993; Taguchi, 1993; Park, 1996)

In relation to the use phase, there are:

- Design for Reliability—Design of components and systems aimed at improving the reliability of the constructional unit (Rao, 1992; Birolini, 1993; Wallace and Stephenson, 1996)
- Design for Serviceability/Maintainability—Design oriented toward facilitating servicing and maintenance, to favor repair or interventions directed at excluding the possibility of failure phenomena (e.g., the replacement of worn parts) (Gershenson and Ishii, 1993; Klement, 1993; Dewhurst and Abbatiello, 1996; Kusiak and Lee, 1997)
- Design for Safety—Design oriented toward controlling safety standards and the prevention of malfunctions during use (Gauthier and Charron, 1995; Wang and Ruxton, 1998)

In relation to the end-of-life phase:

- Design for Product Retirement/Recovery—Design directed at the planning of disposal and recovery strategies at the end of the product's useful life (Ishii et al., 1994; Navin-Chandra, 1994; Zhang et al., 1997; Gungor and Gupta, 1999)

In contrast, other objective properties are applied across the entire DFX system, since they cannot be related to a single phase of the life cycle. Among these, it is worth noting those specifically linked to the containment of costs that serve as the basis for Design for Cost (introduced in Chapter 5), and those associated with environmental protection that are at the root of Design for Environment and Design for Sustainability—techniques of the DFX system already considered in Chapter 1.

7.3.2.3 Choice of Design for X Tools and Their Use in the Design Process

The same premises that led to the introduction and development of DFX are those that render the components of the DFX system completely harmonious with Life Cycle Design. Thus, these components can be operational tools for LCD, each specific for a particular objective property of the product, characterizing it in relation to one or more phases of its life cycle.

The choice of the DFX tools and techniques to use in the design process depends on the requirements defined as the target. Their use must be planned according to not only the phase of the life cycle they affect, but also to the specific design phases where they can be most effective. To aid the designer in identifying the most suitable tool, it is possible to represent graphically the most appropriate fields of application, in relation to the phases of the product's life cycle and to the phases of the design process (Ishii, 1995). Figure 7.7 proposes the contextualization of some DFX techniques, summarizing the product life cycle (horizontal axis) in the phases of component fabrication, assembly, use, and retirement, and the design process (vertical axis) in the phases of conceptual, embodiment, and detail design.

It should be noted that the objective properties of the most commonly used DFX tools are strictly dependent on the design variables typical of the embodiment design phase (distribution of functional units, layout) and the detail design phase (shape of components, dimensions, materials) (Hubka and Eder, 1996; Hubka, 2000). These two phases are, therefore, those where DFX tools can intervene with greatest efficacy, as shown in the figure. However, the possibility of also using these tools in prior phases of the design process (problem specification, conceptual design) should not be ignored. This is true, for example, in Design for Manufacturing (Herrmann et al., 2004).

A more detailed approach to the choice and use of tools for DFX in the product development process consists of classifying the techniques in terms of the sources of information (Watson and Radcliffe, 1998). This approach is associated with the presuppositions according to which each technique of the DFX system must provide designers with a structured method allowing them to manage and elaborate, in the most efficient manner, the information set necessary for the design intervention. Depending on the information that these techniques require for their application, and which they can subsequently give to the development process, it is possible to assess whether each tool can be effectively used in the different phases of the design and development process.

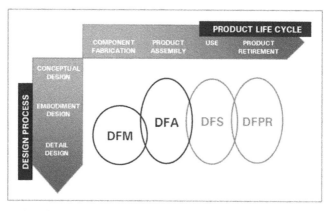

□ DFM - Design for Manufacturing   □ DFS - Design for Serviceability
□ DFA - Design for Assembly   □ DFPR - Design for Product Retirement

FIGURE 7.7 Contextualization of DFX techniques. (Adapted from Ishii, K., Life-cycle engineering design, *Journal of Mechanical Design,* 117, 42–47, 1995, fig. 2. With permission.)

A further question of particular importance for the purposes of using DFX tools is that linked to a possible methodological framework for their development, and specific proposals have been formulated regarding this issue (Huang and Mak, 1997).

7.3.2.4 Design for X and Design-Centered Model in Relation to Other Methodological Approaches

DFX is a system of tools and techniques for design which, if suitably structured and managed, can be used effectively in the product development process, permitting the improvement of a wide range of aspects in industrial production, including product quality, costs, and time to market. Although DFX can thus be considered an approach allowing the implementation of CE (focusing on a limited number of aspects at a time) (Huang, 1996), in the context of CE its tools are not considered essential because their contribution is provided by the direct competence of the multidisciplinary team (Yazdani and Holmes, 1999). The essential difference between the design-centered model (to which the DFX system is grafted) and the concurrent model at the root of CE lies precisely in the fact that they are based on different strengths: in the first case, it is the greater effectiveness of the specifically designed-related phases and of decision making; in the second, it is the greater competence found in the development team.

This same strength of the design-centered model (i.e., the particular emphasis given to the design phases and the scope of the DFX system tools) differentiates it from other widely used methodological approaches that tend to involve the use of DFX tools to aid assessing the effectiveness of the design choices (Ullman, 2003), rather than as an aid to the true and proper development of the design solution. The latter possibility, exploited by the design-centered model, is particularly effective in design for environmental requirements, and is useful for an efficient integration of environmental issues in the product development process, as will be discussed in the next chapter.

7.4 Summary

Referring to concrete industrial products with an engineering content, design activity is a process that transforms resources (cognitive, human, financial, material) in order to translate a set of functional requirements into the description of a physical solution (product) satisfying these requirements. The process of product design and development involves specific design activities but also encompasses a wider arena, which begins with the identification of a market opportunity and finishes with the ramp-up of production. Such a complete process involves all the main company functions (marketing, planning, production) and relates them to customer demands.

The evident complexity of the design process and of the system of factors influencing it in various ways has, in recent years, been further complicated by the need to fulfill an ever-broader spectrum of requirements while at the same time respecting the necessities imposed by an increasingly competitive market. As a direct consequence, the design process and production management have required considerable innovation, principally with the aim of reducing the time and resources used in the design, production, and distribution of products whose performance must meet increasingly diverse demands.

These new exigencies have evidenced the inadequacy of the traditional sequential structuring of the design process, characterized by long development times and a limited capacity to optimize design solutions. The rigidly sequential models have been supplanted by new methodological contexts providing for simultaneous and closely interactive design actions of analysis and synthesis, regarding all the phases of product development. Concurrent Engineering and Design for X are two approaches representing this new simultaneous/integrated methodology. By virtue of their premises and main characteristics, they fully embrace the life cycle approach, which is the ineluctable basis for an efficient and effective design oriented toward environmental protection.

7.5   References

Abdalla, H.S., Concurrent engineering for global manufacturing, *International Journal of Production Economics*, 60–61, 251–260, 1999.

Alger, J.R.M. and Hays, C.V., *Creative Synthesis in Design*, Prentice-Hall, Englewood Cliffs, NJ, 1964.

Allen, C.W., *Simultaneous Engineering: Integrating Manufacturing and Design*, Society of Manufacturing Engineers, Dearborn, MI, 1990.

Alting, L., Life-cycle design of products: A new opportunity for manufacturing enterprises, in *Concurrent Engineering: Automation, Tools and Techniques*, Kusiak, A., Ed., John Wiley & Sons, New York, 1993, 1–17.

Andreasen, M.M. et al., *Design for Assembly*, 2nd ed., Springer-Verlag, Berlin, 1988.

Asimow, M., *Introduction to Design*, Prentice Hall, Englewood Cliffs, NJ, 1962.

Birolini, A., Design for reliability, in *Concurrent Engineering: Automation, Tools and Techniques*, Kusiak, A., Ed., John Wiley & Sons, New York, 1993, 307–348.

Blessing, L.T.M., Comparison of design models proposed in prescriptive literature, in *The Role of Design in the Shaping of Technology*, Perrin, J. and Vincks, D., Eds., Office for Official Publications of the European Communities, Luxembourg, 1996, 187–212.

Boothroyd, G., Dewhurst, P., and Knight, W., *Product Design for Manufacture and Assembly*, 2nd ed., Marcel Dekker, New York, 2001.

Braha, D. and Maimon, O., The design process: Properties, paradigms, and structure, *IEEE Transactions on Systems, Man, and Cybernetics—Part A*, 27(2), 146–166, 1997.

Bralla, J.G., *Design for Excellence*, McGraw-Hill, New York, 1996.

Bralla, J.G., *Design for Manufacturability Handbook*, 2nd ed., McGraw-Hill, New York, 1998.

Bras, B.A. and Mistree, F., Designing design processes in decision-based concurrent engineering, *SAE Transactions—Journal of Materials and Manufacturing*, 100, 451–458, 1991.

Brown, S.L. and Eisenhardt, K.M., Product development: Past research, present findings, and future directions, *Academy of Management Review*, 20(2), 343–378, 1995.

Buhl, H.R., *Creative Engineering Design*, Iowa State University Press, Ames, IA, 1960.

Cooper, R.G., *Winning at New Products: Accelerating the Process from Idea to Launch*, Addison-Wesley, Reading, MA, 1993.

Corbett, J. et al., *Design for Manufacture: Strategies, Principles, Techniques*, Addison-Wesley, Workingham, UK, 1991.

Cross, N., *Development in Design Methodology*, John Wiley & Sons, New York, 1984.

Cross, N., *Engineering Design Methods: Strategies for Product Design*, 2nd ed., John Wiley & Sons, Chichester, UK, 1994.

Devon, R. and van de Poel, I., Design ethics: The social ethics paradigm, *International Journal of Engineering Education*, 20(3), 461–469, 2004.

Dewhurst, P. and Abbatiello, N., Design for serviceability, in *Design for X: Concurrent Engineering Imperatives*, Huang, G.Q., Ed., Chapman & Hall, London, 1996, 298–317.

Dieter, G.E., *Engineering Design: A Materials and Processing Approach*, 3rd ed., McGraw-Hill, Singapore, 2000.

Eppinger, S.D., Model-based approaches to managing concurrent engineering, *Journal of Engineering Design*, 2(4), 283–290, 1991.

Eppinger, S.D. and Salminen, V., Patterns of product development interactions, in *Proceedings of ICED '01—International Conference on Engineering Design*, Glasgow, UK, 2001, 283–290.

Eppinger, S.D. et al., A model-based method for organizing tasks in product development, *Research in Engineering Design*, 6(1), 1–13, 1994.

Erixon, G., Modular function deployment: A method for product modularisation, Ph.D. thesis, The Royal Institute of Technology, Stockholm, 1998.

Ertas, A. and Jones, J.C., *The Engineering Design Process*, 2nd ed., John Wiley & Sons, New York, 1996.

Finger, S. and Dixon, J.R., A review of research in mechanical engineering design—Part I: Descriptive, prescriptive, and computer-based models of design processes, *Research in Engineering Design*, 1(1), 51–67, 1989a.

Finger, S. and Dixon, J.R., A review of research in mechanical engineering design—Part II: Representations, analysis, and design for the life cycle, *Research in Engineering Design*, 1(2), 121–137, 1989b.

Gatenby, D.A. and Foo, G., Design for X: Key to competitive, profitable markets, *AT&T Technical Journal*, 69(3), 2–13, 1990.

Gauthier, F. and Charron, F., Design for health and safety: A simultaneous engineering approach, in *Proceedings of ICED '95—International Conference on Engineering Design*, Prague, Czech Republic, 1995, 891–896.

Gershenson, J. and Ishii, K., Life-cycle serviceability design, in *Concurrent Engineering: Automation, Tools and Techniques*, Kusiak, A., Ed., John Wiley & Sons, New York, 1993, 363–384.

Glegg, G.L., *The Science of Design*, Cambridge University Press, Cambridge, UK, 1973.

Green, P.E. and Srinivasan, V., Conjoint analysis in marketing: New developments with implications for research and practice, *Journal of Marketing*, 54(4), 3–19, 1990.

Gungor, A. and Gupta, S.M., Issues in environmentally conscious manufacturing and product recovery: A survey, *Computers and Industrial Engineering*, 36, 811–853, 1999.

Herrmann, J.W. et al., New directions in Design for Manufacturing, in *Proceedings of ASME Design Engineering Technical Conference*, Salt Lake City, UT, 2004, DETC2004–57770.

Hill, P.H., *The Science of Engineering Design*, Holt, Rinehart and Winston, New York, 1970.

Huang, G.Q., Introduction, in *Design for X: Concurrent Engineering Imperatives*, Huang, G.Q., Ed., Chapman & Hall, London, 1996, 1–18.

Huang, G.Q. and Mak, K.L., The DFX Shell: A generic framework for developing design for X tools, *Robotics and Computer-Integrated Manufacturing*, 13(3), 271–280, 1997.

Hubka, V., Design for X – DFX, in *Proceedings of DESIGN 2000—International Design Conference*, Dubrovnik, Croatia, 2000, 7–14.

Hubka, V. and Eder, W.E., *Engineering Design*, Heurista, Zürich, 1992.

Hubka, V. and Eder, W.E., *Design Science*, Springer-Verlag, London, 1996.

Ishii, K., Life-cycle engineering design, *Journal of Mechanical Design*, 117, 42–47, 1995.

Ishii, K., Eubanks, C.F., and Di Marco, P., Design for product retirement and material life cycle, *Materials and Design*, 15(4), 225–233, 1994.

Jo, H.H., Parsaei, H.R., and Sullivan, W.G., Principles of concurrent engineering, in *Concurrent Engineering: Contemporary Issues and Modern Design Tools*, Parsaei, H.R. and Sullivan, W.G., Eds., Chapman & Hall, London, 1993, 3–23.

Jones, J.C., A method of systematic design, in *Conference on Design Methods*, Jones, J.C. and Thornley, D., Eds., Pergamon Press, Oxford, UK, 1963, 53–73.

Jones, J.C., *Design Methods: Seeds of Human Futures*, John Wiley & Sons, New York, 1970.

Kamrani, A.K. and Salhieh, S.M., *Product Design for Modularity*, Kluwer Academic Publishers, Boston, 2000.

Keoleian, G.A. and Menerey, D., Life Cycle Design Guidance Manual, EPA/600/R-92/226, U.S. Environmental Protection Agency, Office of Research and Development, Cincinnati, OH, 1993.

Klement, M.A., Design for maintainability, in *Concurrent Engineering: Automation, Tools and Techniques*, Kusiak, A., Ed., John Wiley & Sons, New York, 1993, 385–400.

Konda, S. et al., Shared memory in design: A unifying theme for research and practice, *Research in Engineering Design*, 4(1), 23–42, 1992.

Krishnan, V. and Ulrich, K.T., Product development decisions: A review of the literature, *Management Science*, 47(1), 1–21, 2001.

Kuo, T.-C., Huang, S.H., and Zhang, H.-C., Design for manufacture and design for "X": Concepts, applications, and perspectives, *Computers and Industrial Engineering*, 41, 241–260, 2001.

Kusiak, A. and Larson, N., Decomposition and representation methods in mechanical design, *Transaction of the ASME—Special 50th Anniversary Design Issue*, 117, 17–24, 1995.

Kusiak, A. and Lee, G., Design of parts and manufacturing systems for reliability and maintainability, *International Journal of Advanced Manufacturing Technology,* 13, 67–76, 1997.

Kusiak, A. and Wang, J., Decomposition in concurrent design, in *Concurrent Engineering: Automation, Tools and Techniques,* Kusiak, A., Ed., John Wiley & Sons, New York, 1993, 481–507.

Maffin, D., Engineering design models: Context, theory and practice, *Journal of Engineering Design,* 9(4), 315–327, 1998.

Mahajan, V. and Wind, J., New product models: Practice, shortcomings and desired improvements, *Journal of Production Innovation Management,* 9(2), 128–139, 1992.

Martin, M.V. and Ishii, K., Design for variety: Developing standardized and modularized product platform architectures, *Research in Engineering Design,* 13(4), 213–235, 2002.

McGrath, M.E., Anthony, M.T., and Shapiro, A.R., *Product Development: Success through Product and Cycle-Time Excellence,* Butterworth-Heinemann, Boston, 1992.

Miyakawa, S., Ohashi, T., and Iwata, M., The Hitachi new assemblability evaluation method (AEM), *Transactions of the North American Manufacturing Research Institute of SME,* 18, 352–359, 1990.

Navin-Chandra, D., The recovery problem in product design, *Journal of Engineering Design,* 5(1), 67–87, 1994.

Nevins, J.L., Whitney, D.E., and De Fazio, T.L., *Concurrent Design of Products and Processes: A Strategy for the Next Generation in Manufacturing,* McGraw-Hill, New York, 1989.

Oosterman, B.J., Improving product development projects by matching product architecture and organization, Ph.D. thesis, Rijksuniversiteit Groningen, Groningen, The Netherlands, 2001.

Pahl, G. and Beitz, W., *Engineering Design: A Systematic Approach,* 2nd ed., Springer-Verlag, London, 1996.

Park, S.H., *Robust Design and Analysis for Quality Engineering,* Chapman & Hall, London, 1996.

Phadke, M.S., *Quality Engineering Using Robust Design,* Prentice Hall, New York, 1989.

Poli, C., *Design for Manufacturing: A Structured Approach,* Butterworth-Heinemann, Boston, 2001.

Prasad, B., *Concurrent Engineering Fundamentals,* Vol. 1, Prentice Hall, Englewood Cliffs, NJ, 1996.

Pugh, S., *Total Design: Integrated Method for Successful Product Engineering,* Addison-Wesley, Wokingham, UK, 1990.

Rao, S.S., *Reliability-Based Design,* McGraw-Hill, New York, 1992.

Robertson, D. and Ulrich, K., Planning for product platforms, *Sloan Management Review,* 39(4), 19–31, 1998.

Sanchez, S. et al., Quality by design, in *Concurrent Engineering: Automation, Tools and Techniques,* Kusiak, A., Ed., John Wiley & Sons, New York, 1993, 235–286.

Simon, H.A., *The Science of the Artificial,* 2nd ed., MIT Press, Cambridge, MA, 1981.

Smith, R.P., The historical roots of concurrent engineering fundamentals, *IEEE Transactions on Engineering Management,* 44(1), 67–78, 1997.

Smith, R.P. and Eppinger, S.D., A predictive model of sequential iteration in engineering design, *Management Science,* 43(8), 1104–1120, 1997.

Smith, R.P. and Morrow, J.A., Product development process modeling, *Design Studies*, 20(3), 237–261, 1999.

Sohlenius, G., Concurrent engineering, *Annals of the CIRP*, 41(2), 645–656, 1992.

Sriram, D. et al., Knowledge-based system applications in engineering design: Research at MIT, *AI Magazine*, 10(3), 79–96, 1989.

Starr, M.K., *Product Design and Decision Theory*, Prentice Hall, Englewood Cliffs, NJ, 1963.

Steward, D.V., The design structure system: A method for managing the design of complex systems, *IEEE Transactions on Engineering Management*, EM-28(3), 71–74, 1981.

Suh, N.P., *The Principles of Design*, Oxford University Press, New York, 1990.

Syan, C.S. and Menon, U., *Concurrent Engineering: Concepts, Implementation and Practice*, Chapman & Hall, New York, 1994.

Taguchi, G., *Taguchi on Robust Technology Development: Bringing Quality Engineering Upstream*, ASME Press, New York, 1993.

Tipnis, V.A., Evolving issues in product life cycle design: Design for sustainability, in *Handbook of Life Cycle Engineering*, Molina, A., Sanchez, J.M., and Kusiak, A., Eds., Kluwer Academic Publishers, Dordrecht, The Netherlands, 1998, 413–459.

Ullman, D.G., *The Mechanical Design Process*, 3rd ed., McGraw-Hill, New York, 2003.

Ulrich, K.T., The role of product architecture in the manufacturing firm, *Research Policy*, 24, 419–440, 1995.

Ulrich, K.T. and Eppinger, S.D., *Product Design and Development*, 2nd ed., McGraw-Hill, New York, 2000.

Usher, J.M., Roy, U., and Parsaei, H.R., *Integrated Product and Process Development: Methods, Tools, and Technologies*, John Wiley & Sons, New York, 1998.

von Hippel, E., Task partitioning: An innovation process variable, *Research Policy*, 19(5), 407–418, 1990.

Wallace, K.M. and Stephenson, J., Design for reliability, in *Design for X: Concurrent Engineering Imperatives*, Huang, G.Q., Ed., Chapman & Hall, London, 1996, 245–267.

Wang, J. and Ruxton, T., A design for safety methodology for large engineering systems, *Journal of Engineering Design*, 9(2), 159–170, 1998.

Watson, B. and Radcliffe, D., Structuring design for X tool use for improved utilization, *Journal of Engineering Design*, 9(3), 211–223, 1998.

Whitney, D.E., Manufacturing by design, *Harvard Business Review*, 66(4), 83–91, 1988.

Whitney, D.E., Designing the design process, *Research in Engineering Design*, 2(1), 3–13, 1990.

Winner, R.I. et al., The Role of Concurrent Engineering in Weapons System Acquisition, IDA Report R-338, Institute for Defense Analyses, Alexandria, VA, 1988.

Woodson, T.T., *Introduction to Engineering Design*, McGraw-Hill, New York, 1966.

Yan, P., Zhou, M., and Sebastian, D., An integrated product and process development methodology: Concept formulation, *Robotics and Computer-Integrated Manufacturing*, 15, 201–210, 1999.

Yazdani, B. and Holmes, C., Four models of design definition: Sequential, design centered, concurrent and dynamic, *Journal of Engineering Design*, 10(1), 25–37, 1999.

Zhang, H.C. et al., Environmentally conscious design and manufacturing: A state of the art survey, *Journal of Manufacturing Systems*, 16(5), 352–371, 1997.

Chapter 8

Integration of Environmental Aspects in Product Design

One of the most important aspects of Design for Environment (DFE) is that it can act as a connecting bridge between production planning and development and the environmental management of the same, two functions that are usually separate. In order to fulfill this role, the design activity must have several ineluctable features: a product life cycle orientation; the balancing of a wide range of requirements; and a simultaneous and integrated structure of the design intervention. Only on the basis of these premises is it possible to conceive a process of product development that furthers the sustainability of its life cycle, with the ideal objective of obtaining a product whose manufacture, use, and disposal have the least possible effects on the environment. This chapter traces the general picture of how an intervention directed at environmental protection can be integrated in the product design and development process. It will also identify the most appropriate strategies and tools for an integrated design process that considers all the phases of the life cycle, analyzing and reconciling determinant factors such as producibility, requisites for use, cost, and environmental aspects.

8.1 Orientation toward Environmental Aspects in the Design Process

While the more important issues associated with the environmental aspects of industrial production are the subject of much discussion nowadays, manufacturing companies still have difficulty in achieving environmentally sustainable production. One of the crucial factors in this problem is that the principles and methods of designing for the environmental quality of products have not yet been integrated into design and managerial practice (Gutowski et al., 2005). The result is that the success factors in product design still remain limited to those of quality and development costs (i.e., to those that can be understood as factors associated with the product's impact on the business environment).

8.1.1   Premises for the Integration of Environmental Requirements

The life cycle approach can provide a qualitative leap in the statement of product development, "making the product fit its natural environment as much as it fits the business environment" (Krishnan and Ulrich, 2001). This affirmation originates in the recognition that there is a need for a "life cycle thinking approach" to the environmental question. It is confirmed by certain observations regarding determinant factors obstructing the implementation of environmentally oriented product development (Ries et al., 1999):

- Poor understanding of the environmental impacts of products
- Cost-oriented approach to the product development process
- Lack of a homogenous and efficient implementation, within the context of the entire development process, of an approach directed at the environmental requirements of products

Manufacturing companies' limited knowledge of the impacts of products on the environment is historically linked to producers needing to address principally those aspects regarding the impact at production sites (consumption of resources, generation of emissions and waste), not directly attributable to products and limited to the context of the production phase alone. The result has been a lack of primary information that could support a strategy to improve the environmental quality of products—a strategy, as has been repeatedly emphasized in this book, requiring a vision extended over a product's entire life cycle. This problem can be resolved by implementing the techniques used in Life Cycle Assessment (LCA), fully discussed in Chapter 4. Particularly LCA in its simplified form (Streamlined LCA) can overcome the disadvantages of an analysis too detailed to be undertaken in the preliminary phases of product development (Section 4.4).

Traditional cost-oriented formulations of the development process stem from an outdated, defensive approach to the environmental question that considers the environment a restrictive and generally troublesome constraint, without being able to appreciate its potential positive value. This problematic factor becomes particularly significant when one considers the weight that cost planning and marketing functions have in the product development process. The lack of accurate economic analysis and a non-perception of a product's "environmental value" can seriously hamper eco-compatible design. Also in this case, life cycle–oriented techniques can come to the rescue, primarily Life Cycle Cost Analysis (LCCA, treated in Chapter 5) and Environmental Accounting, together with the other techniques integrating economic and environmental analysis of the life cycle introduced in Chapter 6.

The lack of a homogenous, environmentally oriented approach, thoroughly integrated into the entire development process, is one of the crucial factors. It has often been observed that this lack is usually most evident in the preliminary phases of product development (Bhamra et al., 1999; Ries et al., 1999), where there is a scarcity of methods and tools oriented toward environmental aspects. It should be noted how, more generally, design practice lacks an organic approach to environmental aspects in the entire development process, despite such an approach clearly being desirable at the theoretical level.

The life cycle approach, which in the strictly design dimension is represented by Life Cycle Design (LCD), can constitute an effective basis for the integration of environmental aspects into product development. In particular, when LCD is expressly oriented toward environmental requirements (Section 3.2, Chapter 3), it can become an example of a completely environmentally oriented approach to the design process, and provide a reference model to achieve the complete integration of environmental aspects within the development process. As already noted in Chapter 3, Section 3.2 and summarized in Figure 3.3, and as will be further discussed below, the specification of the design objectives and strategies plays a crucial role in this respect.

Another vital role can be played by Design for X (DFX), which provides the tools and techniques for a design directed at specific product requisites so that environmental requirements can be included among the others (Chapter 7, Section 7.3.2). This issue will be considered in greater detail below.

This analysis is summarized in Figure 8.1, showing the instruments with which the life cycle approach can help overcome the factors impeding the implementation of environmentally oriented product development in company practice. The same figure shows another important obstructive factor, the cross-functional character of both design practice and environmental aspects. It is linked to the multidisciplinary nature of the competencies required and to the transversal nature of the correlated activities with respect to the principal

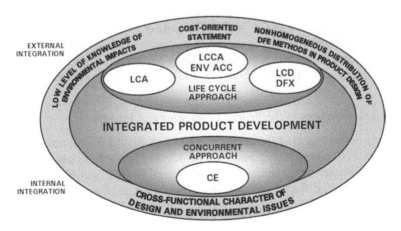

FIGURE 8.1 Approaches to factors impeding the implementation of environmentally oriented product development.

company functions (design, production, marketing). This issue was introduced in the previous chapter in relation to the organization and planning of the product design and development process, explaining how Concurrent Engineering (CE) was conceived precisely in order to address these needs in design practice. Environmental aspects can, therefore, be integrated into product development through implementing the organizational structures of CE.

To sum up, and again referring to Figure 8.1, the full integration of environmental aspects in product development must occur at two different and complementary levels:

- External integration—Concerns the relation between the product development process and factors external to the design team that must be taken into consideration (i.e., customer and market demands, production constraints, and environmental requirements). This integration, as shown in Figure 8.1, is obtained by adopting the life cycle approach and using its tools.

- Internal integration—Concerns the relation between the internal functions and competencies of the design team. This integration is necessary in order to best manage the cross-functional character of design practice and of the environmental aspects, and is obtained through a simultaneous and concurrent approach to product development.

Having achieved the integration on this dual level, it is finally possible to speak of Integrated Product Development (IPD), understood in its most complete sense and including environmental aspects. In this regard, it is interesting to note how IPD can assimilate the general concept of improving the design solution in terms of its response to consumer demands and to market opportunities (Wang, 1997). The life cycle approach extends this perspective, addressing the needs of the consumer as well as of all the other actors involved in the various phases of the product's life cycle (Prudhomme et al., 2003). Further extending the concept underlying IPD to include a response to the needs of the environment thus constitutes the fundamental premise for achieving and integrated product design that also takes into account environmental requirements.

8.1.2   Interventions in the Product Development Process

Referring to the vision of the entire product design and development process described in Chapter 7, Figures 7.2 and 7.3, it is possible to say that the full and homogenous integration of environmental aspects results from a series of interventions, differing according to the different phases of the development process:

- In the preliminary phases (project definition, development process planning, problem specification), this integration is achieved through the extension of the factors conditioning the preliminary structuring of the project, and the definition of product specifications and requisites. These, together with consumer requirements and market opportunities, will also include environmental necessities; the latter are given their due weight in defining company policies and strategies. Referring to Figure 7.2, this involves adding to the information input of the design process a set of information and data, not exclusively environmental, regarding the expected life cycle of the product.

- The definition of the specifically design-related phases (i.e., those comprising the product design process, again referring to Figure 7.2) must be guided by appropriate approaches to the environmental aspects of the product's life cycle. This particular consideration will be analyzed in further detail below.

- In the main phases of the design process, beginning from conceptual design and with particular regard to the phases of embodiment and detail design, the definition of the design intervention must be directed at harmonizing the ever-wider range of design requirements, as envisaged by LCD (Chapter 3, Figure 3.2). On the basis of this statement, the various specifications can be achieved using the tools of the DFX system, each addressing a specific typology of product requisite, giving appropriate emphasis to those oriented toward environmental requirements (Chapter 7, Section 7.3.2).

- The postdesign planning phase must be integrated with the product design phase, which in the general scheme of the product development process precedes it (Figures 7.1 and 7.3), as is established in concurrent design (Section 7.3). This integration must be performed according to the presuppositions already introduced in Section 7.2.4.2 (i.e., extension of postdesign planning to cover the entire life cycle, including the production, distribution, use, and retirement of the product). It is precisely in relation to the planning of the production–consumption–disposal cycle that the most appropriate tools of the DFX system are introduced.

8.2   Environmental Strategies for the Life Cycle Approach

Design strategies play an essential role in the life cycle approach. They allow the environmental requisites demanded of the product to be translated into design practice. It should, therefore, be emphasized that the environmental strategies most appropriate and effective for a specific design problem must be carefully chosen only after the objectives of the project have been accurately translated into product requirements (Keoleian and Menerey, 1993).

In general, strategies oriented toward the environmental efficiency of the life cycle can be defined on the basis of the product's primary impact(s) on the environment, ascribable to exchanges with the ecosphere of the physical–chemical flows involved in the technological processes making up the life cycle:

- Consumption of material resources and saturation of waste disposal sites
- Consumption of energy resources and loss of the energy content of dumped products
- Total direct and indirect emissions of the entire product–system

Thus, for a complete environmental analysis (where it is opportune to refer to the activity model represented in Figure 2.3), it is necessary to identify not only the flows of materials in the life cycle but also those of energy and emissions, in both their explicit and implicit forms. There are numerous environmental strategies directed at reducing this wide spectrum of impacts (Keoleian and Menerey, 1993; Hanssen, 1995; Fiksel, 1996; Bhander et al., 2003). They can be distinguished on the basis of the phase of the life cycle on which they are intended to intervene, as shown in Table 8.1 where some of the more important environmental strategies are reported.

8.2.1   Environmental Strategies in Product Design

Given that the environmental efficiency of a product is directly dependent on its design, it is of fundamental importance that any strategy to be followed be put in relation to the main design parameters (Whitmer et al., 1995). However, not all the strategies reported in Table 8.1 can be likened to true and proper design strategies. In fact, some of them consist of interventions not directly linked to design choices. Summarizing the various strategies presented in the table, it is possible to conclude that a design intervention intended to take account of a product's behavior, in environmental terms, during its life cycle must, in general, have the aim of optimizing the distribution of the flows of resources and emissions by:

- Reducing the volumes of materials used and extending their life span
- Closing the cycles of resource flows through recovery interventions
- Minimizing the emissions and energy consumption in production, use, and disposal

To fully achieve these conditions it is necessary to intervene in the two separate areas of product design and process design. Although process design is of primary importance and (following the principles of Concurrent

TABLE 8.1 Environmental strategies and life cycle phases

LIFE CYCLE PHASES	ENVIRONMENTAL STRATEGIES
Preproduction	Reducing the use of raw materials Choosing plentiful raw materials Reducing toxic substances Increasing the energy efficiency of processes Reducing discards and waste Increasing flows of recovery and recycling
Production	Reducing the intensive use of materials Using materials with low impact Reducing the use of toxic materials Using recycled and recyclable materials Using materials on the basis of their required duration Selecting processes with low impact and high energy efficiency Selecting processes with high technological efficiency Reducing discards and waste
Distribution	Planning the most energy-efficient shipping Reducing the emissions of transport Using containment systems for toxic or dangerous materials Reducing packaging Using packaging with low environmental impact Reusing packaging
Use	Using products under the intended conditions Planning and execution of servicing interventions (diagnostics, maintenance, repair) Reducing energy consumption and emissions during use
Retirement	Facilitating product disassembly at end-of-life Analyzing the condition of materials and their residual life Planning the recovery of components at end of use Planning material recycling at end of use Reducing volumes for disposal

Engineering) more frequently considered to be intimately linked to the product development process, it is not directly relevant to the objectives of this book. Here, attention is focused more on the design of the product understood as a material object—a set of material components designed in such a way that they constitute a functional system that satisfies certain requisites demanded of it. This is the product–entity dimension directly linked to the choices made in the specifically design-related phases of the development process (conceptual, embodiment, detail design), whose parameters are ascribable to precisely the product's physical dimension: materials, component form and dimensions, system architecture, interconnections, and junctions.

This physical dimension of the product–entity is expressed in its life cycle by the flows of material resources. This, therefore, leads back to the first of the three main aspects of a product's impact on the environment, that of the employment and consumption of material resources. This was shown in the overview of the product life cycle and of the resource flows characterizing it, in Figure 2.9.

This partial view of the environmental problem may seem limited, but in reality it is very wide-ranging; the only aspect completely ignored is that of intervening on the various technological processes constituting the life cycle. This view does not exclude the possibility of taking into account the other two aspects of impact (energy consumption and product–system emissions) in environmental evaluations. With regard to how the energy and emission content of the materials in play contribute to the environmental impact, these are clearly ascribable to the volumes of the material flows. Regarding how the energy fueling the process and the direct emissions from it contribute to the environmental impact, these can also be generally ascribed to the volumes being processed or to specific process parameters dependent on the physical properties of the materials or on the geometries. These can all be managed through the choices of product design; the definition of the materials and of the main geometric parameters condition the choice of the processes and how these are performed.

Focusing on the material flows, and therefore on the physical dimension of the product-entity, the environmental performance of the life cycle can be improved through the application of two main types of strategies (Giudice et al., 2002a; Giudice et al., 2002b), as summarized in Figure 8.2:

- Useful Life Extension Strategies, directed at extending the product's useful life and so conferring increased value on the materials used and on all the other resources employed in its manufacture—Product maintenance, repair, upgrading, and adaptation
- End-of-Life Strategies, directed at recovering material at the end of the product's useful life, closing the cycle of materials and recovering, at least in part, the other resources used in its manufacture—Reusing systems and components, recycling materials in the primary production cycle or in external cycles

Although these strategies must already be taken into consideration during the design phase, in order to facilitate their application if this is considered appropriate, clearly they do not have an effect until after the product has been manufactured. As shown in Figure 8.2, however, a third important type of environmental strategy, known as Resource Reduction Strategies, becomes operational before the production phase. Again associated with the product's material dimension, these strategies are directed at reducing the resources

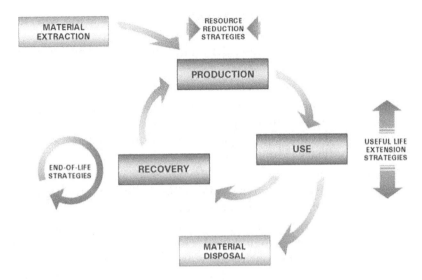

FIGURE 8.2 Environmental strategies for the life cycle of products.

used in its manufacture and include all the interventions and choices that favor a reduction in the use of material and energy resources. Thus, in general terms, they are referable to a wide spectrum of expedients that regard not only product design but also production process planning. They may also include radical strategies, such as "dematerialization" (i.e., the reduction of the quantity of materials necessary to achieve an economic function) (Wernick et al., 1997), promoting the evolution from the sale of products to the sale of services (Tomiyama, 1997), and therefore more properly allocated to the realm of business strategies.

In the sections that follow, attention will be focused on the first two types of design strategies, together with some of the tools available to the designer wishing to implement them. Subsequently, it will be shown how these strategies can be incorporated in a methodological framework for product design, outlining the full integration of environmental aspects. A more detailed description of these environmental strategies, and of the design tools and techniques available to attain them, will be proposed in Chapter 9.

8.2.2 Useful Life Extension Strategies

With reference to the product's useful life (i.e., the period of time over which the product is used while ensuring that it meets the required operating standards), extending this life results in a saving of energy and material resources upstream and a reduction in waste downstream of the use phase. With this intervention, in fact, it is possible to satisfy the same demand with fewer product units.

The extension of a product's useful life may be obtained through four intervention typologies:

- Maintenance—Includes periodic and preventive checking operations. As well as monitoring and diagnostic interventions for the programmed substitution of parts subject to wear, maintenance also includes ordinary cleaning operations.

- Repair—Essentially consists of the removal and substitution of damaged parts in order to reestablish the operational condition and level of performance required of the product.

- Upgrade and adaptation—Similar interventions, in that both are motivated by technological and cultural obsolescence, and by changes in the conditions of the working environment and in the exigencies of the user. They differ in intervention typology, since upgrading provides for the substitution or addition of components, while adaptation involves a reconfiguration of the main components of the product.

8.2.3   End-of-Life Strategies

Recovery interventions at the end of the product's useful life allow the life cycle to become closed, as shown in a simplified manner in Figure 8.2, with consequent environmental benefits: decrease in the raw materials entering the cycle because they are partly substituted by recovered resources; recovery of energy and material resources used in production, and therefore a better exploitation of their use; and decrease in the waste flows.

Some preliminary considerations regarding recovery flows of material resources have already been drawn in Chapter 2, Section 2.5. On the basis of these premises, as suggested by several authors (Dowie, 1994; Ishii et al., 1994; Navin-Chandra, 1994), the strategies for the recovery of resources at the end-of-life can be grouped according to their different recovery levels.

In general, the three main recovery levels are direct reuse, reuse of parts, and recycling of materials. A different potential of environmental benefit corresponds to each of these, depending on the level of the recovery flows in the life cycle (as is evident in the reference model of Figure 2.9):

- Direct reuse—At the end of use, the product can be directly reused, possibly after having been checked and repaired, with consequent savings in energy consumption, any possible emissions, costs relative to the production and assembly of components, and in the volumes of virgin materials.

- Reuse of parts—Components that have not undergone excessive deterioration during use can be recovered, possibly after being regenerated through intermediate processes, as components for reassembly, with savings in energy, possible emissions, costs relative to the process of producing the parts, and in the volumes of virgin materials.

- Recycling materials—The materials of parts that cannot be reused can be recycled by the recovery processes included in the materials' own life cycles, or they can be treated and used in external production cycles to manufacture products with less stringent material property requirements.

8.2.4 Introduction of Environmental Strategies into the Design Process

The environmental strategies for improving the life cycle of a product, introduced above and grouped according to the two typologies proposed, can in practice lead to appropriate design strategies able to guide the designer in the choices that must be made at the different levels of design development. Table 8.2 summarizes the design strategies of greater significance in this respect, classified in relation to the main design parameters. The latter are categorized according to whether they concern the system design (characteristics of the architecture, particularly layout, and relationships between components) or the detailed design of components (materials, shape, geometric parameters). This table also shows the direct correlations between each design strategy proposed and the environmental strategies it can support. This makes it possible to outline a preliminary methodological statement that would allow the integration of environmental aspects into design practice. The statements are schematized in Figure 8.3, and can be summarized in the following points:

- Definition of the environmental requirements to be attained
- Choice of the environmental strategies most appropriate to the desired requisites
- Identification of the design strategies that can enhance the chosen environmental strategies
- Definition of the design parameters to use in interventions at the two design levels (system and component design)

TABLE 8.2 Design parameters, design strategies, and environmental strategies

DESIGN LEVEL	DESIGN PARAMETERS	DESIGN STRATEGIES	ENVIRONMENTAL STRATEGIES					
			USEFUL LIFE EXTENSION			END-OF-LIFE RECOVERY		
			(ES1)	(ES2)	(ES3)	(ES4)	(ES5)	(ES6)
System	LAYOUT	Minimize number of components	✓	✓			✓	✓
		Optimize modularity	✓	✓	✓		✓	✓
		Design multifunctional and upgradable components			✓	✓		
		Plan accessibility to components	✓	✓			✓	✓
	RELATIONS BETWEEN COMPONENTS	Reduce number of connections	✓	✓	✓		✓	✓
		Reduce variety of connecting elements	✓	✓	✓		✓	✓
		Increase ease of disassembly	✓	✓	✓		✓	✓
Component	MATERIALS	Reduce unsustainable and hazardous materials	✓	✓				✓
		Increase biodegradable and low-impact materials		✓				✓
		Reduce material variety						✓
		Increase material compatibility and recyclability						✓
		Specify and label materials						✓
	FORM	Optimize performance, resistance, and reliability	✓		✓	✓		
		Design for easy removal		✓	✓	✓	✓	✓
	DIMENSIONS	Reduce mass	✓	✓	✓	✓	✓	
		Optimize performance, resistance, and reliability	✓		✓	✓		
		Design for easy removal	✓	✓	✓		✓	✓

(ES1) maintenance; (ES2) repair; (ES3) upgrading and adaptation; (ES4) direct reuse; (ES5) reuse of parts; (ES6) recycling materials.

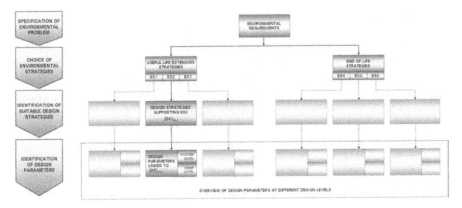

FIGURE 8.3 Introduction of environmental strategies into the design process: Preliminary methodological statement.

In any case, the environmental strategies followed must be in harmony with the entire spectrum of requirements, reconciling any conflicts that may arise from design orientations directed at diverse objectives. Fulfilling criteria not directly referable to environmental benefits, such as the cheapness and quality of the product, becomes in itself part of the concept of "eco-efficiency" of the design solution (Lye et al., 2001).

The result of the preliminary statement outlined above and shown in Figure 8.3, consisting of the overview of a set of design parameters upon which to intervene in order to follow the environmental strategies and achieve the desired requisites, can be extremely useful in the management of conflicts between the strategies themselves. Not infrequently, design interventions directed at different environmental objectives are mutually opposing (Luttropp and Karlsson, 2001). Furthermore, this overview is a first step toward clarifying the links between the choices directed at environmental aspects and those inspired by conventional criteria.

Achieving equilibrium between different necessities (performance, economic, environmental), which is clearly essential for the realization of a full and efficient integration of environmental aspects into design practice, represents an important critical point. This is due to the very nature of the environmental problem and of the design intervention oriented toward environmental necessities. Managing these necessities in a truly effective manner requires radical strategies, potentially conflicting with traditional product requisites (performance, producibility, cost). The environmental strategies introduced in the previous section are developed on the basis of a problem-focused approach to the design process (i.e., an approach initially concentrating on the problem to be resolved and subsequently arriving at the solution to the problem) (Maffin, 1998). This statement, frequently adopted in reference methodology frameworks for the product design process (Pahl and Beitz, 1996), requires considerable freedom in the preliminary phase of structuring the project development and, in effect, it is only possible in the case where there is the opportunity to ideate and develop a new and highly innovative product. More commonly, a product-oriented approach is usually adopted instead (i.e., an approach privileging the analysis of the concepts of preexisting products, subsequently elaborating these and adapting them to any new necessities). This second type of approach is greatly conditioned by previously acquired experience and is expressed in practice through the use of general guidelines and rules of thumb. This latter characteristic, in particular, reveals its inadequacy as an aid in environmentally-effective design. In fact, the design intervention cannot be easily reduced into guidelines and systematic procedures, as noted previously (Chapter 3, Section 3.2.2). Furthermore, the later can be useful only in exploring the potential opportunities of eco-efficient interventions in the product development process (Chapter 1, Section 1.4.4).

The reconciling of environmental strategies with conventional product requirements can be achieved by mediating between these two contrasting approaches. This harmonizing approach can be implemented by focusing on the weight given to the environmental strategies when applied to the design process, as shown in Figure 8.4. Although the environmental requirements must also be clearly defined in the preliminary phase of problem specification, together with all the other product requisites, the application of the product-oriented approach, above all in the early design phases, can ensure the attainment of the conventional requirements precisely through the exercise of experience and established rules. The application of environmental strategies that require a problem-oriented approach can be conditioned by the weight given to the environmental requirements, varying both the extent of the field of application within the design process and the weight given to these strategies in the various phases of design.

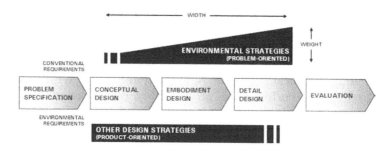

FIGURE 8.4 Introduction of environmental strategies into the design process: Equilibrium between conventional design and environmental aspects.

The problem of reconciling conventional design and environmental aspects is then reduced to the identification of the appropriate state of equilibrium between:

- The preliminary product-oriented statement, ensuring the conventional requisites
- The weight of the environmental strategies, typically problem-oriented, that make it possible to achieve the environmental requirements, and whose effectiveness is the greater the earlier the intervention is in the design process.

8.3 Tools and Techniques for Environmental Requirements of the Life Cycle

As discussed previously, a design intervention directed at harmonizing the ever-greater range of requirements (including those concerning a product's environmental performance during its life cycle) can be achieved when it is based on the methodological structure of LCD. Incorporating the most appropriate DFX tools, each oriented toward a specific typology of product requisite and implemented primarily in the phases of embodiment and detail design, can aid the designer in translating the product concept into detail design.

8.3.1 Role of Design for X

Although DFE is sometimes understood as belonging to the DFX system, it is more appropriate to consider it as a design approach. It is not, therefore, an actual operational design tool but, rather, a design philosophy implying a profound change in the way in which industry relates to the environmental question (Allenby, 1994). As a design approach, it requires operational tools that embody its premises and objectives. Some of the DFX system tools can perform this role effectively (Chapter 7, Section 7.3.2). In this respect, it is interesting to note that by their very nature these tools are based on the problem-oriented approach, confirming that they are particularly suitable with regard to the environmental aspects of product design.

Another very interesting aspect of DFX tools is their specificity, allowing the decomposition of a design problem that is already very wide-ranging and segmented in its conventional dimension, and is further complicated by environmental requirements. Each DFX technique is characterized by methods, procedures, and models, and allows the elaboration of specific data through appropriate analytical functions. A suitable set of DFX tools can, therefore, allow specific parts of the problem to be treated separately, so that each is managed by those members of the design team most skilled in that area.

This approach based on the decomposition of both the problem and the design intervention itself, as shown in Chapter 7, Section 7.2.2, is the cornerstone of modern methods of product design and can constitute an effective resource for achieving the integration between environmental and traditional necessities (Jackson et al., 1997). At the same time, it is important not to overlook the negative effect that excessively specific and separate design actions may have on the design process (Bras, 1997)—delaying or even blocking convergence on a final balanced solution (i.e., one that is effective in the widest sense, feasible and marketable). This dangerous tendency would be remedied by the highly desirable integrated and simultaneous structuring of the design intervention.

8.3.2 DFX Tools for Environmental Strategies

Of the different DFX typologies, several are of particular interest in relation to the two intervention strategies for the environmental quality of the life cycle identified in Section 8.2:

- Those directed at facilitating the continued functionality of the product during the phase of use, in that they favor the extension of its useful life. In this case, one speaks of Design for Maintainability and Design for Serviceability (Makino et al., 1989; Eubanks and Ishii, 1993; Gershenson and Ishii, 1993; Klement, 1993; Subramani and Dewhurst, 1993; Dewhurst and Abbatiello, 1996; Kusiak and Lee, 1997). Considering the necessities associated with the whole set of servicing operations (diagnosis, maintenance, repair), Design for Serviceability usually also encompasses Design for Maintainability.

- Those oriented at the planning of processes at the end-of-life, in that they are directed at the reduction of the impact of disposal and at the recovery of resources. In this case, one speaks in general terms of Design for Product Retirement/Recovery (Ishii et al., 1994; Navin-Chandra, 1994; Zhang et al., 1997; Gungor and Gupta, 1999), or, more specifically, of Design for Remanufacturing (Shu and Flowers, 1993; Amezquita et al., 1995; Bras and McIntosh, 1999) and Design for Recycling (Burke et al., 1992; Beitz, 1993; Kriwet et al., 1995). These distinctions depend on whether greater emphasis is placed on the reuse of components or on the recycling of materials.

In both cases, the tools used to achieve the various objective requisites are those that intervene directly on the most significant design parameters, linked to the product architecture and to the characteristics of the components (exactly as required by the statement proposed in this book).

The same can also be said for the third typology of DFX, directed at the design and planning of the disassembly of constructional systems, known as Design for Disassembly (DFD). DFD is frequently oriented toward interventions of recovery at the end of a product's useful life (Simon, 1991; Jovane et al., 1993; Li et al., 1995; Harjula et al., 1996; Srinivasan et al., 1997), and it is sometimes considered an integral part of Design for Recovery and Recycling (Beitz, 1993; Ishii et al., 1994; Navin-Chandra, 1994; Pnueli and Zussman, 1997). More in general, Design for Disassembly cuts across the two environmental strategies, extension of useful life and recovery at end-of-life (Boothroyd and Alting, 1992; Ishii et al., 1993; Sodhi et al., 2004). Both strategies, in fact, benefit from an intervention directed at achieving a specific and important product characteristic—the ease with which products or components can be disassembled. This is further confirmed by authors who expressly integrate Design for Disassembly with the requirements of servicing operations (Eubanks and Ishii, 1993; Vujosevic et al., 1995).

To integrate environmental aspects into product design, the DFX tools that support environmental strategies must clearly assume a determining role, taking into account a wide variety of product requisites, as described in Chapter 7, Section 7.3.2. Their relationship to and integration with the other tools of the DFX system thus become crucial factors in obtaining a final solution that succeeds in interpreting the different requisites and in balancing the various design strategies. In this respect, it is worth noting some interesting studies on the links between Design for Environment and Design for Manufacture, one of the more widely used DFX techniques, that have reported encouraging results regarding their complementarity and the possibility of conducting interventions of mutual improvement in relation to their respective objectives (Ufford, 1996; Rounds and Cooper, 2002).

The DFX tools supporting the environmental strategies introduced here will be considered in greater detail in Chapter 9.

8.4   Integration in Product Development: Proposed Framework

The choice of fulfilling environmental requirements through the use of the most appropriate DFX techniques implies the adoption of a design-centered product development model, already introduced in Chapter 7, Section 7.3.2. A design-centered model is most appropriate for the introduction of DFX tools due to the particular importance these assume in the phases specifically concerning design.

The integration of environmental aspects can then be summarized according to the scheme shown in Figure 8.5, analogous to that of Figure 7.6 in Chapter 7.

The specifically design-related phases are preceded by the phase consisting of the preliminary definition of the product's fundamental requirements (problem specification). At this level, as noted earlier, the objectives of the

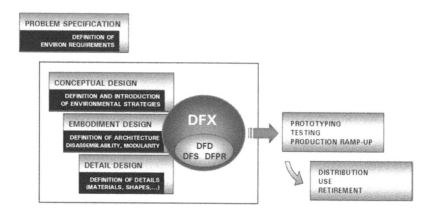

FIGURE 8.5 **Product design and development process: Integration of environmental aspects using DFX tools.**

development process are defined, along with their correlated problems and the design exigencies and constraints. This is the phase wherein the environmental objectives are also defined, identifying the project variables correlated to them and analyzing the problems resulting from their fulfillment. The importance of this preliminary stage has already been emphasized; the most appropriate and effective environmental strategies can be chosen correctly only after the design objectives have been accurately translated into product requirements. Clearly, the life cycle approach must be adopted when formulating the environmental requirements. This approach involves, de facto, an added dimension that greatly enlarges the domain of the design objectives (Keoleian and Menerey, 1993).

The phase of problem specification is followed by the main design phases, not in a rigidly sequential order but largely simultaneous:

- Conceptual Design—Having defined the design objectives and having translated them into product requirements, it is necessary to develop design ideas that can allow these to be obtained. In this phase, a certain number of general proposals are defined through the initial description of the functions and principal characteristics of the product. These are then evaluated in order to determine which of them best meets the intended target. Given that the earlier an intervention oriented toward environmental aspects is implemented in the development process, the greater is its effectiveness, this is the stage where it is necessary to introduce those environmental strategies considered, on the basis of the issues raised in Section 8.2.4, to be the most appropriate with respect to the chosen objectives.

- Embodiment Design—Having introduced the design strategies and having identified the preliminary concept, the next phase consists of producing a more complete description of the product idea. The concepts formulated in the preceding phases are developed, their feasibility is verified, and they are finally translated into an approximation of the product layout that defines the subsystems and functional components. In this phase, DFX techniques supporting the environmental strategies are introduced, applying methods and tools aiding a correct definition of the product architecture. Particular attention is given to the properties of modularity and disassemblability that can favor interventions of both servicing and recovery.

- Detail Design—The product architecture developed in the preceding phase is translated into a complete and detailed solution. The DFX techniques supporting the environmental strategies are introduced again, applying methods and tools aiding a correct definition of the design details. The choice of materials, study of the shape, and the definition of component geometry and junction systems must all be guided not only by performance and economic requirements but also by the environmental requirements.

The design phases, where the most appropriate tools of the DFX system are applied (Design for Disassembly—DFD, Design for Serviceability—DFS, and Design for Product Retirement—DFPR), are followed by the final phases of product development: prototyping, testing, and production ramp-up. However, the definitive verification of the environmental performance of the finished product, which can only be a first approximation if based on various appositely developed tools and metrics (such as those introduced in Section 3.2.3 of Chapter 3), requires that the entire life cycle runs its course through the phases of distribution, use, and disposal. Only then is it possible to evaluate the effective environmental performance of the product and identify the factors to be considered in any future redesign.

8.4.1 Tools and Techniques for Integrated Design: Overview

A brief panorama of the various DFX typologies most widely used in current design practice was given in Chapter 7, Section 7.3.2, to demonstrate how these can constitute a versatile system of tools and techniques for an integrated design. These tools and techniques are directed at a wide variety of product requisites and characteristics for specific phases of the life cycle, spanning the entire cycle. Of these, Design for Manufacturing and Design for Assembly (concerning the necessities of the production phase) and Design for Reliability, Design for Maintainability, and Design for Quality (concerning the correct functioning and quality of products during their use) have now acquired accepted methodological structures and models, as witnessed by the various studies in the literature (Ishii, 1995; Kuo et al., 2001).

As already noted in Chapter 7, Section 7.3.2.3, DFX tools are more effective when implemented in the design phases of embodiment design (operating on the definition of the functional units and layout) and of detail design (operating on the shape of components, dimensions, and materials). The influence these can have on the preceding conceptual design phase is limited to the application of fundamental criteria that can aid the designer in choosing among the initial ideas. This particular phase, by its very nature, offers the greatest opportunity for creativity, and therefore induces the designer to use more appropriate instruments, such as various concept generation methods (Ullman, 2003). Also, with regard to the environmental perspective, tools other than those of DFX can be more effective in this phase. These include techniques based on the analysis of different categories of environmental impact, contextualized with respect to the life cycle phases, which make it possible to formulate the best opportunities for implementing these aspects in product planning (O'Shea, 2002). Industrial Design is also extremely important in this particular phase of design development, as well as in the context of the design of environmentally oriented products, where the industrial designer can make a particularly significant contribution (Lofthouse, 2004).

It is also possible to use specific techniques in the problem specification phase, which precedes the design phases and is central to setting up correct design strategies. For example, Quality Function Deployment (QFD) is one of the techniques most widely used as an aid to understanding the problem correctly and generating project specifications. QFD enables the needs and wants of the consumer to be translated into project specifications and product requirements while taking into account market competition and identifying the relationships between the requirements themselves (Hauser and Clausing, 1988). Derived from this technique, Green QFD constitutes a methodology for the preliminary statement of product development integrating consumer demands with the environmental and economic necessities of the life cycle (Zhang et al., 1999).

This is not the only case of a technique developed for other purposes that has subsequently been reinterpreted from an environmental perspective. Another example is that of Environmental FMEA (E-FMEA), clearly derived from the well-established Failure Mode Effect Analysis (FMEA) technique. This is used to identify, at the design stage, the potential failures that may occur in achieving the required functionalities of the product under development. Based on the same methodological statement, and incorporating the life cycle approach, E-FMEA differs from FMEA in that it focuses on failures in fulfilling the product's environmental performance rather than being directed at the identification of failures in achieving technical functionalities (Lindahl, 1999).

All these instruments must be placed alongside those of traditional engineering design oriented toward the product's primary performance functions (Performance Analysis—PA). Those based on Finite Element Analysis (FEA) are prominent for their wide use (usually in computer software programs) and their high level of evolution. These, together with tools aiding graphical representation, modeling, and simulation (Computer-Aided Design—CAD), comprise what can be defined as a system of tools enabling the designer to create a "virtual prototype" of the product under development, going so far as providing a simulation of its realization and functional behavior. This makes it possible to improve the productivity of designs and the accuracy of analysis, thus limiting the problems met during its manufacture and use and reducing development times and costs (Kim et al., 2002).

Ultimately, in order to attain a complete design that is effective and environmentally oriented, all these tools must be included and managed in an Integrated Product Development process, understood in the sense described in Section 8.1.1 and represented in Figure 8.1, where a determinant role is assumed by life cycle–oriented techniques (LCA, LCCA) and the DFX tools supporting the main environmental strategies (Section 8.3).

With particular regard to the two environmental and economic life cycle analysis techniques, LCA and LCCA, and their respective potential for integration in the product development process, refer to the aspects considered in Chapters 4, 5, and 6, and to the specific contributions of other authors (Weustink et al., 2000; Nielsen and Wenzel, 2002; Bhander et al., 2003; Fixon, 2004).

Many of the diverse tools and techniques described above are compiled in the diagram shown in Figure 8.6. This diagram takes as a starting point that

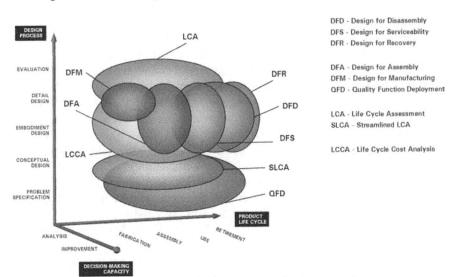

FIGURE 8.6 Tools and techniques for integrated design: Overview.

of Figure 7.7 in Chapter 7, which classifies DFX tools according to the specific design phases where they can be most effective, and to the life cycle phases they can influence. In Figure 8.6, these two dimensions are complemented by a third that characterizes the tools according to the type of information they can provide to the designer, and therefore to their potentiality with respect to decision making.

8.5   Toward an International Standard: The ISO/TR 14062 Technical Report

In response to the growing awareness of consumers and manufacturers with regard to environmental issues associated with industrial production, and to the consequent need to integrate environmental aspects into the product design and development process, in 1998 an ISO technical committee (ISO/TC 207—Environmental Management) set up a working group to study the specific theme of "Design for Environment." The result of their four-year-long proceedings, the technical report ISO/TR 14062 "Environmental Management—Integrating Environmental Aspects into Product Design and Development" (ISO/TR 14062, 2002) has the aim of providing people directly involved in the design and development phase, regardless of the typology or size of the company they work in, with a systematic program for predicting and identifying the possible effects their future products could have on the environment, and for taking effective decisions during the conception and development of these products in order to improve their environmental performance. Some of the main issues treated in the technical report are considered below, showing how it accurately and completely embraces the general perspective of the problems associated with product design for the environment and the methodological statement considered here.

The evolution of this first report toward a standard that can be integrated with the other environmental management norms of the ISO 14000 series would not only favor the realization of ever-more-complete and integrated environmental management systems, but could also clearly stimulate the implementation of DFE in design practice. This is confirmed by the interest generated by the technical report in the specific context of regulatory standards. This is the case in the electromechanical and electronic components production sector, where it is noted that the new editions of the IEC standards, IEC Guide 109 "Environmental Aspects—Inclusion in Electrotechnical Products Standards" (IEC Guide 109, 2003) and ECMA-341 "Environmental Design Considerations for ITC (Information and Communication Technology) and CE (Consumer Electronic) Products" (ECMA-341, 2004), make explicit reference to ISO/TR 14062 and adopt its key concepts, general statement, and design principles.

8.5.1   General Premises and Fundamental Concepts

The process of integrating environmental aspects into product development must interpret, in the most efficient way, the growing interest in the identification, assessment, and minimization of a product's environmental impacts in all the phases of its life cycle, from the extraction of raw materials, through fabrication, packaging and use, to disposal and recycling. It must, therefore, assimilate the life cycle approach and the holistic view of the product–system—recurrent concepts not only in this book but also in the other standards of the ISO 14000 series, to which this technical report makes clear reference.

This integration process is necessary to achieve a proactive intervention (i.e., one that forestalls the environmental impacts of a product by acting preemptively on its life cycle in the ideation and design phases). It must be continuative and sufficiently versatile to allow improvements to be made during the design process, promoting creativity and maximizing innovation and opportunities for the environmental improvement of a wide variety of product typologies, including both material goods and services.

Apart from the life cycle approach, other issues considered to be of strategic importance are:

- Early integration—By integrating the environmental aspects upstream of the design process, it is possible to achieve that versatility necessary in order to intervene and improve products during their development.

- Functionality thinking—In product development, rather than reasoning in terms of technical solutions, it is more appropriate to consider solutions in terms of functionality (i.e., their capacity to satisfy the desires and needs of the consumer). This provides a wider margin for ameliorative interventions.

- Multicriteria concepts—Design solutions must be developed to combine, as much as possible and in the most efficient way, a great variety of diverse criteria, from the conventional (performance, quality, cost) to the environmental. These environmental criteria again differ according to the type of environmental effect to which they are directed.

- Trade-off levels—Design choices must have the aim of realizing an equilibrium condition, achieving an efficient compromise at different levels, from a balance between different environmental requirements, to that between environmental, economic, and social benefits, and between the environmental aspects and technical and/or quality requisites.

Various benefits are to be expected from an integration obtained on the basis of these premises: reduction of costs through the optimization of the use of resources, energy efficiency, and the reduction of waste; greater incentives for innovation in processes and products; new business opportunities through the identification of new products; customer satisfaction, possibly even exceeding customer expectations; improvement of the manufacturer's image and market share; reduction of risks; and improvement of employees' motivation and of relations with regulating and controlling agencies.

8.5.2 Environmental Objectives and Design Strategies

The environmental objectives to be achieved in product design (strategic, product-related, environmental objectives) can be summed up in two principal categories:

- Conservation of resources, recycling, energy recovery—Consists of optimizing the use of resources required to produce a product, with respect to all the other performance requirements.
- Prevention of pollution, waste, other impacts—Consists of eliminating or reducing the causes of pollution and other impacts generated by the product over its entire life cycle.

These objectives can be achieved through an appropriate combination of design strategies, which must be chosen in relation to company policies, product typology, and various other socioeconomic factors. These strategies include improvement of materials and energy efficiency, optimization of functionality, avoidance of hazardous substances, design for cleaner production and use, design for durability, and design for reuse and recycling.

8.5.3 Integration of Environmental Aspects in the Design Process

The integration of environmental aspects into the product development and design process refers to a process model having six distinct phases, largely following the sequential and iterative models discussed in depth in Chapter 7. The possible interventions allowing the integration of environmental aspects are, in relation to each phase:

- Planning—Formulation of environmental requirements, in accordance with company strategies and consumer needs, and choice of the most appropriate environmental strategies for the design approach
- Conceptual design—Development of product concepts able to meet the environmental requirements

- Detailed design—Application of the design strategies and translation of environmental requirements into a detailed solution
- Testing and prototype—Verification of product specifications and environmental performance from the perspective of the entire life cycle
- Production and market launch—Release of documentation regarding the environmental aspects of the product, together with suggestions for its efficient use and disposal
- Product review—Analysis and evaluation of environmental aspects and of impacts associated with the product's real life cycle

Finally, the entire process of product design and development must be monitored to assess its effectiveness and to identify possible margins for improvement. This monitoring can be extended to cover different aspects, such as product functionality, economic efficiency of production, and environmental benefits. It can allow an evaluation of the effectiveness of the tools and techniques used, identify any problems, and define the most appropriate corrective interventions.

8.6 Summary

In order for environmental requirements to become factors of innovation in the development of a product that is, in the widest sense, sustainable, it is necessary to conduct an integrated and simultaneous product design, taking account of an ever-wider range of specifications and requisites. The design intervention must be structured according to the principles of Design for Environment and Life Cycle Design already discussed. Having defined the main phases of a product's life cycle, these must be taken into consideration starting from the definition of the design problem and the development of product requirements, which will subsequently be translated first into the product concept and then into the detailed solution. Only through an accurate definition of the environmental objectives and of the consequent product requirements is it possible to identify the most suitable design strategies. These strategies can be supported by different typologies of techniques, completing the versatile and structured system of design tools for product requisites known as Design for X. Appropriately introduced into an integrated development process that assimilates the life cycle approach and its correlated methodologies, the tools and techniques for environmental requirements can help designers make decisions regarding the most important design parameters, strongly influencing the final solution.

8.7   References

Allenby, B.R., Design for environment, in *Industrial Ecology, U.S.–Japan Perspectives,* Richards, D.J. and Fullerton, A.B., Eds., National Academy Press, Washington, DC, 1994, 32–34.

Amezquita, T., Hammond, R., and Bras, B., Design for remanufacturing, in *Proceedings of ICED 95 10th International Conference on Engineering Design*, Prague, Czech Republic, 1995, 1060–1065.

Beitz, W., Designing for ease of recycling, *Journal of Engineering Design*, 4(1), 12–23, 1993.

Bhamra, T.A. et al., Integrating environmental decisions into the product development process—Part 1: The early stages, in *Proceedings of EcoDesign '99: 1st International Symposium on Environmentally Conscious Design and Inverse Manufacturing*, Tokyo, 1999, 329–333.

Bhander, G.S., Hauschild, M., and McAloone, T., Implementing life cycle assessment in product development, *Environmental Progress*, 22(4), 255–267, 2003.

Boothroyd, G. and Alting, L., Design for assembly and disassembly, *Annals of the CIRP*, 41(2), 625–636, 1992.

Bras, B., Incorporating environmental issues in product design and realization, *Industry and Environment*, 20(1–2), 7–13, 1997.

Bras, B. and McIntosh, M.W., Product, process, and organizational design for remanufacture: An overview of research, *Robotics and Computer Integrated Manufacturing*, 15, 167–178, 1999.

Burke, D., Beiter, K., and Ishii, K., Life-cycle design for recyclability, in *Proceedings of ASME Design Theory and Methodology Conference*, Scottsdale, AZ, 1992, 325–332.

Dewhurst, P. and Abbatiello, N., Design for serviceability, in *Design for X: Concurrent Engineering Imperatives*, Huang, G.Q., Ed., Chapman & Hall, London, 1996, 298–317.

Dowie, T., Green design, *World Class Design to Manufacture*, 1(4), 32–38, 1994.

ECMA-341, Environmental Design Considerations for ICT and CE Products, Standard ECMA-341, Ecma International, Geneva, 2004.

Eubanks, C.F. and Ishii, K., AI methods for life-cycle serviceability design of mechanical systems, *Artificial Intelligence in Engineering*, 8(2), 127–140, 1993.

Fiksel, J., *Design for the Environment: Creating Eco-Efficient Products and Processes*, McGraw-Hill, New York, 1996.

Fixson, S.K., Assessing product architecture costing: Product life cycles, allocation rules, and cost models, in *Proceedings of ASME Design Engineering Technical Conference*, Salt Lake City, UT, 2004, DETC2004–57458.

Gershenson, J. and Ishii, K., Life-cycle serviceability design, in *Concurrent Engineering: Automation, Tools and Techniques*, Kusiak, A., Ed., John Wiley & Sons, New York, 1993, 363–384.

Giudice, F., La Rosa, G., and Risitano, A., An ecodesign method for product architecture definition based on optimal life-cycle strategies, in *Proceedings of Design 2002—7th International Design Conference*, Dubrovnik, Croatia, 2002a, 1311–1322.

Giudice, F., La Rosa, G., and Risitano, A., Prodotti eco-compatibili, *Progettare*, 257, 69–73, 2002b.

Gungor, A. and Gupta, S.M., Issues in environmentally conscious manufacturing and product recovery: A survey, *Computers and Industrial Engineering*, 36, 811–853, 1999.

Gutowski, T. et al., Environmentally benign manufacturing: Observations from Japan, Europe and the United States, *Journal of Cleaner Production*, 13, 1–17, 2005.

Hanssen, O.J., Preventive environmental strategies for product systems, *Journal of Cleaner Production*, 3(4), 181–187, 1995.

Harjula, T. et al., Design for disassembly and the environment, *Annals of the CIRP*, 45(1), 109–114, 1996.

Hauser, J.R. and Clausing, D., The house of quality, *Harvard Business Review*, 66(3), 63–73, 1988.

IEC Guide 109, Environmental Aspects—Inclusion in Electromechanical Product Standards, Guide 109—IEC: 2003(E), International Electrotechnical Commission, Geneva, 2003.

Ishii, K., Life-cycle engineering design, *Journal of Mechanical Design*, 117, 42–47, 1995.

Ishii, K., Eubanks, C.F., and Di Marco, P., Design for product retirement and material life cycle, *Materials and Design*, 15(4), 225–233, 1994.

Ishii, K., Eubanks, C.F., and Marks, M., Evaluation methodology for post manufacturing issues in life-cycle design, *Concurrent Engineering: Research and Applications*, 1(1), 61–68, 1993.

ISO/TR 14062, Environmental Management—Integrating Environmental Aspects into Product Design and Development, ISO/TR 14062:2002(E), International Organization for Standardization, Geneva, 2002.

Jackson, P. et al., An analytical method for integrating environmental and traditional design considerations, *Annals of the CIRP*, 46(1), 355–360, 1997.

Jovane, F. et al., A key issue in product life cycle: Disassembly, *Annals of the CIRP*, 42(2), 651–658, 1993.

Keoleian, G.A. and Menerey, D., Life Cycle Design Guidance Manual, EPA/600/R-92/226, U.S. Environmental Protection Agency, Office of Research and Development, Cincinnati, OH, 1993.

Kim, H., Querin, O.M., and Steven, G.P., On the development of structural optimization and its relevance in engineering design, *Design Studies*, 23, 85–102, 2002.

Klement, M.A., Design for maintainability, in *Concurrent Engineering: Automation, Tools and Techniques*, Kusiak, A., Ed., John Wiley & Sons, New York, 1993, 385–400.

Krishnan, V. and Ulrich, K.T., Product development decisions: A review of the literature, *Management Science*, 47(1), 1–21, 2001.

Kriwet, A., Zussman, E., and Seliger, G., Systematic integration of design for recycling into product design, *International Journal of Production Economics*, 38, 15–22, 1995.

Kuo, T.-C., Huang, S.H., and Zhang, H.-C., Design for manufacture and design for 'X': Concepts, applications, and perspectives, *Computers and Industrial Engineering*, 41, 241–260, 2001.

Kusiak, A. and Lee, G., Design of parts and manufacturing systems for reliability and maintainability, *International Journal of Advanced Manufacturing Technology*, 13, 67–76, 1997.

Li, W. et al., Design for disassembly analysis for environmentally conscious design and manufacturing, in *Proceedings of ASME International Mechanical Engineering Congress and Exposition*, San Francisco, 1995, 969–976.

Lindahl, M., E-FMEA: A new promising tool for efficient design for environment, in *Proceedings of EcoDesign '99: 1st International Symposium on Environmentally Conscious Design and Inverse Manufacturing*, Tokyo, 1999, 734–739.

Lofthouse, V., Investigation into the role of core industrial designers in ecodesign projects, *Design Studies*, 25, 215–227, 2004.

Luttropp, C. and Karlsson, R., The conflict of contradictory environmental targets, in *Proceedings of EcoDesign 2001: 2nd International Symposium on Environmentally Conscious Design and Inverse Manufacturing*, Tokyo, 2001, 43–48.

Lye, S.W., Lee, S.G., and Khoo, M.K., A design methodology for the strategic assessment of a product's eco-efficiency, *International Journal of Production Research*, 39(11), 2453–2474, 2001.

Maffin, D., Engineering design models: Context, theory and practice, *Journal of Engineering Design*, 9(4), 315–327, 1998.

Makino, A., Barkan, P., and Pfaff, R., Design for serviceability, in *Proceedings of ASME Winter Annual Meeting*, San Francisco, 1989, 117–120.

Navin-Chandra, D., The recovery problem in product design, *Journal of Engineering Design*, 5(1), 67–87, 1994.

Nielsen, P.H. and Wenzel, H., Integration of environmental aspects in product development: A stepwise procedure based on quantitative life cycle assessment, *Journal of Cleaner Production*, 10, 227–257, 2002.

O'Shea, M.A., Design for environment in conceptual product design: A decision model to reflect environmental issues of all life-cycle phases, *Journal of Sustainable Product Design*, 2, 11–28, 2002.

Pahl, G. and Beitz, W., *Engineering Design: A Systematic Approach*, 2nd ed., Springer-Verlag, London, 1996.

Pnueli, Y. and Zussman, E., Evaluating the end-of-life value of a product and improving it by redesign, *International Journal of Production Research*, 35(4), 921–942, 1997.

Prudhomme, G., Zwolinski, P., and Brissaud, D., Integrating into the design process the needs of those involved in the product life-cycle, *Journal of Engineering Design*, 14(3), 333–353, 2003.

Ries, G., Winkler, R., and Zust, R., Barriers for a successful integration of environmental aspects in product design, in *Proceedings of EcoDesign '99: 1st International Symposium on Environmentally Conscious Design and Inverse Manufacturing*, Tokyo, 1999, 527–532.

Rounds, K.S. and Cooper, J.S., Development of product design requirements using taxonomies of environmental issues, *Research in Engineering Design*, 13, 94–108, 2002.

Shu, L. and Flowers, W., A structured approach to design for remanufacture, in *Proceedings of ASME Winter Annual Meeting*, New Orleans, LA, 1993, 13–19.

Simon, M., Design for dismantling, *Professional Engineering*, 4, 20–22, 1991.

Sodhi, R., Sonnenberg, M., and Das, S., Evaluating the unfastening effort in design for disassembly and serviceability, *Journal of Engineering Design*, 15(1), 69–90, 2004.

Srinivasan, H., Shyamsundar, N., and Gadh, R., A framework for virtual disassembly analysis, *Journal of Intelligent Manufacturing*, 8, 277–295, 1997.

Subramani, A. and Dewhurst, P., Efficient design for service considerations, *Manufacturing Review*, 6(1), 40–47, 1993.

Tomiyama, T., A manufacturing paradigm towards the 21st century, *Integrated Computer Aided Engineering*, 4, 159–178, 1997.

Ufford, D.A., Leveraging commonalities between DFE and DFM/A, in *Proceedings of IEEE International Symposium on Electronics and the Environment*, Dallas, TX, 1996, 197–200.

Ullman, D.G., *The Mechanical Design Process*, 3rd ed., McGraw-Hill, New York, 2003.

Vujosevic, R., et al., Simulation, animation, and analysis of design disassembly for maintainability analysis, *International Journal of Production Research*, 33(11), 2999–3022, 1995.

Wang, B., *Integrated Product, Process and Enterprise Design*, Chapman & Hall, London, 1997.

Wernick, I.K. et al., Materialization and dematerialization: Measures and trends, in *Technological Trajectories and the Human Environment*, Ausubel, J.H. and Langford, H.D., Eds., National Academy Press, Washington, DC, 1997, 135–156.

Weustink, I.F. et al., A generic framework for cost estimation and cost control in product design, *Journal of Materials Processing Technology*, 103, 141–148, 2000.

Whitmer, C.I., Olson, W.W., and Sutherland, J.W., Methodology for including environmental concerns in concurrent engineering, in *Proceedings of 1st World Conference on Integrated Design and Process Technology*, Austin, TX, 1995, 8–13.

Zhang, H.C. et al., Environmentally conscious design and manufacturing: A state of the art survey, *Journal of Manufacturing Systems*, 16(5), 352–371, 1997.

Zhang, Y., Wang, H.P., and Zhang, C., Green QFD-II: A life cycle approach for environmentally conscious manufacturing by integrating LCA and LCC into QFD matrices, *International Journal of Production Research*, 37(5), 1075–1091, 1999.

Chapter 9

Life Cycle Environmental Strategies and Considerations for Product Design

To implement a process of integrated design that takes into consideration all the phases of the life cycle, from the definition of product specifications to its disposal, harmonizing a wide range of factors and including environmental aspects, it is necessary to use opportune strategies allowing environmental requirements to be incorporated in design practice. Chapter 8 outlined a general view of how an intervention oriented toward environmental protection can be integrated in the design and development process through suitable design strategies for the environmental performance of a product's life cycle.

In this chapter, these same environmental strategies are considered in greater detail, highlighting any aspects of particular interest to the designer and describing the tools and techniques aiding their application. For each of the strategies, particular emphasis will be placed on the definition of the determinant factors conditioning their applicability and effectiveness, and on the connection between these strategies and certain product properties (reliability, durability)—the traditional subjects of engineering design.

9.1 Strategies for Improving Resources Exploitation and Determinant Factors

A general overview of what may be considered the most effective environmental strategies in the life cycle approach was presented in Chapter 8. Underscoring the need to assimilate such strategies in the product design and development process, it is opportune to focus on those which can be directly linked to design choices, and thus become true and proper design strategies (Section 8.2.1).

As highlighted before, the material dimension of the product–entity is directly connected to choices made in the specifically designed-related phases of the development process—conceptual, embodiment, and detail design. The important parameters of such choices are referable to precisely this physical dimension of the product (system architecture, materials, component shapes and dimensions, interconnections, and junctions) and it was suggested how an improvement in the environmental performance of its life cycle could be achieved. Figure 9.1 highlights two main strategy typologies that, as well as expedients for reducing the resources used in product manufacture, could achieve this improvement:

- Useful Life Extension Strategies—Maintenance, repair, upgrading, and adaptation of the product
- End-of-Life Strategies—Reuse and remanufacturing of systems and components, recycling of materials in the primary production cycle or in external cycles

Both of these types of intervention strategies have great potential in relation to the environmental optimization of resource flows throughout the product's life cycle. Extending the useful life allows greater exploitation of the resources used in production, avoiding the consumption of additional resources in the manufacture of replacement products. Intervening to recover the product, or parts of it, at the end-of-life allows its constituent components or materials to be reused in the production of new products, thus reducing

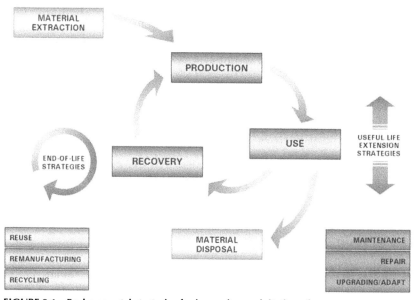

FIGURE 9.1   Environmental strategies for improving exploitation of resources.

the consumption of virgin material resources, conserving all or part of the energy resources used, and reducing the volumes of waste. In the final analysis, these strategies favor an increase in the intensity with which the resources employed in manufacturing the product are used, thereby improving their exploitation. This aspect constitutes a substantial difference with respect to those strategies directed at reducing the resources used in production (Chapter 8, Section 8.2).

An effective integration between these two types of strategies for improving exploitation of resources is both desirable and necessary, given that the actions of interventions for extending the useful life can strongly condition the opportunities for recovery, since these depend on the level of use to which a product, its components, and constituent materials have been subjected during the phases preceding recovery. Products that have been used for long periods of time usually show a significant deterioration in their functional performance, and consequently allow a lower level of recovery than products that are little used. The latter often still exhibit highly efficient performance and are, therefore, suitable for reuse.

9.1.1 Influence of External Factors and Product Durability

Applying strategies for the extension of the useful life and recovery at end-of-life is, in general, conditioned by a wide range of factors determining its effectiveness (Rose et al., 1998; van Nes et al., 1999). The evaluation of these factors is, therefore, essential for a correct implementation of these strategies in product development. In this respect, external factors conditioning the life expectation of a product are of particular importance (Woodward, 1997):

- Functional Life—The period of time for which need for the product is predicted to last
- Technological Life—The period of time that ends when the product is so technologically obsolete that it must be replaced by another based on superior technology
- Economic Life—The period of time that ends when the product's economic obsolescence is such it must be replaced by another characterized by analogous performance but costing less
- Social and Legal Life—The period of time that ends when changes in the desires of the consumer or in normative standards require the product to be replaced

All these factors, which can be considered external to the context of design choices linked to the product's physical dimension, must be considered along with a final, internal factor, which can be identified as the "physical life." This is the period of time for which the product is expected to physically last, maintaining its functional performance. This is, therefore, the factor directly linking the main design choices (architecture, materials, shapes, and geometries) with the predicted lifespan of the product.

In the design phase, the possibility of providing for the extension of the product's useful life and the reuse of parts depends precisely on the length of a product's physical life, which in turn is strictly bound to the durability of its components (Giudice et al., 2003), understood in general terms as the capacity to maintain the functional performance required of them. However, this property should not be maximized indiscriminately since, for example, in product sectors with a high level of technological innovation (and therefore with rapid obsolescence), excessive duration has a negative environmental value, guaranteeing a useless extension of product and component life that uses more resources in the production phase.

9.1.2 Identification of Optimal Strategies

The definition of both strategies for improving resource exploitation must, therefore, be subordinate to an evaluation of the external factors noted above, linked to the market reality and to regulatory standards, company policies, technological innovation, and aesthetic–cultural conditioning—all factors that vary widely according to the product typology. Having quantified the main external factors, it is possible to identify the strategies most appropriate to the product for varying its durability and that of its components.

A series of significant evaluations can be made by comparing the physical life with the "replacement life," defined as the period of time for which the product is effectively usable. This is comparable to the period of time the product is present on the market up to its definitive replacement, thus incorporating all the external conditioning considered above. Physical life represents the predicted duration of a product's full efficiency (its potential lifespan), while replacement life represents its effective lifespan, conditioned by factors such as technological and economic obsolescence and other external factors.

On the basis of this comparison, a distinction can be made between two types of useful life extension strategies:

- Maintenance, repair, and (more generally) service operations constitute strategies intervening on the physical life (Physical life extension strategies).
- Upgrading and adaptation of the product constitute strategies intervening on the replacement life (Replacement life extension strategies).

Figure 9.2 shows how, depending on the product typology, it is possible to identify the favorable conditions for extending the useful life by using the two different types of strategies. For physical life extension strategies, these conditions correspond to a long replacement life (indicating that the product may be used for a long time), and a short physical life (revealing the brief duration of some components, and therefore the limited capacity of the entire system to guarantee the required performance). Conversely, for replacement life extension strategies, these conditions correspond to a short replacement life and a long physical life. These conditions not only indicate the inappropriateness of planning maintenance and service interventions (pointlessly prolonging the product's life), but also reveal a poor design, unsuited to the predicted short span of effective use. Examples of the latter are over-dimensioning or using unnecessarily high-performance materials. This highly ineffective situation can be remedied through the upgrading or adaptation of the product.

The other areas of the graph indicate conditions of equilibrium between the two factors, representing the result of good design wherein the design choices were such that the physical duration of the system was calibrated on its expected effective useful life. This particular condition is generally referred to as a condition of environmental efficiency, where the resources used in manufacturing the product are gauged on the basis of the effective exigencies, avoiding over-dimensioning and consequent pointless wastage.

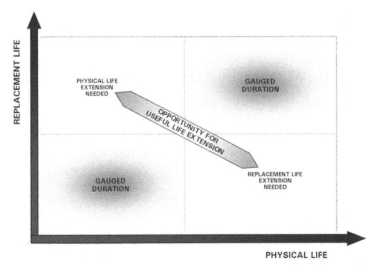

FIGURE 9.2 Identification of optimal strategies: Extension of product useful life.

In an analogous way, Figure 9.3 shows the conditions favoring different recovery strategies, from low-level recovery (recycling of materials) to a higher level (reuse of the entire product) where:

- Area 1 represents the condition where the product, whose effective useful life and presence on the market is expected to be short, is composed of rapidly deteriorating components. This represents a condition of gauged duration and is therefore, in principle, eco-efficient. At the end-of-life, any integral components are potentially reusable in other products, but the most probable recovery strategy is that of recycling the materials where possible.

- Area 2 represents the condition where the product, whose effective useful life and presence on the market is again expected to be short, is composed of long-lasting and functionally efficient components. At the end-of-life, the product may still be fully efficient but, as a result of external factors, cannot easily be reused because it is obsolete. Many of the components can potentially be reused as spare parts or in other products. The most probable recovery strategy is again the recycling of materials where possible.

- Area 3 represents the condition where the product, whose effective useful life and market presence is expected to be long, is composed of rapidly deteriorating components, so that its performance is not long-lasting enough. At the end-of-life, components that are still efficient

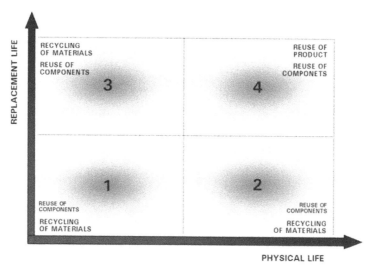

FIGURE 9.3   Identification of optimal strategies: Recovery at product end-of-life.

can be reused in the manufacture of a product of the same type. For the remaining components, only the recycling of materials is possible.

- Area 4 represents the condition where the product, whose effective useful life and market presence is again expected to be long, is composed of long-lasting and functionally efficient components. Like Area 1, this represents, in principle, an eco-efficient condition of gauged duration. If at the end-of-life the entire product is fully efficient and if the length of the replacement life will allow it, in this case it is possible to directly reuse the entire product. This would, in theory, represent the most efficient strategy for environmental protection unless the use of the product involved a significant environmental impact that could be avoided by using a new, more efficient product in its stead. As an alternative to direct reuse, it is possible to reuse some of the components. Finally, it is always possible to resort to recycling the materials if feasible.

9.1.3 Use Process Modeling

The factors conditioning the strategies for improving exploitation of resources, described above, are strictly dependent on how the product is used and on the context in which this use takes place. To best define the most effective design strategies, it is therefore necessary to have a clear vision of the way in which the product's use process (understood as the phase of the life cycle where the product performs its intended function) may develop. This depends on how the product behavior (i.e., the way in which it executes its function) interacts with the behavior of the user, and on how this interaction is configured in the context of the environment where it takes place. Ultimately, the use process is to be understood as an evolutionary process of the product–user–environment system and must be anticipated at the design stage. This requires a modeling of the use process, allowing the designer to simulate the way this process develops so that it is possible to make well-founded projections regarding the various factors that may condition the design strategies, ensuring a sound and truly effective product development. Studies of life cycle simulations, already discussed in Chapter 3, Section 3.2.4, provide meaningful examples of how this problem can be approached—limiting it to particular aspects and then using models to simulate the functional behavior of products and variations in the level of quality (Hata et al., 2000) or user behavior (Sakai et al., 2003).

The complex dynamics of the product–user–environment system determine the various factors, internal and external to the product, influencing both its replacement life and its physical life. Therefore, these dynamics should be analyzed for a meaningful estimation of the two important parameters. However, while the estimation of the physical life can be based on well-established modeling methods (such as finite element modeling that is able to forecast the behavior of a product's performance in relation to the conditions of use with reliability), the estimation of the environmental conditions and the interaction with the user, manifested in the various factors conditioning the replacement life (functional, technological, economic, social) is still a subject of research today (van der Vegte and Horvath, 2002).

9.2   Strategies for Extension of Useful Life and Design Considerations

During their use, products can be subjected to servicing operations such as maintenance and repair of worn or damaged components. The opportuneness of these operations can be assessed after an accurate estimation of the environmental implications; if the maintenance and/or repair results in a significant environmental impact, it may be more appropriate to retire the product unless its substitution requires the manufacture of a product with an even greater environmental cost.

The requisites of durability (the capacity to maintain initial performance levels over time) and maintainability (suitability for maintenance interventions aimed at restoring performance levels to their initial values) are of particular importance in the context of design. In reality, from a complete perspective of environmental efficiency these requisites must not be maximized indiscriminately. As noted in the case of durability, strengthening these properties is positive only up to a certain point, beyond which they begin to generate an impact greater than that caused by the replacement of the product (due, for example, to a higher consumption of resources in the use phase than that of a new, more efficient product).

Maintenance (i.e., the set of activities regarding periodic prevention and minor replacement interventions) is extremely important in limiting the environmental and economic costs of repair, as well as the impacts of dumping in waste disposal sites and of manufacturing a replacement product. To facilitate maintenance it is necessary to perform cleaning operations during use, ensure the accessibility of parts, arrange for the use of adequate equipment, and provide for systems to monitor the condition of parts and components. Other aspects that can extend the useful life of products are upgradability (in relation to various phenomena of technological evolution and modification) and adaptability (for products rapidly becoming obsolete and composed of more reconfigurable components).

The strategies aimed at extending a product's useful life were noted in Chapter 8, Section 8.2.2, and can be summarized as:

- Maintenance—Cleaning components; monitoring and diagnosis; substitution of parts subject to wear

- Repair—Regeneration or replacement of damaged and worn parts
- Upgrading and adaptation—Substitution of obsolete parts; reconfiguration of components, adapting them to user requirements or changes in the operating environment

It is appropriate to specify that in this book "adaptation" is meant as "reconfiguration" (i.e., an operation to change behaviors or states of the product resulting in a new function) (Tomiyama, 1999). This meaning differs from other more general definitions used in the literature, where "adaptation" is described as a process aiming at transferring products or components in additional usage phases, including maintenance, repair, remanufacturing, upgrading, or rearrangement (Seliger et al., 1998).

With regard to the factors that render a product predisposed to the application of one or more of these useful life extension strategies, it is appropriate to make the following distinction:

- The determinant factors making product upgradability and adaptability appropriate can only be analyzed on the basis of purely qualitative assessments.
- The determinant factors making the provision and planning of maintenance and repair interventions appropriate or necessary are quantifiable on the basis of well-established engineering techniques, as is the level of product maintainability and reparability.

This distinction, which is clearly a direct consequence of the different effects these strategies have on the replacement life and physical life (Section 9.1.2), has direct implications for the design tools and techniques, which are adequately developed only in relation to the strategies of servicing and maintenance (i.e., those that can intervene on the product's physical life). In fact, as was noted in Chapter 8 (particularly in Section 8.3.2) with regard to these strategies, the most appropriate DFX components are those aimed at maintaining performance during the phase of use—Design for Serviceability (DFS) (Makino et al., 1989; Gershenson and Ishii, 1993; Subramani and Dewhurst, 1993) and Design for Maintainability (Klement, 1993; Kusiak and Lee, 1997). These approaches to the problem of extending the product's useful life are closely interconnected and substantially complementary. It is necessary, however, to define the relationships between them more fully.

The term "service" includes interventions of diagnosis, maintenance, repair, and whatever else may be necessary to guarantee that the system functions correctly (Gershenson and Ishii, 1993). Design for Serviceability therefore includes, in general terms, Design for Maintainability. In this context, Design for Reliability (Rao, 1992; Birolini, 1993; Wallace and Stephenson, 1996) also plays a leading role. The reliability of the system is generally defined as the measure of its capacity to maintain functionality over a certain period of time, and is expressed by the probability that the system will maintain this functionality (Blanchard et al., 1995). It can also be quantified by the percentage of the time during which the system operates correctly, or by the frequency of malfunctions (Gershenson and Ishii, 1993).

Reliability is thus a product requisite that determines the necessity for interventions of maintenance or repair, predicted and favored by DFS. Ultimately, the latter is the design approach that best combines the most appropriate characteristics for achieving strategies for extending the product's physical life. It must be conducted in strict correlation with the design of the system's reliability, as confirmed by studies on the development and use of design rules formulated to take into account both reliability and maintainability (Kusiak and Lee, 1997).

9.2.1 Design for Serviceability

Using the term "service" for the set of diagnosis, maintenance, and repair interventions, together with any other intervention aimed at maintaining the functionality of a system, the term "serviceability" is understood as the facility with which a system can be subjected to these interventions, expressed by evaluation of:

- How easy it is to perform these interventions
- How much time they require and how much they cost

Design for Serviceability thus has the objective of aiding the designer in making choices promoting the development of products prearranged for service interventions (diagnosis, maintenance, repair) (Makino et al., 1989).

9.2.1.1 Main Aspects of Serviceability

On the basis of the definition of service given above, the main aspects of serviceability can be summarized as follows (Gershenson and Ishii, 1993; Klement, 1993):

- Diagnosability—A property of the constructional system making it possible to identify the causes of malfunction and define the consequent service interventions necessary. Arranging the components in groups according to function can favor diagnosability.
- Maintainability—A property of the constructional system making it possible to operate planned or required maintenance interventions. It renders the components requiring maintenance accessible to the

operator and, by simplifying the maintenance procedures, favors product maintainability.

- Reparability—A property of the constructional system allowing the removal and substitution of parts or components, beyond the replacement interventions foreseen by ordinary maintenance. It can be favored by a correct modularization of the constructional system's architecture, which minimizes the costs of disassembly and replacement.

9.2.1.2 Parameters of Constructional System Reliability

The reliability of a system is generally defined as the measure (in probabilistic terms) of its capacity to maintain its functionality over a certain period of time. An ideal constructional system of unlimited duration would have an unvarying reliability of 1 (i.e., maximum probability of maintaining the performance level). Since, in reality, no constructional system is completely failure-free, it can be appropriate to deal with its functionality in terms of reliability, so that while being characterized by a functionality of limited duration it nevertheless offers a high probability of successfully completing the functional task within a preset time span. Reliability is, therefore, a product requisite determining the necessity for the maintenance or repair interventions foreseen and favored by Design for Serviceability.

For further details regarding reliability parameters (failure rate, reliability of a component or system, reliable life, and reliability frequency), refer to the specialist literature (Bazovsky, 1961 and 2004; Rao, 1992; Ireson et al., 1995).

9.2.2 Quantitative Evaluation of Serviceability Properties

In order to evaluate the effect design choices may have on the product in terms of its suitability for service interventions, it is possible to use some indicators aimed at quantifying the costs and times involved in such interventions. A cost-based measure can aid the designer in decision making. Considering service modes as the ways in which a system may be serviced, Service Mode Analysis is the method of describing which service modes will impact a particular design and in what manner (Bryan et al., 1992; Gershenson and Ishii, 1993).

According to the approach called Component-based Service Mode, malfunctions are directly referable to single components or groups of components, and the required service interventions consist of repairing the components responsible for the failure. Having set the duration of functionality, it is possible to evaluate the cost of service over the life cycle in relation to the reliability of the single components. The service cost of the life cycle defined in this way provides a first evaluation of this property, limited to repair interventions on malfunctioning components.

An alternative and more complete expression can be defined using a different approach, called Phenomena-based Service Mode. This approach takes into consideration all the types of service interventions (repair, preventive and corrective maintenance, diagnostics), treating them as generic service phenomena (service mode phenomena). In this case, it is possible to arrive at a function of the life cycle service cost that evidences the dependence of serviceability properties on the main constructional characteristics of the product and on the reliability of the system, since this determines the number and typology of intervention phenomena and the factor of the frequency of operations, expressed by appropriate reliability parameters (failure rates, reliability frequencies).

Some of these functions allowing the assessment of the life cycle service cost are introduced and used in Chapter 16. As an alternative to evaluating the costs of service, it is possible to refer to service efficiency, and this can also be quantified as a function of the frequency of interventions (Subramani and Dewhurst, 1993).

Maintainability is defined as the probability of repairing a damaged system or component within a certain time interval (Rao, 1992). Maintenance interventions can be grouped according to two typologies:

- Corrective maintenance—Consisting of a set of unplanned interventions required to repair or substitute malfunctioning parts in order to restore functionality that was unexpectedly interrupted
- Preventive maintenance—Consisting of a set of interventions of inspection and possible repair or substitution aimed at preventing malfunctions and the deterioration of performance

For indicators that in some way quantify the maintainability of the constructional system (i.e., its property of allowing and favoring maintenance interventions), certain specialist literature (Blanchard et al., 1995) proposes functions able to express the mean duration of maintenance interventions and the mean time between them. This is called Mean Time Between Maintenance (MTBM), the mean value of the length of time between the beginning of two successive maintenance interventions; it can be expressed by reliability parameters.

9.2.3 Specific Determinant Factors for Useful Life Extension Strategies

With reference to the strategies considered here (maintenance, repair, upgrading, and adaptation), it is worth noting which specific factors determine the suitability of a product or component for each strategy (i.e., the factors conditioning the opportuneness or necessity of following a particular strategy). For example, a component that requires frequent cleaning, is particularly liable to deterioration phenomena and is characterized by low reliability and durability; it is predisposed to the application of maintenance. Thus, the need for cleaning, the sensitivity to phenomena of physical deterioration, and the reliability and durability properties can be considered specific determinant factors for the maintenance strategy.

Table 9.1 (upper part) summarizes the primary specific factors for each strategy:

- Maintenance—Necessity for cleaning; physical or mechanical deterioration (wear, aging, corrosion); reliability and durability of components and system.

TABLE 9.1 Environmental strategies for improving exploitation of resources and determinant factors

ENVIRONMENTAL STRATEGIES		SPECIFIC DETERMINANT FACTORS
Useful Life Extension	Maintenance	Necessity for cleaning
		Physical or mechanical deterioration
		Reliability and durability of components and system
	Repair	Physical or mechanical deterioration
		Risk of damage
		Reliability and durability of components and system
	Upgrading and Adaptation	Obsolescence (technological, cultural, functional)
		Changes in the use mode and environment
End-of-Life Recovery	Reuse	Physical or mechanical deterioration
		Risk of damage
		Technological obsolescence (also other types)
		Reliability and durability of components and system
		Ease of disassembly
	Remanufacturing	Physical or mechanical deterioration
		Technological obsolescence (also other types)
		Reliability and durability of components and system
		Ease of disassembly and cheapness of remanufacturing processes
	Recycling	Physical or mechanical deterioration
		Technological obsolescence (also other types)
		Reliability and durability of components and system
		Recyclability and value of materials
		Ease of material separation and cheapness of recycling processes

- Repair—Physical or mechanical deterioration; risk of damage; reliability and durability of components and system.
- Upgrading and adaptation—Technological, cultural (change in aesthetic values and modes of use), and functional (change in exigencies of user) obsolescence; changes in the use mode and environment.

9.2.4   Design Expedients

Some design expedients that facilitate maintenance, repair, upgrading, and adaptation include:

- Maintenance—Facilitate the accessibility of parts to be cleaned; prearrange and facilitate access to and substitution of the parts deteriorating most rapidly; group components according to their physical and mechanical properties, levels of reliability, and shared functions; prearrange systems for the diagnosis of parts requiring maintenance and provide for the application of diagnostic instruments on critical components.
- Repair—Prearrange and favor the removal and reassembly of critical components subject to deterioration and damage; design standardized parts and components.
- Upgrading and adaptation—Design modular and reconfigurable architecture for adaptation to different environments; design multifunctional products for adaptation to the evolutions of the user.

9.2.5   Design Variables

From an analysis of the specific factors determinant for the strategies and of the design expedients directed at achieving the requisites for extending the product's useful life, it is possible to identify the main design variables upon which interventions can be made in order to follow the strategies in question:

- Choice of materials and geometric properties—Influences the physical and mechanical performance of components over time, and their reliability and durability.
- Layout of product—Determines the reliability of the system, influences the component accessibility, and can allow the grouping of components according to their functionality or other performance characteristics.

- Modularity of the architecture—Influences the ease of access, removal and substitution of critical components, and reconfiguration of the system.

9.3 Strategies for Recovery at End-of-Life and Design Considerations

At the end of a product's useful life there are various opportunities for exploiting the resources used in its production; the functionalities of the entire product or some of its parts can be recovered and re-employed for the same task or other tasks (after collection and transport), or its original functionality can be restored and the product used as though new (after reprocessing). Also, its material and energy content can be exploited through recycling, composting, or incinerating its constituent materials. Some of these activities, such as reusing the product or its components, can be considered as extending of the product's useful life; this will be further discussed in Chapter 15. In this preliminary formalization, however, it is preferred to maintain the distinction between strategies intervening during the phase of use (e.g., maintenance) and those intervening after this phase. Reuse can be understood as one of the latter, to the point where some authors consider it a phase in itself, following that of use (Roozenburg and Eekels, 1995).

From this perspective, it becomes important to conceive and design products that are easily disassembled in order to favor the rapid and economic separation of parts or materials and the reuse of components, or to facilitate the separation of materials for recycling (when they are heterogeneous and mutually incompatible), for isolation (when they are toxic or dangerous), or for composting or incineration.

Products destined for reuse, in their entirety or only in part, can be collected and directed at the same original use or, alternatively, at another use demanding lower performance standards and having less-stringent requisites. The operations required may be limited to the cleaning or disassembly of some parts and their reassembly in new products. With regard to the possibility of remanufacturing (i.e., the process of reconditioning products worn through use and returning them to their almost original condition), it is especially important to facilitate the removal and substitution of parts and components and favor their interchangeability within the same line of products.

When the objective of disassembly is recycling, the condition of equilibrium defining the profit margin and economic sustainability is sensible to variations in the price of virgin materials and in the costs of collection and recycling. In this case, therefore, economic efficiency results from minimizing the time and costs of the necessary processes and in both the preservation and the exploitation of the materials recovered (the greater the purity of the materials, the better they preserve their performance characteristics and the greater their market value). As an alternative to disassembly, it is possible to consider a fragmentation process, followed by the separation and cleaning of the pieces.

Apart from the economic aspect, not all materials are equally recyclable. It is possible that the performance characteristics of the recycled material are substantially different than those of the initial virgin material, or that a material is easy to recycle at the technological level but requires a great expenditure of energy in the application of the necessary processes, outweighing the environmental advantages offered by the recycling itself.

As an alternative to recycling, another way of exploiting materials is that of recovering their energy content through incineration or other suitable processes. In either case, the environmental advantage is twofold: first, the impact due to the disposal of waste materials is avoided and, second, nonvirgin resources are made available, thus avoiding the impacts due to the production of a corresponding quantity of energy obtained from natural resources. Therefore, the criteria for the choice of materials direct the designer toward materials allowing their original performance characteristics to be recovered easily (avoiding the use of composites and additives) or those allowing the efficient recovery of their energy content, with little impact on the environment.

9.3.1 Definition of Recovery Strategies at End-of-Life

The strategies for the recovery of resources at the end of a product's useful life can be divided into various typologies (Overby, 1979; Ishii et al., 1994; Navin-Chandra, 1994; Zang et al., 1997; Gungor e Gupta, 1999; Rose et al., 2002). In Chapter 8, a preliminary distinction was drawn between strategies of direct reuse, reuse of parts, and the recycling of materials (Section 8.2.3). This discussion showed how each of these corresponds to a different potential environmental benefit, in turn depending on the level at which the recovery flow reenters the life cycle (this aspect is illustrated in Chapter 2, Figure 2.9). Another distinction that must be made is that between the recycling flows which, as was anticipated in Chapter 2, Section 2.5.2, can be closed (closed loop) or open (open loop), leading to recycling internal or external to the primary life cycle.

A complete panorama of the entire range of possible recovery strategies can be structured by clearly distinguishing between strategies for the recovery of the entire product and those for its parts or components (Dowie, 1994). Starting from this presupposition, at the end of a product's useful life the recovery intervention can follow one of the following different routes (Figure 9.4):

- If the product is in optimum working condition and guarantees the functional standards, it can be directly reused (in the figure, Reuse of Product).

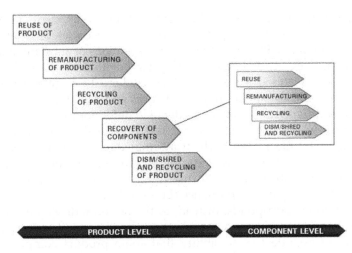

FIGURE 9.4 End-of-life strategies: Recovery options.

- If it is in good condition it can be reprocessed, reconditioned, and reused (Remanufacturing of Product).

- If it is entirely composed of recyclable and compatible materials, it can be fully recycled (Recycling of Product).

- If the product contains parts that can be reused, reprocessed, or recycled, it must be disassembled. This operation can, in turn, lead to four different possibilities of recovery: reusable parts can be detached from the product, returned to the production process, tested and reassembled in new products (Reuse of Components); other parts can be reprocessed before being reassembled in new products (Remanufacturing of Components); parts unsuitable for the two previous solutions and composed of recyclable materials can be directly routed to material recycling (Recycling of Components); parts containing volumes of recyclable materials can be dismantled and the resulting materials separated into volumes for recycling and volumes for dumping (Dismantling, Shredding, and Recycling of Components).

- If the product contains small volumes of recyclable materials, it can be dismantled to allow the subsequent selection and recycling of these materials (Dismantling, Shredding, and Recycling of Product).

- Finally, if the product does not contain reusable or recyclable parts it is directly routed to the waste disposal site (Disposal).

In this clearly hierarchical demarcation of the possible options at the product's end-of-life, some ambiguity may be found in the interpretation of the term "remanufacturing." It is appropriate, therefore, to specify that in this book "remanufacturing" of either product or component means an intervention of reuse that requires intermediate processes of regeneration and reconditioning, without resorting to substitutions or substantial repairs. This meaning is embedded in the more general definitions, where "remanufacturing" is understood as a process of reuse where the original forms of the product and components are preserved (Bras and McIntosh, 1999). At the same time it differs from other, more detailed definitions frequently used in the literature, where "remanufacturing" is defined as the process by which a product at end-of-life is completely disassembled in order to recover components that are still usable. After cleaning and inspection, these components are reassembled with others, both used and new, in order to recreate the product with its original performance (Lund, 1984).

This definition corresponds, instead, to the option indicated as "reuse of components" in the scheme reported above. In all cases, it was considered preferable to exclude the possibility that some product components are replaced or repaired, an intervention typology that is included within the useful life extension strategies, although other authors consider it assimilable to remanufacturing (Pnueli and Zussman, 1997). For the same reason, interventions of upgrading are also excluded in this book, despite the fact that some authors consider it part of remanufacturing (Thierry et al., 1995).

9.3.2 Management and Optimization of Recovery Strategies

The goal of applying the possible recovery strategies described above is to realize an ideal condition of a closed loop resource flow, where materials, components, and products circulate in a qualitatively and economically efficient way (Giudice et al., 2001; Umeda, 2001). With this perspective, the environmental strategies of recovery achieve this good cycling of resources and product design puts these strategies into action. Life Cycle Management (Chapter 3) must then perform the important role of maintaining this active and efficient flow of resources. With particular reference to the management of recovery strategies, Product Recovery Management (PRM) is understood as the management of the entire ensemble of products, components, and materials (used or discarded) that are the responsibility of a manufacturing company (Thierry et al., 1995).

The overview of the possible recovery strategies presented above follows a hierarchical order, according to a decreasing potential for environmental benefit. However, this criterion of hierarchical distinction between recovery levels allows the definition of the most effective strategy only in relation to the environmental aspect of the recovery question. In this case, the optimal solution is one that harmonizes environmental requirements with the times and costs of the processes under consideration. On the other hand, even when considering only the environmental aspect of the problem, the overall plan for recovery at end-of-life must be such that it maximizes the recovery of resources while simultaneously minimizing the use of resources and the production of emission phenomena by the recovery processes themselves. This confirms the validity of the systemic approach, adopted from the very beginning of this book, according to which the correct analysis of the environmental issues associated with the life cycle of a product requires an evaluation of all the flows of resources and emissions involved in each phase of the cycle.

According to this comprehensive perspective of the question, the problem of recovery can then be formulated as: Given a product (or a design proposal), determine the recovery plan able to effectively balance the costs of the recovery processes and the corresponding profits, understood in terms of resources used and recovered (Navin-Chandra, 1994).

Figure 9.5 represents the problem defined in this way, making it possible to identify the most favorable conditions. The trends of the curves are typical of the functions they express. The costs of recovery (disassembly, testing, remanufacturing) become prohibitive with the increasing depth of the recovery process. On the other hand, the revenues, beginning from a certain depth of recovery, tend to stabilize. As a consequence, the profit curve (given by the difference between the curves of revenue and cost) has a trend of the type shown in the figure.

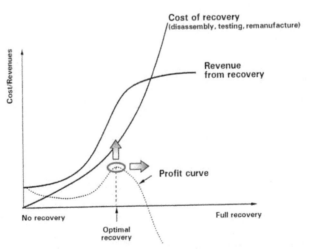

FIGURE 9.5 Recovery curves and optimization of recovery planning. (Adapted from Navin-Chandra, D., The recovery problem in product design, *Journal of Engineering Design*, 5(1), 67–87, 1994, Fig. 3-1.)

The optimization of recovery, therefore, consists of designing the product and its recovery plan in order to:

- Shift the peak of the profit curve toward a greater depth of the recovery intervention
- Increase the value of the profit at the peak

9.3.3 Approaches and Tools for Design

As was noted in Chapter 8, Section 8.3.2, various DFX tools are oriented toward the planning of processes at the end-of-life. They belong to the general area of Design for Product Retirement/Recovery (Ishii et al., 1994; Navin-Chandra, 1994; Kriwet et al., 1995; Zhang et al., 1997; Gungor and Gupta, 1999). Using more specific terminology, reference is frequently made to Design for Remanufacturing (Shu and Flowers, 1993; Bras and McIntosh, 1999) and Design for Recycling (Burke et al., 1992; Beitz, 1993).

Another DFX technique, known as Design for Disassembly, is oriented toward design and planning for the disassembly of systems (Boothroyd and Alting, 1992; Jovane et al., 1993; Scheuring et al., 1994; Harjula et al., 1996). As noted previously, although Design for Disassembly is frequently directed at recovery interventions at the end-of-life, in reality it is a tool that cuts across the two environmental strategies (extension of useful life and recovery at end-of-life) since it is targeted at a product characteristic—the ease of disassembly—that facilitates both strategies. Some important considerations regarding Design for Disassembly and Design for Recycling that represent (at the current state of the art) the most well-established design approaches, are discussed below.

9.3.3.1 *Design for Disassembly*

The disassembly of a used product is necessary whenever it is advantageous to proceed with the recovery of a product's subunits or single components. Disassembly can be defined as a systematic removal of the desired parts from an assembly, with the condition that the disassembly process does not result in any damage to the parts (Brennan et al., 1994). Therefore, it differs from "dismantling" due to its reversible and nondestructive character.

Disassembly is essential in the context of recovery strategies and also has great importance for useful life extension strategies because it can favor the properties of product serviceability discussed in Section 9.2.1. The objectives of disassembly are, therefore, to facilitate (Lambert, 1997):

- Recovery of parts, components, and subassemblies reusable in new products

- Recovery of recyclable materials
- Removal of dangerous or toxic components and materials
- Accessibility to parts or components that may be subjected to service operations

These objectives may also be extended to removal and replacement of components to be upgraded, and separation of parts and components to be reconfigured.

Design for Disassembly (DFD) can thus be defined as a design approach whose objective is that of optimizing product architecture and the other design parameters in relation to the following requisites:

- Simple and rapid separability of parts to be serviced, replaced, or recovered
- Possibility of separating fractions of a component composed of recyclable materials without compromising their potential for recycling
- Limitation of disassembly costs

Understood as a design tool, DFD can, therefore, target the harmonization between product layout, component geometries, and materials and junction systems in relation to the disassemblability of the system.

With this objective, the product design intervention can be interpreted on different levels (Yamagiwa et al., 1999): Frame Design (study of product layout); Part Design (study of the geometry and materials of parts); and Joint Design (study of the junction systems). Some design guidelines for an intervention directed at facilitating the disassembly of the constructional system can be associated with each level of design action, as shown in Table 9.2. These can be summarized on the basis of certain constructional standards (GE Plastics, 1992; ICER, 1993; VDI 2243, 2002) and contributions found in the literature (Beitz, 1993; Jovane et al., 1993; Chen et al., 1994; Dowie and Simon, 1994).

For more details regarding the disassembly of products, refer to Chapters 13 and 14, where the issues of the disassembly depth of the components of a constructional system and the optimal planning of disassembly processes are examined.

9.3.3.2 Design for Recycling

In general, recycling can be defined as the set of processes for the recovery of materials or components from a used product in order to render them usable in new products (Jovane et al., 1993; Kriwet et al., 1995). Design for Recycling thus consists of the design intervention intended to provide for and to facilitate the operations of recovery and recycling at the end of a product's life.

TABLE 9.2   Design for Disassembly: Guidelines, expedients, and requisites

DESIGN LEVELS	GUIDELINES
Frame Design (Product Architecture and Layout)	Separability of toxic or harmful parts and materials
	Separability of high-value parts and materials
	Subdivision into easily separable subunits
	Modularity of architecture
	Simplification of the hierarchy of connections between parts
	Prearrangement of accessible and recognizable pathways for disassembly operations
Part Design (Geometries and Materials)	Less variety and incompatibility of materials
	Fewer parts and components which are asymmetrical or difficult to handle
	Presence of flat surfaces and standardized handholds
	Arrangement of handholds near the center of gravity
	Provision of lines or areas of preferential breakage (elimination of incompatible inserts)
	Provision of cutting or fracture paths along the interfaces of incompatible materials
	Highlighting breakage points to facilitate identifying and reaching them
Joint Design (Junction Systems)	Use reversible junction systems
	Use of junction elements that can be destroyed physically or chemically
	Less variety of fasteners and fewer types of fasteners that are difficult to remove
	Fewer fastening systems that to be opened require simultaneous actions

The problems involved are numerous, and some are closely linked to those typical of disassembly regarding the separation of materials and components (Table 9.2). The essential requirements for a design oriented toward recycling (provided that they are limited to those strictly tied to the properties of the materials, the subject of recycling interventions) are briefly summarized below:

- Distribution of the materials in the product architecture in relation to their intrinsic properties of recyclability, their compatibility in terms of the recycling processes, and to the heterogeneity of the components or main subunits

- Absence of environmentally dangerous materials (toxic for humans, other organisms, or the environment in relation to the phases of production, use, and recycling)
- Clearly identifiable parts and their constituent materials

Recycling can also be understood as a strategy for the extension of the useful life of materials, given that it allows some kind of reuse of the materials recovered. Extending the life of materials, therefore, means making them function beyond the duration of the product they are part of. As noted earlier, the resulting environmental advantage is twofold:

- Avoiding the environmental impact due to the disposal of waste materials
- Making available nonvirgin resources for the production of materials or energy, according to the type of recycling, with a consequent reduction in the impact due to the production of corresponding quantities of materials and energy obtained from virgin natural resources

With regard to the typologies of recycling processes, these generally consist of a series of activities directed at the regeneration of material and its subsequent transformation. These activities include collection, separation, and cleaning, together with the processing of materials recovered from the waste flow and reintroduced into the original or external lifecycle. These typologies are grouped on the basis of the type of processing used on the materials (chemical recycling, mechanical recycling). Given that chemical recycling is not, at present, very practical because of its complexity and cost, the physical–mechanical recycling typology is the most commonly used on both homogenous and heterogeneous materials.

A design intervention with the objective of defining a recyclable product must evaluate two possible alternatives:

- The use of materials homogenous from a physical-chemical point of view, which can be directly recycled, eliminating even the operations of disassembly (this alternative is suited to the case where the product is characterized by limited functional complexity and, therefore, does not require the use of materials with widely differing performances)
- The use of materials that are mutually incompatible but easily separable (this alternative is suited to products characterized by a functional complexity requiring materials with different performances)

In economic terms, the recycling strategy is characterized by costs and revenues dependent on factors of various kinds:

- Recycling costs—The costs of the various phases of the recycling processes (cost of the operations of collection and transport, disassembly and separation, cleaning and reprocessing). They largely depend on the materials themselves, on the product architecture, and on the other constructional characteristics.

- Recycling revenues—Depend on various kinds of economic factors, such as the cost of virgin materials (if this increases, it raises the value of the recycled materials) and the efficiency of the product architecture and of the recycling processes (easily disassembled materials tend to be less contaminated, and the purity of the material influences its characteristics and, therefore, its market value).

These simple considerations demonstrate the importance of a design intervention that defines a product architecture and component geometry aimed at the economically feasible recycling of materials.

9.3.4   Quantitative Evaluation of the Potential for Recovery

In order to evaluate the effect that design choices can have in terms of disassemblability and recyclability, certain indices can be used that quantify various characteristics of the constructional system that are ascribable to these properties (Navin-Chandra, 1991; Simon and Dowie, 1993; Takata et al., 2003). The more important indices are based on certain primary considerations. For the purposes of their definition, it is possible to represent the disassembly process with two primary models (for a complete overview of disassembly process modeling, see the literature) (Tang et al., 2000; Lambert, 2003).

The first model, evidencing the depth of the disassembly levels (see also Chapter 13), correlates the successive phases of disassembly with the corresponding parts that become separated, and is represented in tree diagrams (Figure 9.6). Its first level corresponds to the product still assembled, the second level corresponds to main subassemblies, and so on down to the single components. A graphical representation of this kind is clearly helpful for simplifying the architecture because, if the diagram is structured on numerous levels, the product is characterized by great structural depth and, consequently, by inefficient disassembly.

In the second model, the process of disassembly is represented by a diagram of the order of precedence, commonly used for assembly planning. When applied to the inverse process, it evidences the complexity of the disassembly process. Figure 9.7 shows an example of a diagram of precedence wherein

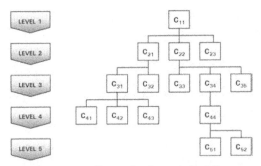

FIGURE 9.6 Tree diagram for the analysis of disassembly depth.

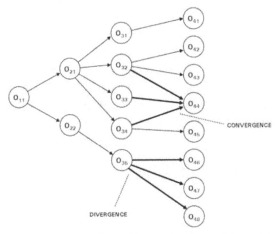

FIGURE 9.7 Precedence diagram for disassembly modeling.

each node of the diagram represents an operation of disassembly; the arrows indicate the direction of the process and thus the sequence of the operations.

In effect, two possible relations are created:

- Relation of divergence—Represents the condition where the process of disassembly offers a choice of several operations that can be performed in parallel, reducing the time required to complete the operation and, therefore, the cost.

- Relation of convergence—Represents the condition where the disassembly process decreases the levels of freedom with which it can be performed, because two or more processes must be completed before it is possible to begin the next operation. Relations of convergence normally imply an increase in the times required for disassembly because they reduce the possibility of conducting disassembly operations in parallel.

If the term "part" is used to indicate a single component or subassembly with the potential for independent recovery, and "connection" (or "junction system") indicates a system of connections between the parts consisting of one or more fasteners, and each junction is not considered a part of the system, it is possible to define the following quantitative evaluation indices:

- Depth of structure—Number of parts at each disassembly level (from the tree diagram)
- Dependence of disassembly—Number of parallelisms present in the disassembly process (from the diagram of precedence)
- Simplicity of structure—Number of parts connected by each connection
- Compatibility of materials—Number of connections uniting materials compatible for the purposes of recycling
- Product recyclability—Fraction of recyclable material present in the product
- Number of junctions—Number of junctions making up each connection
- Types of junction—Number of different types of junctions present in the constructional system
- Compatibility of junctions—Number of connections consisting entirely of junctions compatible for the purposes of recovery

These indices are independent of each other, so different indices do not indirectly evaluate the same property. They can be normalized in order to compare different architectures and design solutions and, if necessary, they can be unified in a single indicator by introducing a weighting coefficient.

9.3.5 Specific Determinant Factors for End-of-Life Strategies

In relation to the main strategies of recovery (reuse, remanufacturing, and recycling), it is worth noting which specific factors determine the suitability of a product or component for each strategy (i.e., the factors conditioning the advantage or necessity of following one strategy rather than another, as in the case of useful life extension strategies discussed in Section 9.2.3). Table 9.1 (lower part) summarizes the specific determinant factors for each end-of-life strategy:

- Reuse—Physical or mechanical deterioration; risk of damage; technological obsolescence; reliability and durability of components and system; ease of disassembly
- Remanufacturing—Physical or mechanical deterioration; technological obsolescence; reliability and durability of components and system; ease of disassembly and cheapness of remanufacturing processes
- Recycling—Physical or mechanical deterioration; technological obsolescence; reliability and durability of components and system; recyclability and value of materials; ease of material separation and cheapness of recycling processes

9.3.6 Design Expedients

Various considerations proposed in this chapter outline a set of expedients for a design intervention directed at favoring the strategies of recovery at the end-of-life, facilitating the disassembly of the constructional system and the possible levels of recovery of resources. These can be summarized, on the basis of well-known constructional standards and the main contributions present in the literature (Section 9.3.3.1), with regard to the main design factors defining the construction system: materials, junction systems, and layout of product architecture:

- Choice of materials—Minimize the number of different materials; make subassemblies and parts irreversibly connected from a single material or from compatible materials; use both recyclable and recycled materials; avoid the use of toxic or dangerous materials; avoid metal inserts or reinforcing in plastic components.
- Junction systems—Minimize the number of junctions and the number of different types of junctions; use rapidly reversible junctions, especially for the connections of high-value components or incompatible components; make points of junction easily accessible; use junctions made from materials compatible with the connected parts; provide for separation by breaking, to accelerate their disassembly process; simplify and standardize the junctions between components.
- Layout of product—Minimize the number of parts; make the architecture as modular as possible, separating the functions of

components; locate nonrecyclable parts so that they are easily detachable; locate parts of high value so that they are easily accessible and easily recoverable.

9.3.7 Design Variables

In conclusion, from the analysis of the specific determinant factors for the end-of-life strategies, and of the design expedients directed at achieving the requisites of product recoverability at the end of its useful life, it is possible to identify the main design variables upon which interventions can be made in order to follow the strategies in question:

- Choice of materials—Depending on their properties of durability and recyclability, materials influence the opportuneness of recovery.
- Geometries of the parts—Influence the ease of disassembly, and also the characteristics of durability.
- Layout of product—Influences component accessibility; can allow the grouping of components according to their functionality or other performance characteristics in relation to the recovery strategies.
- Modularity of the architecture—Influences the ease of access, separation, and recovery of components or subassemblies.

9.4 Product Modularity as a Key Concept for the Application of Environmental Strategies

One of the essential properties of product architecture is the level of dependence between the constructional units. Depending on this level, the architecture can be modular or integral (Chapter 7, Section 7.2.2.1). In accordance with the design principle that states "[A]n optimal design always maintains the independence of functional requirements" (Suh, 1990), the constructional units of a modular product implement only one, or a few, functional elements and are linked together by a limited number of well-defined interactions. To realize this condition, a connection between functional dependence and physical dependence is required. Therefore, minimizing interactions between physical components becomes an important requirement in product design for modularity (Ulrich and Tung, 1991).

With these premises, the implications deriving from product modularity have been examined by various authors (Ulrich, 1995; Erixon, 1998; Kusiak, 2002). The benefits of product modularity include component economies of scale, increased product variety, design standardization, ease of testing and production, and reduction in product development time. These benefits result from subdividing the system into modules with clear definition of the interfaces; this permits the design problem to be decomposed and the design tasks to be decoupled (Kamrani and Salhieh, 2000). As a consequence, design complexity is reduced and design tasks may be performed concurrently, reducing product development time and finally implementing the Concurrent Engineering approach (Chapter 7, Section 7.3).

In the context of the life cycle approach, some benefits of modularity are particularly interesting (Ulrich and Tung, 1991; Gershenson et al., 1999):

- Flexibility to meet changes in processes or requirements
- Ability to reduce life cycle costs by reducing the number of processes
- Ease of service due to differential consumption
- Ease of product updating due to the functional modules

A "module" is defined as a component or group of components that provides a basic main function and that can be removed from the system nondestructively as a unit (Allen and Carlson-Skalak, 1998). Thus, product modularity becomes a key concept for the application of the environmental strategies treated in this chapter (Gu and Sosale, 1999):

- By grouping components into detachable modules, fault analysis of product is facilitated and critical modules can be easily identified and replaced; serviceability (ease of maintenance and repair) is thus improved.
- By changing the arrangement of modules or upgrading one or more modules, new required functions and performances may be realized by an existing product; adaptation and upgrading are thus facilitated.
- By grouping components with similar life duration and compatible materials into detachable modules, they can be easily reused and recycled.

It is clear that Design for Modularity could become a useful technique of the DFX system, aimed at the development of complex products using the modularity concept. It can aid the designer in implementing an integrated product design and development approach (Chapter 8, Section 8.1.1), matching the criteria set by design for functionality and manufacturing, and those related to the environmental requirements.

Some implications of modularity on product design for the life cycle are treated in Chapter 11. For further details, refer to other resources in the literature (Gershenson and Prasad, 1997; Ishii, 1998; Newcomb et al., 1998; Kimura et al., 2001; Giudice et al., 2002; Zhang and Gershenson, 2003).

9.5   Summary

The environmental strategies enabling the extension of a product's useful life and its recovery at the end-of-life favor an increase in the intensity with which the resources employed in its manufacture are used. They can, therefore, be understood as appropriate strategies for the improvement of resource exploitation, and they differ substantially from strategies for the limitation or reduction of the resources used to manufacture a product.

Both typologies (useful life extension and recovery at end-of-life) are characterized by their high potential in relation to the environmental optimization of the flows of resources within the life cycle. Extending the useful life allows better exploitation of the resources used in production, avoiding the consumption of additional resources in the manufacture of replacement products. Intervening to recover all or parts of the product at the end of its life allows the reuse of its constituent components or materials in the production of new products, thus reducing the consumption of virgin materials, conserving the energy resources used, and reducing the volumes of waste.

For a truly effective application of these strategies, it is necessary to operate on different levels. In the preliminary phase of identifying those most appropriate to the case in question, an accurate analysis must be made of a wide range of factors that greatly condition their applicability and efficacy. In the implementation of these strategies, achieved through the use of specific tools and design approaches, it is necessary to take account of the relations linking these factors with the requirements of conventional engineering design, especially those relevant to the product's physical life. In managing the strategies themselves, they must be integrated in an effective manner, given that the actions of interventions extending the useful life can have considerable effect on the opportunities for recovery, which depend on the level of use to which the product, its components, and constituent materials have been subjected during the phases preceding recovery.

9.6   References

Allen, K.R. and Carlson-Skalak, S., Defining product architecture during conceptual design, in *Proceedings of 1998 ASME Design Engineering Technical Conference*, Atlanta, GA, 1998, DETC98/DTM-5650.

Bazovsky, I., *Reliability Theory and Practice*, Prentice Hall, Englewood Cliffs, NJ, 1961; Dover Publications, New York, 2004.

Beitz, W., Designing for ease of recycling, *Journal of Engineering Design*, 4(1), 12–23, 1993.

Birolini, A., Design for reliability, in *Concurrent Engineering: Automation, Tools and Techniques*, Kusiak, A., Ed., John Wiley & Sons, New York, 1993, 307–348.

Blanchard, B.S., Verma, D.C., and Peterson, E.L., *Maintainability: A Key to Effective Serviceability and Maintenance Management*, 2nd ed., John Wiley & Sons, New York, 1995.

Boothroyd, G. and Alting, L., Design for assembly and disassembly, *Annals of the CIRP*, 41(2), 625–636, 1992.

Bras, B. and McIntosh, M.W., Product, process, and organizational design for remanufacture: An overview of research, *Robotics and Computer Integrated Manufacturing*, 15, 167–178, 1999.

Brennan, L., Gupta, S.M., and Taleb, K.N., Operations planning issues in an assembly/ disassembly environment, *International Journal of Operations and Production Management*, 14(9), 57–67, 1994.

Bryan, C., Eubanks, C.F., and Ishii, K., Design for serviceability expert system, in *Proceedings of ASME Conference on Computers in Engineering*, San Francisco, 1992, 91–98.

Burke, D., Beiter, K., and Ishii, K., Life-cycle design for recyclability, in *Proceedings of ASME Design Theory and Methodology Conference*, Scottsdale, AZ, 1992, 325–332.

Chen, R.W., Navin-Chandra, D., and Prinz, F.B., Product design for recyclability: A cost-benefit analysis model and its application, *IEEE Transactions on Components, Packaging, and Manufacturing Technology*, 17(4), 502–507, 1994.

Dowie, T. and Simon, M., Guidelines for Designing for Disassembly and Recycling, Report DDR/TR18, Manchester Metropolitan University, Manchester, UK, 1994.

Dowie, T., Green design, *World Class Design to Manufacture*, 1(4), 32–38, 1994.

Erixon, G., Modular function deployment: A method for product modularisation, Ph.D. thesis, The Royal Institute of Technology, Stockholm, 1998.

GE Plastics, Design for Recycling, General Electric Plastics, Pittsfield, MA, 1992.

Gershenson, J. and Ishii, K., Life-cycle serviceability design, in *Concurrent Engineering: Automation, Tools and Techniques*, Kusiak, A., Ed., John Wiley & Sons, New York, 1993, 363–384.

Gershenson, J.K. and Prasad, G.J., Modularity in product design for manufacturability, *International Journal for Agile Manufacturing*, 1(1), 99–110, 1997.

Gershenson, J.K., Prasad, G.J., and Allamneni, S., Modular product design: A life-cycle view, *Journal of Integrated Design and Process Science*, 3(4), 13–26, 1999.

Giudice, F., La Rosa, G., and Risitano, A., An ecodesign method for product architecture definition based on optimal life-cycle strategies, in *Proceedings of Design 2002—7th International Design Conference*, Dubrovnik, Croatia, 2002, 1311–1322.

Giudice, F., La Rosa, G., and Risitano, A., Optimisation and cost-benefit analysis of product recovery-cycles, in *Proceedings of ICED01—13th International Conference on Engineering Design: Design Applications in Industry and Education—Design Methods for Performance and Sustainability* Cotley, S. et al., Eds. Professional Engineering Publishing, London, 2001, 629–636.

Giudice, F., La Rosa, G., and Risitano, A., Product recovery-cycles design: Extension of useful life, in *Feature-Based Product Life-Cycle Modelling*, Soenen, R. and

Olling, G., Eds., Kluwer Academic Publishers, Dordrecht, The Netherlands, 2003, 165–185.

Gu, P. and Sosale, S., Product modularization for life cycle engineering, *Robotics and Computer Integrated Manufacturing*, 15, 387–401, 1999.

Gungor, A. and Gupta, S.M., Issues in environmentally conscious manufacturing and product recovery: A survey, *Computers and Industrial Engineering*, 36, 811–853, 1999.

Harjula, T. et al., Design for disassembly and the environment, *Annals of the CIRP*, 45(1), 109–114, 1996.

Hata, T. et al., Product life cycle simulation with quality model, in *Proceedings of 7th CIRP International Seminar on Life Cycle Engineering*, Tokyo, 2000, 60–67.

ICER, Design for Recycling: General Principles, Industry Council for Electronic Equipment Recycling, London, 1993.

Ireson, W.G., Coombs, C.F., and Moss, R.Y., *Handbook of Reliability Engineering and Management*, 2nd ed., McGraw-Hill, New York, 1995.

Ishii, K., Modularity: A key concept in product life-cycle engineering, in *Handbook of Life Cycle Engineering*, Molina, A., Sanchez, J.M., and Kusiak, A., Eds., Kluwer Academic Publishers, Dordrecht, The Netherlands, 1998, 511–531.

Ishii, K., Eubanks, C.F., and Di Marco, P., Design for product retirement and material life cycle, *Materials and Design*, 15(4), 225–233, 1994.

Jovane, F. et al., A key issue in product life cycle: Disassembly, *Annals of the CIRP*, 42(2), 651–658, 1993.

Kamrani, A.K. and Salhieh, S.M., *Product Design for Modularity*, Kluwer Academic Publishers, Boston, 2000.

Kimura, F. et al., Product modularization for parts reuse in inverse manufacturing, *Annals of the CIRP*, 50(1), 89–92, 2001.

Klement, M.A., Design for maintainability, in *Concurrent Engineering: Automation, Tools and Techniques*, Kusiak, A., Ed., John Wiley & Sons, New York, 1993, 385–400.

Kriwet, A., Zussman, E., and Seliger, G., Systematic integration of design for recycling into product design, *International Journal of Production Economics*, 38, 15–22, 1995.

Kusiak, A., Integrated product and process design: A modularity perspective, *Journal of Engineering Design*, 13(3), 223–231, 2002.

Kusiak, A. and Lee, G., Design of parts and manufacturing systems for reliability and maintainability, *International Journal of Advanced Manufacturing Technology*, 13, 67–76, 1997.

Lambert, A.J.D., Optimal disassembly of complex products, *International Journal of Production Research*, 35(9), 2509–2523, 1997.

Lambert, A.J.D., Disassembly sequencing: A survey, *International Journal of Production Research*, 41(16), 3721–3759, 2003.

Lund, R.T., Remanufacturing, *Technology Review*, 87(2), 18–29, 1984.

Makino, A., Barkan, P., and Pfaff, R., Design for serviceability, in *Proceedings of ASME Winter Annual Meeting*, San Francisco, 1989, 117–120.

Navin-Chandra, D., Design for environmentability, in *Proceedings of ASME Conference on Design Theory and Methodology*, Miami, FL, 1991, DE-31, 119–125.

Navin-Chandra, D., The recovery problem in product design, *Journal of Engineering Design*, 5(1), 67–87, 1994.

Newcomb, P.J., Bras, B., and Rosen, D.W., Implications of modularity on product design for the life cycle, *Journal of Mechanical Design*, 120, 483–490, 1998.

Overby, C., Product design for recyclability and life extension, in *Proceedings of American Society of Engineering Education Annual Conference*, Baton Rouge, LA, 1979, 181–196.

Pnueli, Y. and Zussman, E., Evaluating the end-of-life value of a product and improving it by redesign, *International Journal of Production Research*, 35(4), 921–942, 1997.

Rao, S.S., *Reliability-Based Design*, McGraw-Hill, New York, 1992.

Roozenburg, N.F.M. and Eekels, J., *Product Design: Fundamentals and Methods*, John Wiley & Sons, Chichester, UK, 1995.

Rose, C.M., Ishii, K., and Stevels, A., Influencing design to improve product end-of-life stage, *Research in Engineering Design*, 13, 83–93, 2002.

Rose, C.M., Masui, K., and Ishii, K., How product characteristics determine end-of-life strategies, in *Proceedings of IEEE International Symposium on Electronics and the Environment*, Oak Brook, IL, 1998, 322–327.

Sakai, N., Tanaka, G., and Shimomura, Y., Product life cycle design based on product life control, in *Proceedings of EcoDesign 2003: 3rd International Symposium on Environmentally Conscious Design and Inverse Manufacturing*, Tokyo, 2003, 102–108.

Scheuring, J., Bras, B., and Lee, K.M., Significance of design for disassembly on integrated disassembly and assembly processes, *International Journal of Environmentally Conscious Design and Manufacturing*, 3(2), 21–33, 1994.

Seliger, G., Grudzien, W., and Muller, K., The acquiring and handling of devaluation, in *Proceedings of 5th CIRP Seminar on Life Cycle Engineering*, Stockholm, 1998, 99–108.

Shu, L. and Flowers, W., A structured approach to design for remanufacture, in *Proceedings of ASME Winter Annual Meeting*, New Orleans, LA, 1993, 13–19.

Simon, M. and Dowie, T., Quantitative Assessment of Design Recyclability, Report DDR/TR8, Manchester Metropolitan University, Manchester, UK, 1993.

Subramani, A. and Dewhurst, P., Efficient design for service considerations, *Manufacturing Review*, 6(1), 40–47, 1993.

Suh, N.P., *The Principles of Design*, Oxford University Press, New York, 1990.

Takata, S. et al., Framework for systematic evaluation of life cycle strategies by means of life cycle simulation, in *Proceedings of EcoDesign 2003: 3rd International Symposium on Environmentally Conscious Design and Inverse Manufacturing*, Tokyo, 2003, 198–205.

Tang, Y. et al., Disassembly modelling, planning, and application: A review, in *Proceedings of IEEE International Conference on Robotics and Automation*, San Francisco, 2000, 2197–2202.

Thierry, M. et al., Strategic issues in product recovery management, *California Management Review*, 37(2), 114–135, 1995.

Tomyiama, T., Reversible reconfiguration: A key for reuse, in *Proceedings of EcoDesign '99: 1st International Symposium on Environmentally Conscious Design and Inverse Manufacturing*, Tokyo, 1999, 310–315.

Ulrich, K.T., The role of product architecture in the manufacturing firm, *Research Policy*, 24, 419–440, 1995.

Ulrich, K.T. and Tung, K., Fundamentals of product modularity, in *Proceedings of ASME Winter Annual Meeting Symposium on Design and Manufacturing Integration*, Atlanta, GA, 1991, 73–79.

Umeda, Y., Toward a life cycle guideline for inverse manufacturing, in *Proceedings of EcoDesign 2001: 2nd International Symposium on Environmentally Conscious Design and Inverse Manufacturing*, Tokyo, 2001, 143–148.

van der Vegte, W.F. and Horvath, I., Consideration and modeling of use processes in computer-aided conceptual design: A state of the art review, *Journal of Integrated Design and Process Science*, 6(2), 25–59, 2002.

van Nes, N., Cramer, J., and Stevels, A., A practical approach to the ecological lifetime optimization of electronic products, in *Proceedings of EcoDesign '99: 1st International Symposium on Environmentally Conscious Design and Inverse Manufacturing*, Tokyo, 1999, 108–111.

VDI 2243, Recycling-Oriented Product Development, VDI 2243 German Standards, Verein Deutscher Ingenieure, Düsseldorf, Germany, 2002.

Wallace, K.M. and Stephenson, J., Design for reliability, in *Design for X: Concurrent Engineering Imperatives*, Huang, G.Q., Ed., Chapman & Hall, London, 1996, 245–267.

Woodward, D.G., Life cycle costing: Theory, information acquisition and application, *International Journal of Project Management*, 15(6), 335–344, 1997.

Yamagiwa, Y., Negishi, T., and Takeda, K., Life cycle design achieving a balance between economic considerations and environmental impact with assembly-disassembly evaluation/design, in *Proceedings of EcoDesign '99: 1st International Symposium on Environmentally Conscious Design and Inverse Manufacturing*, Tokyo, 1999, 760–765.

Zhang, H.C. et al., Environmentally conscious design and manufacturing: A state of the art survey, *Journal of Manufacturing Systems*, 16(5), 352–371, 1997.

Zhang, Y. and Gershenson, J.K., An initial study of direct relationships between life-cycle modularity and life-cycle cost, *Concurrent Engineering: Research and Applications*, 11(2), 121–128, 2003.

Chapter 10

Engineering Methods for Product Duration Design and Evaluation

One of the primary tasks of product design for the environment consists of harmonizing the requisites of environmental performance with those of conventional design (functionality, safety, duration). To do this, methods and tools must be available to the designer that allow the evaluation and optimization of design parameters determining a product's performance (conventional and environmental) over its entire life cycle.

The definition of strategies for extension of useful life and recovery at end-of-life is conditioned by several factors that limit their effectiveness. The evaluation of these factors is essential for a correct implementation of these strategies in product development. Being able to predict, in the design phase, the extension of a product's useful life and the reuse or remanufacture of parts of it depends on the expected duration of components and on their residual life. As a consequence, any study of the environmental aspects of a product must include the consideration of parameters such as the predicted duration of a component, its resistance to the operating load, and the estimated damage suffered by it.

Accordingly, this chapter briefly treats certain significant aspects of conventional design. In particular, after a short review of material fatigue and damage phenomena, attention is focused on the rapid methods currently used for the fatigue characterization of materials.

10.1 Durability of Products and Components

Deterioration in the functional performance of products and their components, which greatly affects the possibility of applying the environmental strategies for extension of useful life and recovery at end-of-life, is principally due to phenomena conditioning the properties of duration over time:

- Structural obsolescence determined by the physical–mechanical deterioration of materials

- Damage due to improper use or accidental events
- Deterioration due to external factors and operating environment

Contrary to what might be supposed, the durability of components and the constructional system (understood as their capacity to maintain the required operating performance) must not be maximized indiscriminately, but optimized in relation to the feasibility of using the product or reusing its components. Graphs such as those shown in Chapter 9, Figures 9.2 and 9.3 can allow the judicious calibration of product durability (which determines the span of its physical life) in relation to:

- The limits imposed on the effective useful life by the external factors previously defined (expressed by the span of replacement life)
- The range and typology of intervention to be operated through useful life extension and end-of-life strategies

A complete structural durability analysis directed toward the prediction of physical life of components requires the integration of several engineering tools and techniques, and large amounts of data collection and computation (Youn et al., 2005). Nevertheless, the durability of components and systems can be defined and quantified with good approximation in the design stage, using established methods and mathematical tools for design for durability, the result of exhaustive studies on phenomena such as fatigue and damage.

In this context, there are clear and simple rules of design for appropriate durability:

- Design equal duration for components similar in terms of functionality and intensity of use.
- Design duration as a function of the product's effective useful life.
- Design heightened duration for components difficult to repair and maintain, and for those intended for reuse.
- Design limited duration (as close as possible to the effective life required) for components needing substitution during use, and for those intended for recycling or disposal.

With these premises, it could be appropriate to consider some aspects of conventional engineering design, paying particular attention to phenomena of performance deterioration (fatigue and damage), design for component durability, and methods for the evaluation of residual life. These are the basis of the modern computer-aided engineering design processes, developed to carry out design optimization for structural durability and aimed at realizing durable, manufacturable, and cost-effective products.

10.2 Fatigue of Materials

Studies on material fatigue began in the nineteenth century, when, with the daily use of machines, tools, and vehicles, it was observed that working parts subjected to loads that varied over time were damaged and eventually broke, despite the fact that at no time during their use did the stresses reach the safety values determined using normal techniques for studying the resistance of materials. In particular, the earliest scientific investigations on fatigue behavior concerned railway structures. In his first paper, the German engineer August Wöhler reported on the fatigue resistance of railway tracks, the first attempt at a quantitative description of fatigue with the introduction of the concept of fatigue limit. The research undertaken by Wöhler between 1852 and 1870 produced an enormous quantity of data that he presented in graphical form, known as the Wöhler curve and still frequently used today (Wöhler, 1870).

Researchers agree in describing fatigue as a localized phenomenon evolving in four distinct phases:

- Nucleation
- Subcritical propagation of the defect
- Critical propagation of the crack, which can be characterized using the theories of elastic, elastic–plastic, or completely plastic fracture mechanics
- Unstable propagation

The nucleation of the crack occurs in a critical zone of the component or specimen, characterized by an elevated value of local stress different from the stress value measured macroscopically on the same component. This is due to the presence of discontinuities in the material at the structural level (nonhomogeneities, microcracks) or geometric level (notches, irregularities). At the apex of the crack, the material is subjected to a localized plastic deformation. As the dimensions of the crack increase, there is a resulting decrease in the resisting cross-section with a consequent increase in the stress on the material. Large zones of plasticization lead to a decrease in ductility and a reduction of resistance. Thus, fatigue failure always has its origins in plastic deformations occurring at the microscopic level.

According to the American Society for Testing and Materials, the phenomenon of fatigue can be defined as that process that "triggers a progressive and localized permanent structural transformation in the material, whenever it is subjected to loading conditions that produce, in some points of the material, cyclical variations in the stresses or strains" (ASTM E606–92, 2004). These cyclical variations, after a certain number of applications, can culminate in the presence of cracks or in the failure of the component. To study the fatigue behavior of a component it is, therefore, necessary to know the loading history, the characteristics of the material comprising the component, and the geometry of the component itself.

10.2.1   Loading History

Design for component fatigue requires information on the time history of the loading the element will undergo. These loading histories are obtained using experimental techniques on preexisting components or on scale specimens. The stresses σ measured in this way must be representative of those to which the element under examination will actually be subjected. The time histories can be classified as periodic or aleatory, following the scheme proposed in Figure 10.1.

In general, actual loading histories are treated by arranging them in constant amplitude sinusoidal cycles using the Fourier series. Sinusoidal loading is described using the variables in Figure 10.2:

- σ_{max}, maximum stress
- σ_{min}, minimum stress

- $\sigma_m = \dfrac{\sigma_{max} + \sigma_{min}}{2}$, mean stress

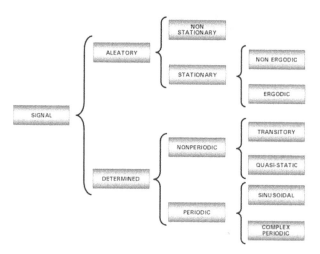

FIGURE 10.1   Classification of signals.

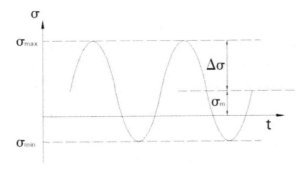

FIGURE 10.2   Representation of a dynamic load.

- $\Delta\sigma = \dfrac{\sigma_{max} - \sigma_{min}}{2}$, load amplitude
- N, number of cycles

10.2.2   Design for Fatigue

The theories of fatigue can be applied using three distinct approaches:

- Design for infinite life
- Design for finite life
- Design for critical dimensions of defects

Of the three approaches, the first is based on Wöhler's theories. The underlying hypothesis is that of the perfect integrity of the material (i.e., the absence of defects or cracks before loading) and that nucleation occurs after the application of the load. It is commonly used for metals, particularly steel, but it is not always applicable to other types of material. Using appropriate damage hypotheses, it is also possible to determine the residual life.

In the 1940s and 1950s, there was considerable development in the design of machines for fatigue testing. By allowing the application of greater loads, such devices made it possible to investigate the behavior of materials under more extensive regimes of plasticization. Since the phenomenon of fatigue is essentially expressed at a local level, it seemed more appropriate to describe this phenomenon through the use of strains rather than stresses. Experimental data were, therefore, represented in terms of stain ε versus number of cycles N. With this approach (design for finite life), it is possible to consider the effects of plasticity, and it is also more adaptable to variations in the test parameters. It is also more suited for application on different materials and different component geometries. However, it is more complicated to apply than the previous approach and requires greater processing power for the elaboration of the data. Furthermore, given its more recent introduction, there is less data available in the literature. Also, here it is assumed that the material subjected to loading is perfectly integral with no initial defects and that the end of its useful life coincides with the formation of a crack.

Conversely, the third approach (design for critical dimensions of defects) assumes that there are always internal defects present in every material and that their characteristic dimensions increase following the application of load. Therefore, a component's useful life does not end when a defect arises but, rather, when this defect reaches critical dimensions. This approach, developed in the 1960s, led to the introduction of complex variables referring to fracture mechanics, such as the stress intensity factor (ΔK_I). This factor is a function of the orientation of the defects and of the dimensions and geometry of the part containing the defect. The growth of the crack under a variable load is usually described using diagrams of the type da/dN (velocity of crack growth) versus ΔK_I. Clearly, it is a considerable advantage to be able to assess components already damaged; however, this approach has the disadvantages of increased calculation times in that it requires nondestructive testing (NDT) in order to evaluate the effective dimensions of the defects present in the component.

10.2.3   Infinite Life Approach

Design for infinite life developed between the end of the nineteenth and beginning of the twentieth century as a result of the Industrial Revolution giving rise to greater complexity of machinery subjected to dynamic loading and, therefore, susceptible to fracture. Often called Design for High Cycle Fatigue (DHCF), design for infinite life is directed at ensuring that the specimen, component, or subassembly under examination remains inside the elastic region throughout its useful life. More explicitly, in a component designed for infinite life the applied loading always remains below the fatigue limit, defined by Wöhler as: "That stress value which does not result in the failure of the component in question whatever the number of application cycles."

In the Wöhler diagram, this value corresponds to the slope of the curve σ versus N, also known as the "elbow" of the curve (Figure 10.3). In fact, for each value of dynamic load it is possible to determine the number of cycles that will lead to failure. The number of cycles to failure N_r increases when the applied load decreases, to arrive at a given value σ_0 corresponding to a number of cycles of infinite life. In testing, since it would be impossible to conduct a test for an infinite number of cycles, it is possible to define a number of cycles (corresponding to the elbow of the Wöhler curve) after which the material can be considered to have an infinite residual life. This number is a characteristic of the type of material. In the case of steel, the elbow is well-defined by the

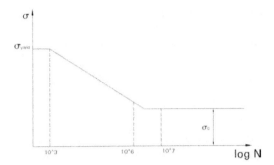

FIGURE 10.3   Wöhler diagram for a steel.

asymptotic trend of the curve σ versus N, beginning from the fatigue limit at around 10^6 cycles. Because of this characteristic, steels are particularly suited to this approach. Conversely, many other materials do not present such a clear trend and even at high numbers of cycles (from 10^6 to 10^9), the σ versus N curves continue to exhibit steep slopes.

Wöhler curves are obtained from controlled loading tests, typically plotting the number of cycles along the x-axis and the load (maximum load σ_{max}, or load amplitude Δσ) along the y-axis. In order to interpret the diagrams correctly, other load characteristics are specified (e.g., the cycle ratio $R = \sigma_{min}/\sigma_{max}$).

The data obtained from experimental tests are highly dispersed, so the construction of the curve requires a large number of specimens for each loading level. Furthermore, this dispersion gradually increases as the load nears the fatigue limit. Interpolating the points with the same probability of failure at different load levels gives the "different probability of failure" curve. The highest curve of the diagram represents 95% probability of failure within the corresponding number of cycles, while the lowest curve represents 5% probability of failure. Wöhler diagrams allow an infinite life component to be dimensioned in terms of resistance to fatigue, referring to the values of the fatigue limit or, in temporal terms, referring to the number of cycles to failure relative to the stress considered.

With regard to the frequency of the applied loads, experience has shown that this has a negligible effect on the relation between the stresses and the number of cycles. In experimental trials on specimens under rotating bending load, with frequencies up to 170 Hz, the value of frequency had no effect. Higher frequencies, up to 500 Hz, produced an increase in fatigue resistance varying between 3% and 13%. It should be noted that the frequency has no effect only when the material under examination does not reach temperatures high enough to alter its structure.

Given that experimental trials are generally performed on simple specimens, to determine the actual fatigue resistance of the component to be designed it is necessary to take into account its shape, surface finishing, heat treatment, (Shigley and Misehke, 1989 and so on). To do so, coefficients are used that evaluate the reduction in resistance due to:

- The type of loading applied
- The stress concentration
- The surface finishing
- The dimensions (scale effect)

These factors are usually evaluated experimentally, as summarized in Table 10.1.

The effective fatigue limit σ_0 is lower than that obtained on a specimen σ_0^I:

$$\sigma_0 = \sigma_o^I C_L C_G C_S \tag{10.1}$$

The effect of the dimensions, or scale effect, is associated with the probability of finding a critical defect in the material; the greater the volume of material subjected to fatigue forces, the higher this probability will be. Also, the type of loading must be seen in terms of the probability of creating conditions of microplasticization in the material. In the case of traction, where all the points of the specimen are subjected to the same stress, a point of discontinuity would reach plasticization and trigger a crack. In the case of torsion, the points with greatest stress are on the external surface of the specimen, and there is, therefore, a lower probability that conditions of microplasticization are generated. The phenomenon is less probable under bending loads, where points of greatest stress are those along the opposite generatrices of a cylindrical specimen.

The surface finishing of parts is extremely important in elements subjected to fatigue. It is possible to show the coefficient of decreased fatigue resistance in relation to the failure load R, for various degrees of surface finishing. From

TABLE 10.1 Fatigue limit reduction factors

		BENDING	TRACTION	TORSION
C_L Load Factor		1	1	0.58
C_G Size Factor				
	Diameter < 10 mm	1	from 0.7 to 0.9	1
	10 mm < Diameter < 50 mm	0.9	from 0.7 to 0.9	0.9
C_S Surface Finishing Factor		See (Shigley and Mischke, 1989, pp. 282–286)		

diagrams like these, it can be seen that the various curves show a decrease on the y-axis for an increase on the x-axis, and thus steels with the highest failure loads are more susceptible to the effect of surface irregularities. This effect can be explained by considering the phenomenon used to determine fatigue failure. Given that the existence of microscopic cracks is inevitable in a mechanical element, all processes that can lead to an increase in their extension will lower the fatigue limit, while those limiting their extension will raise this limit. In general, those processes that generate residual compression stresses in the element are those that increase the fatigue limit, while those that generate residual traction stresses result in a decrease in the fatigue limit. Heat treatments improve, to a greater or lesser extent, the fatigue resistance of the element.

Finally, it is necessary to take into account the effects produced by a variation in the cross-section of the component in question (e.g., coves, notches, or holes near which there is a very steep stress gradient and a maximum stress peak, as shown in Figure 10.4). This phenomenon is defined as Stress Concentration and is more marked as the size of the radius of curvature of the cove, notch, or hole decreases.

The application of St. Venant's torsion theory can only give approximate values of the maximum stresses. To determine the actual stress in each point of the material requires, therefore, the direct application of the general elasticity equations. In the case of moderately simple geometric shapes, Neuber provided some solutions of the stress state along the entire contour, evaluating the maximum stress value (Neuber, 1958).

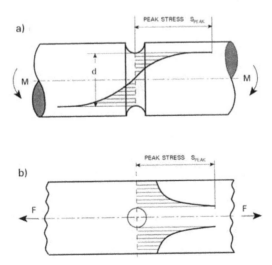

FIGURE 10.4 Stress gradients corresponding to (a) coves and (b) notches.

The value of the nominal stress acting on the component is thus increased by a force concentration factor K_t:

$$K_t = \sigma_{theor}/\sigma_{nom} \tag{10.2}$$

K_t is calculated using the theory of elasticity and the results are presented in Peterson diagrams (Peterson, 1959).

The coefficient K_t is theoretical because the effect of a notch also depends on the type of material and on the type of static or fatigue loading applied on the notched element. If this element is composed of a ductile material and subjected to fatigue loading, there is a redistribution of the stresses due to the plasticity of the material and to metallurgical instability caused by the fatigue process itself. In the fatigue characterization of a material, this effect is taken into account by introducing, at the experimental level, a dynamic or fatigue stress concentration factor.

The fatigue notch factor K_f is defined as:

$$K_f = \sigma_{eff}/\sigma_{nom} \tag{10.3}$$

where σ_{eff} takes account of the distribution of the stresses within the material at the microplasticizations forming in the proximity of zones with concentrated stresses. The two factors are interrelated: $1 \leq K_f \leq K_t$.

When the material is perfectly fragile, the stresses are not redistributed and the preceding inequality becomes $K_f \quad K_t$. The factor K_f can be calculated using empirical relations that take into account the radius of curvature and the properties of the material (e.g., Heywood's equation):

$$K_f = 1 + \frac{K_t - 1}{1 + \sqrt{\dfrac{a}{r}}} \tag{10.4}$$

where r is the radius of curvature and a is a constant, function of the properties of the material, with the magnitude of one length. In practice, the value of K_f can be obtained as a ratio between the high cycle fatigue resistance of the material determined on an unnotched specimen and that on a notched specimen.

In conclusion, it is possible to define the notch sensitivity factor q, by the ratio between the increase ineffective stress due to notch and that in theoretical stress due to notch:

$$q = \frac{K_f - 1}{K_t - 1} = \frac{\sigma_{eff} - \sigma_{nom}}{\sigma_{theor} - \sigma_{nom}} \tag{10.5}$$

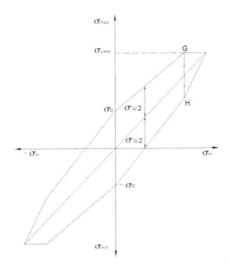

FIGURE 10.5 Goodman–Smith diagram.

as a consequence

$$K_f = 1 + q (K_t - 1) \tag{10.6}$$

The factor q is the ratio between the increase in effective stress due to notch and that in theoretical stress due to notch.

Finally, it is also necessary to consider the influence of the mean stress σ_m. It can be said that with increasing static traction stress σ_m, to ensure the same lifespan (in this case, infinite), the amplitude of the alternate stress $\Delta\sigma$ must decrease. Different models have been proposed to evaluate the influence of mean stress. The most commonly used is the Goodman–Smith diagram, shown in Figure 10.5.

10.2.4 Design for Finite Life

The finite life approach, introduced around 1950, is often referred to as Low Cycle Fatigue (LCF). In this case, the intention is not to impart an infinite life to a component but, rather, to determine the maximum admissible loading depending on what the component's useful life should be. Instead of σ versus N, bilogarithmic ε versus N graphs are used (Figure 10.6), plotting the points of total strain and cycles to failure corresponding to the results obtained in a given test. The total strains, reported on the y-axis, can be separated into plastic and elastic components.

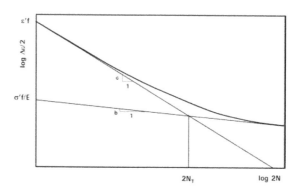

FIGURE 10.6 Amplitude of total strain—cycles of life.

The experimental trials required to determine the strains are more time-consuming in that they involve the use of strain gauges requiring continuous monitoring of the strain force relations, and also control of other parameters affecting the execution of the tests. Combining the outputs from the load cell and strain gauges, it is possible to obtain the hysteresis loop. In general, the hysteresis curve varies with the number of cycles. Maintaining the strain value constant, $\Delta\sigma$ can increase or decrease. When $\Delta\sigma$ increases, it is said that the material undergoes cyclic hardening; when $\Delta\sigma$ decreases, it undergoes cyclic softening.

The tendency of a material to harden or soften is determined by the structure of the material itself. Generally, it is observed that soft materials tend to harden, whereas materials already hardened (e.g., by previous machining) tend to soften.

The area of the hysteresis loop represents the energy of plastic strain expended in the movement of the dislocations. The variation of $\Delta\sigma$ tends to decrease with the number of cycles until, having passed the transition phase, it assumes a stable value. Once they pass this transitory phase, these curves can be used to evaluate the plastic and elastic components of the strain imposed. The total strain amplitude $\frac{\Delta\varepsilon}{2}$ can be divided into two components, one elastic and one plastic, as follows (Figure 10.7):

$$\frac{\Delta\varepsilon}{2} = \frac{\Delta\varepsilon_e}{2} + \frac{\Delta\varepsilon_p}{2} = \frac{\sigma_f^I}{E} \cdot (2N)^b + \varepsilon_f^I \cdot (2N)^c \qquad (10.7)$$

where σ_f^I is the coefficient of resistance to fatigue, b is the exponential of fatigue resistance, ε_f^I and c the coefficient and exponential of fatigue ductility, respectively, and 2N represents the alternations to failure (twice the number of cycles).

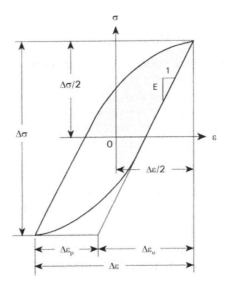

FIGURE 10.7 Hysteresis loop.

One alternation does not imply passing from R = −1, but simply a change in the loading direction so that each cycle consists of two alternations. This relation, known as Manson's equation (Manson, 1954), can be considered a generalized equation of fatigue in that it takes account of both the elastic and plastic components. The formulation of the elastic components was performed by Basquin (Basquin, 1910), and the formulation of the plastic components was performed by Coffin and Manson (Coffin, 1954). σ_f^I and ε_f^I represent the fatigue resistance and ductility, respectively, in the case of a single alternation. In a bilogarithmic diagram, the equation above is represented by the sum of two straight lines, representing the elastic and plastic contributions. Ductile materials, with elevated plastic deformation where the contribution of the second term predominates, show better fatigue behavior than fragile materials.

As noted in the infinite life approach, it is also necessary here to take account of the effects of the mean stress. With this aim, Morrow proposed a modification to the Manson equation (Morrow, 1965):

$$\frac{\Delta\varepsilon}{2} = \frac{\Delta\varepsilon_e}{2} + \frac{\Delta\varepsilon_p}{2} = \frac{\sigma_f^I - \sigma_m}{E} \cdot (2N)^b + \varepsilon_f^I \cdot (2N)^C \qquad (10.8)$$

Considering the effect of the mean stress value allows a better generalization of this approach. Given that, in reality, the components are subjected to a history of aleatory loading, it is therefore often necessary to apply this relation regardless of what the cycle ratio R is.

10.3 Damage Evolution Modeling

Damage is a phenomenon leading to the failure of the material in a more or less progressive manner, depending on the characteristics of the material and on the way in which it is stressed or strained. The gradualness with which this occurs implies that even a component that is apparently integral and able to function correctly, may in effect be damaged and therefore close to failure. The separation into two or more parts that "announces" the failure of a ductile material at the macroscopic level, is caused by the usually extremely rapid propagation of a crack that, in turn, derives from the growth and coalescence of cavities or porosities. These may already be present in the virgin material as it leaves the foundry, or be formed (nucleated) later as a result of strain.

10.3.1   Definition of Ductile Damage and Damage Parameter

The parameter used for the analytical measurement of damage is the percentage ratio of the area (or volume) of the cavities within the elementary cell to its nominal area (volume). The value of this parameter grows in each part of the material during its strain-history due to the effect of the two contributions noted above: the growth of preexisting cavities and the nucleation of new cavities that in turn begin to grow. The formation of a certain number of gas bubbles within a material is, in fact, typical of foundry processes and determines its initial porosity.

Further, any metallic material always contains, dispersed within it, a certain amount of impurities under the form of flakes of material of a different consistency (inclusions) embedded in the surrounding material (matrix). When the stresses or strains exceed certain values, the cohesion between inclusions and matrix is no longer sufficient to guarantee the continuity between the two micromaterials, so that the surface of separation between the inclusion and the matrix becomes the surface of a microcavity within which, possibly, the inclusion is free to move. Nucleation is precisely this phenomenon leading to the formation of cavities that are formed when certain stress or strain values are reached.

Then, when some contiguous microcavities grow large enough, the thin layer of material separating them (ligament) undergoes a sort of small-scale necking and collapses, allowing the microcavities to unite and form one large cavity. The condition of cavity coalescence is when this occurs widely in some of the zones of the material. It is this phase provokes the formation of the microcrack (the result of the coalescence of numerous cavities) which then rapidly degenerates into the fracture of the material.

Clearly, therefore, for any component of finite dimensions, the damage function also will assume diverse values from point to point, and it will always be just one of these points that reaches the condition of nucleation before the others and triggers the crack, involving neighboring points and extending, almost instantaneously, the fracture over the entire surface of the break.

10.3.1.1 Evolution of Cavities

The analytical reconstruction of the behavior of ductile metal materials satisfactorily reproduces the real situation only in those cases where it is not necessary to take account of the fracture phenomenon.

The principal characteristic of materials that is not taken into account is the relatively largescale discontinuity due to the presence of cavities that confer a certain porosity on all metals produced with normal foundry techniques. Further, it is certain that the cavities constituting the porosity of the material grow in number and dimension when the material is subjected to plastic strains, and it is precisely this growth in porosity that triggers the instantaneous and catastrophic fracture of the damaged object. It can be said, therefore, that by ignoring the initial presence and subsequent growth of a characteristic material porosity it becomes impossible to make hypotheses regarding the times and manners of ductile failure, or to accurately assess the material's capacity to respond to loads outside the elastic field.

A more precise understanding of the plastic behavior and above all of its limit in the phenomenon of failure, would require a specific investigation into the mechanisms of the growth of cavities within the material. In the late 1960s, this stimulated the first studies into this aspect (McClintock, 1968; Rice and Tracey, 1969).

10.3.1.2 Continuous Damage Mechanics and Lemaitre's Model

Following the seminal analysis conducted by Lemaitre, based on a representative volume element (RVE) of a damaged body, it is possible to consider here some of the main results obtained (Lemaitre, 1996). In the simple one-dimensional case (force F along the normal to the resisting cross section) and homogeneous damage, defining S and S_D as the area of the normal section and the area of the "void" section respectively, the damage variable can be defined as:

$$D = \frac{S_D}{S} \qquad (10.9)$$

From this definition of the damage variable, it follows that the stress acting at the various points of the elementary resistant section is no longer equal to the macroscopic stress F/S. In a first approximation, however, it is possible to assume that the internal cavities constitute a reduction in cross-section not

accompanied by the stress concentrations characteristic of every discontinuity, so that the effective stress becomes:

$$\breve{\sigma} = \frac{F}{S - S_D} = \frac{F}{S\left(1 - \dfrac{S_D}{S}\right)} = \frac{F}{S(1 - D)} = \frac{\sigma}{1 - D} \tag{10.10}$$

One useful consequence of how these variables are defined is that, from
measurements of the "apparent" elastic modulus in traction trials, is possible to obtain the value of damage according to the criterion:

$$\varepsilon^{el} = \frac{\breve{\sigma}}{E} = \frac{\sigma}{E(1 - D)} = \frac{\sigma}{\breve{E}} \tag{10.11}$$

$$D = 1 - \frac{\breve{E}}{E} \tag{10.12}$$

To determine the relationship between the variable D and the other variables
characterizing the material's behavior, it is necessary to identify a potential that
connects all the thermodynamic variables of the phenomenon. In this respect, it
is worth noting the distinction made by Lemaitre between observable variables
(ε, T), internal variables (ε^e, ε^p, r, α, D) and associated variables (σ, S, R, X, Y:
respectively, σ to ε, ε^e and ε^p, S to T, R to r, X to α, and Y to D), where:

- σ is the tensor of the stresses
- D is the damage
- ε^e elastic strain tensor, is the elastic component of the total strain
  tensor
- ε^p plastic strain tensor, is the plastic component of the total strain
  tensor
- $\varepsilon = \varepsilon^e + \varepsilon^p$ is the total strain tensor
- r is the cumulative plastic strain, dimensionless, piloting the evolution of isotropic hardening
- α is the backstrain tensor, and represents the strain piloting the
  evolution of kinematic hardening
- R is the isotropic hardening stress, scalar [MPa]
- X is the backstress, kinematic hardening tensor [MPa]
- Y is the power density of released strain [J], and corresponds to the
  quantity of energy liberated by the elementary volume as a result of
  the loss of stiffness due to increasing damage

- T is the temperature of the material point
- S is the entropy of the material point

The total potential F_T that, in "State Kinetic Coupling Theory," is used to determine the elastic–plastic constitutive relationship of a material with isotropic and linear kinematic hardening and subjected to damage, has the form:

$$F_T = \left(\breve{\sigma}^D - X \right)_{eq} - R - \sigma_y + F_D \qquad (10.13)$$

being σ^D the deviatoric part of and F_D the damage potential.

Considering that a preliminary hypothesis regarding the term F_D is that it does not explicitly contain the terms σ, X, and R, the duality relationship between the internal variables and associated variables determined by the potential considered is also a function of the variable D:

$$\dot{\varepsilon}_{ij}^P = \frac{\partial F}{\partial \sigma_{ij}} \dot{\lambda} \qquad (10.14)$$

$$\dot{r} = \dot{p} = \frac{\dot{\lambda}}{(1-D)} \qquad (10.15)$$

$$\dot{\alpha}_{ij} = \dot{\varepsilon}_{ij}^P \left(1-D\right) \qquad (10.16)$$

with λ multiplier of plasticity (Lemaitre, 1996).

To obtain the law of evolution of damage D, it is still necessary to define the variable Y associated with the damage at the potential F_D. The term Y is given by the relation:

$$Y = -\frac{1}{2} C_{ijkl} \varepsilon_{ij}^e \varepsilon_{kl}^e \qquad (10.17)$$

being C the elastic stiffness matrix.

Considering the expression of the energy of elastic strain in the damaged material:

$$\omega_e = \int \sigma_{ij} d\varepsilon_{ij}^e = \int C_{ijkl} \varepsilon_{kl}^e (1-D) d\varepsilon_{ij}^e = \frac{1}{2} C_{ijkl} \epsilon_{kl}^e \varepsilon_{ij}^e (1-D) \qquad (10.18)$$

gives the relation:

$$Y = \frac{1}{2}\frac{\partial \omega_e}{\partial D}\bigg|_{\sigma=\text{const}}$$

(10.19)

that is, the variable Y is equal to the reduction in plastic energy occurring in the material subjected to a constant stress and undergoing an infinitesimal increase in damage.

To construct the potential F_D, therefore, it is necessary to keep in mind that the generic form of the law of damage evolution is:

$$\dot{D} = \frac{\partial F_D}{\partial Y}\dot{\lambda} = \frac{\partial F_D}{\partial Y}\dot{p}(1-D)$$

(10.20)

On the basis of the following practical considerations, Lemaitre constructed the first functional form able to elicit the damage variable:

- The total damage is always correlated to a form of irreversibly accumulated strain, already taken into account with the term p.
- When the equivalent plastic strain begins to increase, it is reasonable to assume that the porosity of the material and the correlated damage do not increase until a strain threshold p_0 is reached. This aspect can be reproduced by introducing a step function or "Heavyside Function" of the type $H\,|\,p_0$ into F_D.
- The velocity of damage growth is strongly dependent on the triaxiality factor of the acting load, defined as in the relationship between hydrostatic stress σ_H and equivalent stress σ_{eq}. This dependence is already present in the term Y. In fact, breaking down the generic tensor of the stresses into its hydrostatic σ_H and deviatoric σ^D components gives:

$$Y = \frac{1}{1-D}\int \sigma_{ij}d\varepsilon^e_{ij} = \frac{\sigma_{eq}^2}{2E(1-D)^2}R_v$$

(10.21)

The term R_v is called the triaxiality function, given that it contains the triaxiality factor (σ_H/σ_{eq}) defined above.
- A generic and qualitative relation between the damage velocity and the energy released can be obtained considering their relationship to be linear, so that the potential will be quadratic with respect to Y.

Taking all these considerations into account, the potential proposed by Lemaitre is:

$$F_D = \frac{Y^2}{2S(1-D)} H\big|_{p_0} \tag{10.22}$$

where the term at the numerator 2S was chosen as a scale factor.

Following the law of damage evolution proposed by Lemaitre, numerous variations have been, and continue to be, developed, each offering major or minor improvements aimed at freeing the treatment from the simplified hypotheses of the idealized model:

$$\dot{D} = \frac{Y}{S}\dot{p}H(p-p_0) = \frac{\sigma_{eq}^2}{2ES(1-D)^2} R_v \dot{p}H\big|_{p_0} \tag{10.23}$$

The parameters that appear, S and p_0, characterize the material with regard to the effects of the damage and must be determined experimentally: the term S, for example, is obtained through measurements of the elastic modulus during the unloading phases during the a tensile test.

The aliquot of "plastic power" dissipated from a point in the form of heat is equal to the product of the various types of stress (stresses and hardenings) for the dual strains, that is:

$$\Phi = \sigma_{ij}\dot{\varepsilon}_{ij}^p - R\dot{p} - X_{ij}\dot{a}_{ij} = \sigma_y \dot{p} \tag{10.24}$$

The damage triggering strain is that at which, in the generic situation of triaxiality, the following succession of events occurs:

- The load increases from zero, the material accumulates exclusively elastic energy.
- The fatigue limit is reached and, under continuing loading, in very localized zones, the material also begins to internally absorb plastic energy that cannot be returned. The microcavities intrinsically present in the virgin material are not yet modified in form, size or number, and the temperature at points within the material begins to increase imperceptibly.
- The yield point is reached, the absorbed elastic energy has grown to the level corresponding to a very widespread movement of dislocations, to the extent where, even at the macroscopic level, the irreversible plastic strains begin to affect an entire resistant cross-section and the surrounding zone. The microcavities still remain in their initial

state, and the temperature at the macroscopic level has not yet increased appreciably.

- The plastic energy accumulated in the elementary volume has continued to increase along with the plastic strain that has reached the value p_0: from this moment on, further increases in the stress, that is in the work of plastic strain, will not simply result in increases in the energy irreversibly conserved within the material, but will also be transformed into externally dissipated heat because of the marked increase in temperature. Further, from now on, increments in plastic work will be accompanied by the growth of existing cavities, the nucleation of new cavities, and their coalescence until they reach the critical damage value. Soon after the onset of the increase in damage, the phenomenon of necking begins to appear in specimens subjected to tensile stress.

For an ideal plastic material, because the threshold value of this energy is constant, by measuring the value experimentally for the one-dimensional case it is possible to determine the damage triggering strain for any other value of triaxiality by imposing that the plastic energy not dissipated as heat has a single common value.

10.3.2 Cumulative Damage Fatigue and Theories of Lifespan Prediction

In general, fatigue damage is an incremental phenomenon, increasing with the number of cycles applied and possibly leading to failure. The first theories, proposed by Palmer, were expressed mathematically in 1945 by Miner (Miner, 1945):

$$D = \sum \left(n_i / N_{fi} \right) \tag{10.25}$$

where D represents the cumulative damage, and n_i and N_{fi} are, respectively, the number of cycles applied and the number of cycles to failure for an i-th load of constant amplitude.

Subsequently, numerous authors sought to develop theories of damage. In particular, a distinction can be made between theories formulated before and after the 1970s. The former are based on a more phenomenological approach, the latter on an analytical treatment.

10.3.2.1 *Phenomenological Approach*

The phenomenological approach is based on three main concepts:

- Damages produced at different loading levels are summed linearly.
- The reduction of the fatigue limit due to stress concentration can be a measure of damage.

- The process of fatigue damage can be subdivided into two phases, the nucleation of a fracture and its propagation.

The first concept (Palmgren, 1924) was subsequently translated into mathematical form by Miner, according to the law:

$$D = \sum r_i = \sum \frac{n_i}{N_{fi}} \qquad (10.26)$$

This is a Linear Damage Rule (LDR), based on the principle that for every loading cycle there is a constant absorption of energy and that every material has a characteristic value of absorbed energy to reach failure. According to Miner's hypothesis, each cycle consumes a part of the residual life of the material (n_i/N_{fi}), even though it does not directly cause failure. When the sum of the individual damages reaches the value of 1 (i.e., $\sum r_i = 1$), all the residual life of the component has been consumed and it breaks.

This law can be demonstrated as follows. Knowing the Wöhler curve of a given material, a sample of this material is subjected to fatigue loading for a number n_1 of symmetrical alternating cycles with oscillation semiamplitude greater than the fatigue limit. If at this loading level the life of the sample, evaluated from the Wöhler curve, is equal to N_1, the residual lifespan of the sample is given by the difference $N_1 - n_1$, the percentage of life consumed being equal to the ratio n_1/N_1. Subjecting the same sample to a second loading of different amplitude, with which the life of the virgin sample would be N_2, failure is reached after a number of cycles n_2. If the percentage of residual life was $N_1 - n_1/N_1$, this should equal n_2/N_2 and, therefore:

$$\frac{n_1}{N_1} + \frac{n_2}{N_2} = 1 \qquad (10.27)$$

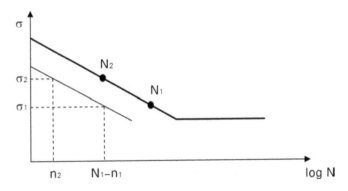

FIGURE 10.8 Damage curve according to Miner.

In a graph of σ versus N, it is possible to plot the curve of residual life. As shown in Figure 10.8, this will essentially be parallel to the original curve.

The main limitations to this theory are that it is independent of the loading level and sequence, as well as the loss of interaction between the different loads. Over time, numerous corrections to this theory have contributed to the improvement of design tools directed at determining the residual life of components. A complete survey of damage and life prediction models has been proposed by Fatemi and Yang (1998). For loading sequences of increasing amplitude (L–H, Low to High), a value of $\Sigma r_i > 1$ is expected and, vice versa, $\Sigma r_i < 1$ for sequences with decreasing amplitude (H–L, High to Low). This was demonstrated in 1954 by Marco and Starkey, who were the first to propose a theory of damage that was nonlinear and dependent on the load, governed by an exponential relation (Marco and Starkey, 1954):

$$D = \sum r_i^{x_i} \tag{10.28}$$

where x_i is a function of the i-th applied load. In a plane D–r, this relation can be represented by a curve parameterized with the stress σ, as shown in Figure 10.9, where the principal diagonal represents the Miner law and the other curves correspond to the Marco and Starkey laws.

Some authors contend that the reduction in the fatigue limit due to an initial preloading could be used as a measure of damage. The theories proposed are all of a nonlinear type and all take account of the actual sequence

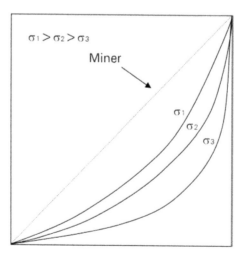

FIGURE 10.9   Representation of Marco–Starkey damage law.

of the loads applied. Some of these can also be used to determine the fatigue limit when the loading history is known. Of special relevance are the Corten–Dolan Hypothesis of Rotation and that of Freudenthal–Heller who, from the observation of experimental data and in order to obtain a model in agreement with this data, proposed the application of a clockwise rotation to the curve σ versus N around a point on the curve (Freudenthal and Heller, 1959). In the first approach (Corten and Dolan, 1956), this point coincides with the highest load represented (yield or failure), while in the second it corresponds to the fatigue limit for 10^3 to 10^4 cycles.

Subsequently, a further improvement of these two theories was proposed, suggesting the construction of the σ versus N curve given by the mean of the results obtained from two-stepped loading tests. Figures 10.10 and 10.11 show the curves for the two loading sequences, H–L and L–H, modified in this way. For comparison, the curves directly representing the Miner law are also shown. It can be seen that the rotation method is much more efficient than the Miner law in taking account of the interaction between the applied loads.

An improvement to the linear models of damage is the two-stepped linear approach, wherein two phases of damage propagation are considered:

- Damage due to the nucleation of the fracture $N_I = \alpha N_f$
- Damage due to the propagation of the fracture $N_{II} = (1-\alpha)N_f$

where α is a reduction factor and a function of the state of initialization.

In a further development of this approach, Manson proposed that the two steps be expressed by (Manson, 1965):

$$N_I = N_f - PN_f^{0.6}$$
$$N_{II} = PN_f^{0.6}$$

(10.29)

where P is a coefficient of the second stage of the fatigue life.

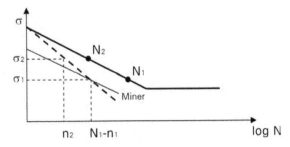

FIGURE 10.10 Rotation hypothesis: H–L loading sequence.

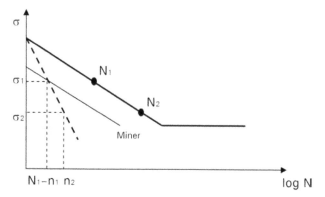

FIGURE 10.11   Rotation hypothesis: L–H loading sequence.

10.3.2.2   Theories Based on Fracture Growth

In the 1950s and 1960s, the introduction of instruments that can reveal microcracks on the order of 1 μm led to the development and acceptance of theories based on fracture growth. The models developed were based on the correlation between the delay in the growth of the fracture and the overloading produced by variable amplitude loading conditions.

One of the most popular of the many models proposed is that of Wheeler, who assumed that the increase in the growth of the fracture is correlated to the residual compression load produced by overloading at the apex of the crack (Wheeler, 1972). This model introduces the use of a delaying factor C_i in the law regulating fracture growth:

$$\frac{da}{dN} = C_i \left[A (\Delta K)^n \right] \tag{10.30}$$

where

$$C_i = \left(\frac{r_{pi}}{r_{max}} \right)^p \tag{10.31}$$

r_{pi} is the radius of the plastic zone associated with the i-th applied load, r_{max} is the maximum distance between the apex of the fracture and the largest adjacent elastic–plastic zone due to overloading, and p is an empirical factor, a function of the properties of the material and of the loading spectrum.

A similar model was also proposed by Willenborg, based on the reduction of the stress intensity factor ΔK (Willenborg et al., 1971). This reduction is due to the instantaneous dimension of the plastic zone at the i-th load and to the maximum dimension of the plastic zone due to overloading. This model introduces the use of an effective stress intensity factor (ΔK_{eff})$_i$ and has the advantage over the previous method of not requiring the calculation of empirical parameters:

Statistical models of the propagation of macrocracks have also been developed. Here, the velocity of crack development is linked to the amplitude of the intensification factor of the effective stresses, based on the curve of the probability density of the load spectrum. The amplitude of the effective stress intensification factor, described in terms of the mean square shift in amplitude of the stress intensification factor proposed by Barsom, is represented by (Barsom, 1971):

$$\Delta K_{rms} = \frac{\sqrt{\sum_{i=1}^{n} \Delta K_i^2}}{n} \tag{10.32}$$

where ΔK_i is the stress intensity factor in the i-th cycle, for loading sequences of n cycles. These models are empirical and do not take account of the effects of the loading sequence.

To evaluate the accumulation of fatigue damage in the initial phase of crack propagation, Miller and Zachariah introduced an exponential relation between the length of the crack and the life consumed (Miller and Zachariah, 1977). In the calculation model, damage is normalized as:

$$D = \frac{a}{a_f} \tag{10.33}$$

where a and a_f are, respectively, the instantaneous and final length of the crack. The model of Later and Ibrahim, based on the propagation mechanism of very small cracks, is described mathematically by:

$$\frac{da}{dN} = \Phi\left(\Delta\gamma_p\right)^\alpha a \tag{10.34}$$

where Φ and α are constants of the material and $\Delta\gamma_p$ is the amplitude of the tangential plastic strain. In subsequent studies, Ibrahim and Miller correlated the parameters N_i and α_i to values of the amplitude of the tangential plastic strain $\Delta\gamma_p$, using an exponential function (Ibrahim and Miller, 1980). Thus, the exponential of damage for the first stage of propagation can be written as:

$$D = \frac{a}{a_f} = \left(\frac{a_I}{a_f}\right)^{\frac{(1-r)}{(1-\eta)}} \tag{10.35}$$

10.3.2.3   *Energy Theories of Damage*

Since the first investigations of fatigue properties and the energy of the hysteresis curve, many researchers have studied this correlation and proposed various theories. In particular, those of the last 20 years have shown that damage parameters based on energy can correlate different damages (thermal, creep, and fatigue) caused by different types of applied load.

Craig, Kujawski, and Ellyin developed a model of preliminary damage using a parameter that takes into account the energy density of plastic formation Δw_p, calculated by integrating the area of the hysteresis loop (Craig et al., 1995). Subsequently, this approach was found to present some inefficiencies, given that Δw_p can be very small for materials subjected to high cycle fatigue. In some cases, even though the behavior of the material at the macroscopic level remains in the elastic and quasi-elastic field, at the macroscopic level plastic deformations may be present due to the nonhomogeneity of the material itself, and to the nonuniform distribution of the stresses and strains. Later, Golos and Ellyin proposed substituting Δw_p with the density of the total energy Δw_t, combining both the elastic Δw_e and plastic Δw_p components, where the elastic contribution can be calculated as (Golos and Ellyin, 1987):

$$\Delta w_e = \frac{1}{2E}\left(\frac{\Delta\sigma}{2} + \sigma_m\right)^2 \tag{10.36}$$

where σ_m is a hydrostatic stress.

To describe the damage, it is possible to introduce a relation of the type $\sigma - N$, which can be represented on a bilogarithmic plane by straight lines. As shown

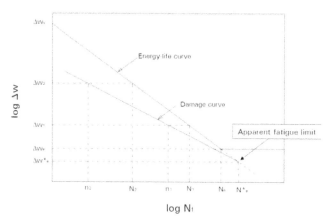

FIGURE 10.12   Damage curve passing through apparent fatigue limit $(N_e{}^*, \Delta w_e{}^*)$.

in Figure 10.12 for a test at two loading levels, the isodamage curve does not pass through the original point of the fatigue limit but, rather, through a point on the extension curve at an initial time corresponding to $(N_e^*, \Delta w_e^*)$, a point defined as the "apparent fatigue limit." This point can be determined using one of the models for fatigue limit prediction or by determining the critical damage curve that delimits the boundary between the nucleation phase and that of defect propagation. Having fixed this point, the isodamage curve branching from this point corresponds to different levels of damage.

Another energy-based approach was proposed by Leis, with a model of energy damage not linearly dependent on the loading history (Leis, 1977):

$$D = \frac{4\sigma_f^I}{E}(2N_f)^{2b_1} + 4\sigma_f^I \varepsilon_f^I (2N_f)^{b_1 + c_1} \qquad (10.37)$$

where σ_F^I and ε_F^I are the coefficients of fatigue stress and fatigue ductility, respectively, while the exponentials b_1 and c_1 are two variables associated with strain-hardening n_1:

$$c_1 = \frac{-1}{1 + 5n_1} \text{ and } b_1 = \frac{-n_1}{1 + 5n_1} \qquad (10.38)$$

According to Leis, the parameter n_1 can be considered a function of the accumulated plastic strain $\Sigma\Delta\varepsilon_p$. Thus, the model represented by Equation (10.37) includes the dependence on strain history.

10.4 Thermography and the Risitano Method

The study of fatigue in materials through the application of Wöhler curves is, as discussed above, one of the classic approaches most widely employed in design, particularly in the mechanical field. For this application it is necessary to determine the fatigue limit and plot the Wöhler curve, entailing time-consuming laboratory trials on a large number of specimens taken to failure. The problem is further complicated when the same data are to be determined for the dimensioning or verification of machine components since, in the industrial context, the production of such a large number of specimens destined for destruction would lead to an unacceptable increase in costs.

These difficulties stimulated the development of several rapid methods, based on certain hypotheses or empiric observations, that make it possible to determine the fatigue limit and, more rarely, the entire curve in a shorter time and with the destruction of fewer samples. From the early 1900s until today, numerous rapid methods have been developed, often arousing interest when proposed but thereafter little-used. However, some have been readopted in recent years; of these, the most important are:

- Methods based on the heat developed by the specimen (thermal methods). These refer to the observation that fatigue failure is preceded by phenomena of plastic deformation and, therefore, by the development of heat that can be measured using thermal detection equipment.

- Methods based on damping. Here, measurements are made of the energy dissipated under different loading conditions. Different techniques are used for detecting the energy, with the simplest being that performed on a Schenk rotating bending machine where the engine is mounted on a pendulum system allowing the measurement of the torque. The curve of the energy dissipated versus loading is analogous to the temperature curve, and a similar method of intersecting the tangent is used to determine a force value related to the resistance to fatigue.

- A method based on variations in the magnetic characteristics of a specimen subject to variable loading. This requires the measurement of the magnetic induction, the magnetic–elastic effect, or the loss due to parasitic currents. When permanent deformations occur, there is a more or less marked variation in these parameters. This very rapid test requires the use of only one specimen.

- Methods based on measuring the strain. The specimen is subjected to a small, symmetrical, alternating axial load, with an increasing static load superimposed, and the mean strain is measured. This increases with the load, and the corresponding graph is a straight line until, at a certain load, it presents an "elbow" after which the strain increases more rapidly than before. The load at this elbow is defined as the limit of dynamic sliding and coincides with the fatigue limit. Another method is based on evaluating the deflection on a specimen subjected to an increasing rotating bending load. In the resulting stress versus deflection graph, the deflection increases proportionally to the loading until a certain value. Above this load, the increase is no longer proportional and the load value is very close to the fatigue limit.

Subsequently, rapid methods were developed to evaluate the fatigue limit with only a few trials. In particular, it is worth examining the methods of Prot and Locati in greater detail.

The Prot method consists of subjecting specimens to fatigue loading using appropriate test machines (called Prot machines) to apply increasing unitary loads (Prot, 1951). During the test there is a uniform increase in the stress (with a fixed increment of MPa) for a certain number of cycles N (i.e., with a value of the rate of increase $\frac{d\sigma}{dN}$). Different specimens are tested with different rates of increase and then a graph is drawn, plotting the failure loads of the specimens tested as a function of the square roots of the rates of increase adopted in the corresponding trials. This gives a series of points aligned along a straight line that intersects the y-axis at a value corresponding to the usual fatigue resistance (in fact, in this case there is a zero rate of increase).

With the Locati method it is possible to determine the fatigue limit of an element using only one specimen (Locati, 1935). In a graph of σ versus N, three presupposed Wöhler curves are drawn in such a way that the horizontal section of the middle curve is equidistant (in terms of stresses) from the horizontal sections of the upper and lower curves. A single specimen is then subjected to oscillating loads for a fixed number of cycles. Subsequently, the load is increased (maintaining, however, the upper and lower limits between which the opposing loads oscillate) and the specimen is subjected to an additional identical number of cycles. Testing proceeds in this way until a specimen breaks. Connecting the three points (cumulative damage–load amplitude) relative to the three Wöhler curves, the fatigue limit is determined as the value of the curve corresponding to a damage value of 1. In fact, according to Miner's hypothesis, failure occurs when the cumulative damage assumes a value of 1.

The principal limitation of this method is its need to presuppose a priori the trend of three Wöhler curves in a complete manner (i.e., also with regard to the section of resistance over time). Furthermore, given that it is an application of the hypothesis of linear damage, the damage is subjected to the same uncertainties as Miner's law, and in subsequent applications the results obtained were unreliable. Finally, it is important to note that both these methods involve destructive testing.

Of the many methods proposed, one of the most efficient is certainly the Risitano method. This allows the very rapid (on the order of a few hours) determination of not only the fatigue limit but also of the time curve using, at best, only one sample or component.

The Risitano method (Geraci et al., 1992; Geraci et al., 1995; La Rosa and Risitano, 2000; Fargione et al., 2002) is based on the correlation that exists between the increase in surface temperature and the effective applied stress on the loaded specimen. It has been observed that the temperature in each point of the specimen progressively increases until, for each load levels, it reaches a given stable value. In the initial phases, the higher the loading level the specimen is subjected to, the faster the temperature rises. Measuring the surface temperature of the specimen, it is possible to derive the component's fatigue limit. The surface temperature of the loaded specimen is mapped using thermography, briefly described below in terms of theory and application in order to better clarify the Risitano method.

10.4.1   Thermography

Thermography, from the Greek *thermos* (heat) and *graph* (write), is the measurement of the infrared radiation emitted by an object. The term commonly refers to a visual, photographic, or graphic representation, obtained using appropriate methods and tools, of the natural or reflected radiations emitted by a physical entity in the infrared spectrum. However, it also often means the graphic representation of the thermal state of a physical entity (the pictorial representation of the temperature map of a physical entity).

This is possible due to the fact that all structures, animate and inanimate, with a temperature above absolute zero emit and absorb infrared radiation until they reach thermal equilibrium. In terms of infrared radiation, different physical entities behave differently, ranging from those with a high level of absorption to those which, conversely, almost completely reflect thermal radiation. The latter condition is exemplified by an infrared mirror whose surfaces reflect infrared radiation, while an example of the former is the so-called black body which, once in thermal equilibrium, irradiates as much energy as it absorbs.

The surface thermal mapping of a physical entity can be performed continuously or at certain stages, either by measuring the direct conduction of heat between the physical entity under examination and the sensor (contact method) or by capturing the radiation transmitted or reflected from the physical entity using heat sensors (the remote method).

The techniques commonly used in thermographic investigations can be divided into two basic groups: those exploiting the emission of internal energy from the object, and those that require external thermal stimulation. In general, above 800 K an unlit object becomes visible to the naked eye because a significant fraction of its emitted radiation lies in the visible part of the spectrum. At lower temperatures, instruments sensible to infrared radiation (IR) can be used to obtain black and white or false color images, where the gray or color scale is directly linked to the temperature and the surface properties of the object under examination.

A thermographic system essentially consists of a thermal scanner with an infrared sensor, a monitor, and a computer to elaborate the images. This method of analysis, best known in military applications, offers a wide variety of civilian uses in fields ranging from agriculture to geology, meteorology, and medicine. Particularly relevant here are applications in the fields of industrial manufacturing; in the preventive maintenance of plants and machinery employed in energy production and utilization; in the quality control of production processes; and, more generally, in the ambit of nondestructive testing (NDT). In NDT, thermography assumes particular importance given that the distribution of the surface temperature of a component can provide useful information about the presence of surface and/or subsurface defects, especially in those materials particularly difficult to inspect with other nondestructive methods (e.g., dielectric materials, composite materials, etc.).

Within the wide electromagnetic radiation spectrum, the term "thermal radiation" is used for wavelengths between 10^{-7} and 10^{-3} m, and the thermal infrared band is further divided into:

- Near infrared, 0.75–3 μm
- Middle infrared, 3–6 μm
- Far infrared, 6–15 μm
- Extreme infrared, 15–100 μm

The total energy emitted by a physical entity is termed its "global emission" and the quantity of energy irradiated by a unit of surface in a unit of time is termed the "global emissive power," indicated by the symbol $\dot{q}$ and measured in W/m^2. The global emissive power depends principally on the surface state of the physical entity and its temperature, and contains all the contributions $\Delta_\lambda q$ relative to all the intervals of wavelength $\Delta\lambda$ included in the spectrum.

A black body is able to absorb all incident energy, regardless of the wavelength. In practice, such an ideally defined object can be represented by a hollow structure with opaque walls and a very small opening; all incident radiation is almost completely absorbed and only a minute fraction will be reflected out through this opening.

Then, defining the coefficients (a, absorption; r, reflection; and t, transmission), the relation between these components of heat becomes:

$$1 = a + r + t \tag{10.39}$$

In general, a, r, and t are functions of λ and of T.

Thus, a black body will have $a = 1$ and $r = 0$. For "nonblack bodies," it is possible to define the emissivity coefficient ε as the ratio between the integral emissive power of the nonblack body $\dot{q}(T)$ in question and that of a black body $\dot{q}_0(T)$:

$$\varepsilon = \frac{\dot{q}(T)}{\dot{q}_0(T)} \tag{10.40}$$

The expression given by Stefan–Boltzmann for a black body is modified for the case of any body as:

$$q = \varepsilon \cdot \sigma_0 \cdot T^4 \tag{10.41}$$

where σ_0 is the Boltzmann's constant.

The thermal equipment is able to detect the temperature of the physical entity under analysis, measuring the intensity of radiation (I) and, consequently, the amount of heat (q) emitted by the physical entity.

There are numerous problems that must be resolved for correct interpretation of the data acquired by the instrument. First, considering the nonlinearity of the Stefan–Boltzmann equation, it is clear that the system is imprecise at low T, because discrete variations in T are associated with only slightly different values of q. A second problem is that the q measured by the equipment is not that which the physical entity exchanges with the environment but, rather, that which the physical entity itself emits (i.e., the T of the analyzed physical entity is falsified by the T of other physical entities irradiating it).

As represented schematically in Figure 10.13, physical entities interact through thermal exchange by radiation occurring in every direction. Centering the thermal scanner on physical entity "A" measures a greater quantity, falsified by the presence of physical entity "B":

$$q_{measured} = q_1 + (1-a)q'_2 \qquad (10.42)$$

where a represents the coefficient of absorption of physical entity "A."

To resolve the situation, physical entities with values of ε as close as possible to 1 are used. When examining physical entities with very low ε, it is necessary to coat them with particular paints in order to obtain high values of ε. Test specimens are usually painted an opaque black.

Remote sensing systems are designed to operate in the thermal infrared field at wavelengths between 2 and 5.4 μm, or between 8 and 14 μm. The scanner converts the infrared radiation into digital signals that the system interprets and returns in the form of color or gray-scale images. The radiation received by the thermal scanner is not only that emitted by the object under examination, but includes the following contributions:

- Radiation emitted by the object
- Radiation emitted by the environment and reflected on the object
- Radiation emitted by the atmosphere

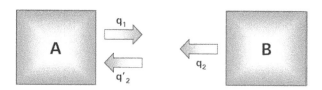

FIGURE 10.13   Interaction between emitting bodies.

- Radiation emitted by the sensor itself
- Radiation emitted by the lenses

In calculating the temperature of the object, it is therefore necessary to take into consideration all these components, as well as the fading the radiation undergoes due to the distance traveled through the atmosphere. These collateral effects are satisfactorily eliminated by the analysis software and internal compensation systems.

The emissivity factor ε is a parameter that must be taken into consideration in the measurement of the infrared temperature, since the actual objects analyzed are rarely black bodies. The values of emissivity can be measured or obtained from appropriate tables on the basis of the material and surface treatment of the object; these values vary between 0.1 for mirrored surfaces and 0.95 for painted surfaces.

Nonoxidized metals represent an extreme case of almost perfect opacity and highly mirrored reflection, where the emissivity does not vary excessively with the wavelength. Consequently, the ε of metals is low and varies with the temperature. For nonmetal materials, the emissivity tends to be high and to decrease with the temperature.

The reflection factor of an opaque object can be written as:

$$\rho = 1 - \varepsilon \qquad (10.43)$$

A low emissivity factor indicates not only that the emission of the object is lower than that of a black body, but also the existence of undesired radiation emitted from the surrounding environment and reflected by the object.

10.4.2   The Risitano Method

10.4.2.1   Fatigue Limit Determination

Traditional methods of evaluating the fatigue limit require many lengthy trials and are both time-consuming and expensive. The development of thermographic techniques for checking defects in components and materials has broadened the opportunities for application in fatigue testing.

The Risitano method, proposed for the first time in 1984 at Catania University in Italy, originated with the empirical observation of the temperature distribution on fatigue-loaded specimens or components (Curti et al., 1986). The essential advantages of this methodology are:

- The possibility of localizing, in a relatively short time, the critical zone where the fatigue fracture originates
- The evaluation of the fatigue limit without requiring the destruction of the object under examination

- The low dispersion of the results
- The construction of the whole Wöhler curve

It is clear that this technique can be useful in the study of the fatigue behavior of prototypes, and in the rapid analysis of the effect of all those parameters that can lead to variation in the fatigue limit. Anyone who has conducted fatigue testing, even simply bending a metal wire back and forth until it breaks, will have observed that the failure of the component involves the production of heat. Starting from this observation, the authors of the method conducted a series of trials that allowed them to evaluate the correlation between the typical parameters of fatigue (σ, N) and the trend of the temperature, and to derive a law that allowed the determination of the fatigue limit σ_0 and, subsequently, the construction of the time curve.

First, it was observed that for values of $\sigma > \sigma_0$, the trend of the temperature exhibits three distinct phases, as shown in Figure 10.14. In the initial phase, the increase in temperature is more rapid as the load applied to the specimen increases. In practice, for loads over the fatigue limit, the thermal variation increases during the first part of the trial (phase 1) and then remains constant (phase 2) until just before failure, near which there is a further increase in temperature (phase 3). Phase 1 is usually limited to a relatively low number of cycles compared to the number required for failure. The stable temperature phase (phase 2) is usually longer, in terms of time, than the other two, and it varies relative to $\Delta\sigma$, becoming shorter as the load level increases. In particular, this phase is extremely short for loads close to the yield load, while it extends to almost the entire life of the specimen for loads close to the fatigue limit. In phase 3, corresponding to failure (complete plasticization in one part of the specimen), the temperature increases rapidly within a limited number of cycles.

The increase in temperature is merely the observable phenomenon resulting from the application of fatigue loads; it physically expresses a structural change within the material due to damage provoked by the load and, in

FIGURE 10.14   Theoretical increase in temperature.

particular, the microplasticization produced by the stresses at local discontinuities. As a result of this observation, a new definition of fatigue limit was
proposed: the maximum value of the stress applicable to the material that
does not induce damage to it.

Taking advantage of this relationship, it is possible to determine the fatigue
limit by finding the stress value that does not induce an increase in temperature. It is not necessary to take the specimen to failure; it is sufficient to note
when (the number of cycles) the maximum temperature of the specimen
becomes stable. Through the set of parameters (maximum temperature, load
level, and number of cycles), it is possible to determine the fatigue limit. In
practice, the method prose is the construction of a diagram correlating the
applied stress $\Delta\sigma$ to the gradient $\Delta T/\Delta N$ of phase 1. The points plotted are
interpolated linearly to obtain a straight line that intersects the abscissa at a
point representing the stress value for which there would be no increase in
temperature and, therefore, indicating the fatigue limit (Figure 10.15).

To obtain a sufficient number of points on the $\sigma - \Delta T/\Delta N$ graph, it is advisable
to program no more than five loading levels, as shown in Figure 10.16. The trials

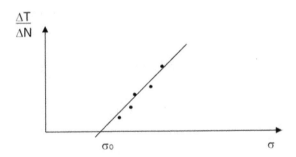

FIGURE 10.15 Interpolation of points $\Delta T/\Delta N$–σ for the
determination of σ_0.

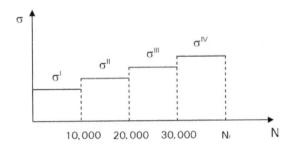

FIGURE 10.16 Trial with stepped increase in applied load.

can be conducted by taking at least one specimen to failure for each loading level. In order to have sufficient data to obtain a mean value, it is usual to perform at least three trials for each load level. Alternatively, different load levels can be applied on the same specimen, increasing σ with predetermined or variable time intervals (number of cycles).

It should be noted that this method, when applied under particular conditions, can be considered nondestructive. In order to determine the fatigue limit of the material, it is sufficient to determine only the gradient of the ΔT–N curve for several loads without necessarily taking the element to failure (although, of course, damage will be induced in the component).

10.4.2.2   Construction of the Wöhler Curve

Failure is a heat-dissipating process during which heat is produced by plastic work and at the same time by the fracture. When the process is adiabatic, this release of heat results in a local rise in temperature that alters the characteristics of the material itself. The theories of thermodynamics can be applied to this process to determine the relation between the mechanical and thermal energies.

The first principle of thermodynamics can be written (for an enclosed system) as:

$$\oint (dQ - dW) = 0 \tag{10.44}$$

where the term $dQ - dW$ is the total internal energy:

$$dQ - dW = dU_t \tag{10.45}$$

The internal energy is given by the difference between the work applied to the system and heat yielded to the surrounding environment. In the case considered here, the work applied to the system is mechanical, given by the sum of an elastic component W_e and a plastic component W_p. However, the two components contribute thermally at different levels. In fact, areas affected by plasticity are usually defined as hot zones while those subjected to plastic deformation only are defined as cold zones. In particular, the contribution of the elastic component is regulated by the thermoelastic law and is negligible respect to the plastic contribution. The global equilibrium between the mechanical energy supplied in elastic and plastic form and the heat dissipated is given by the heat equilibrium equation:

$$\dot{W} + K\nabla^2 T + (3\lambda + 2\mu)\alpha T_0 \dot{\varepsilon}^e = \rho c_E \dot{T} \tag{10.46}$$

where K is the thermal conductivity, ρ the density, c_E the specific heat, and T the temperature. β is the fraction of released thermal power on mechanical potential (ignoring the thermoplastic component), λ and μ are the Lamè constants, α the coefficient of thermal expansion, and W represents the energy dissipated per unit volume. The latter coincides with the area of the hysteresis loop and can be calculated by the expression:

$$W = \int_0^T \sigma d\varepsilon^P \tag{10.47}$$

$\dot{W}$ is, therefore, equal to the product of the deviatoric component of the force and the increase in plastic deformation in one cycle.

For the materials most commonly used in industrial components (steel and cast iron), β can be considered constant with respect to the stresses and strains and, from the literature, has an indicative value of 0.9. Furthermore, when a high frequency of alternate loading is used in the trials it is possible to ignore the thermoelastic component, so that:

$$\beta\dot{W} = \rho c_E \dot{T} \tag{10.48}$$

The surface temperature reached under these conditions is, therefore, directly proportional to the mechanical energy of deformation (Figure 10.17).

According to one of the most accepted theories of physics, the fatigue failure of materials occurs when the stored internal energy E_i per cycle reaches a

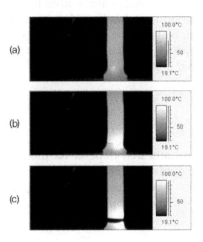

FIGURE 10.17 Localization of fracture point in the last 3,000 load cycles.

constant value that is characteristic of each material and is dependent on the mechanical and thermal characteristics of the material itself:

$$\int E_i dN = \text{constant} \tag{10.49}$$

From the first principle of thermodynamics (excluding the transitory process), the level of the internal energy per cycle can be expressed as the difference between the energy supplied E_W and that converted to heat Q_e:

$$E_i = E_W - Q_e \tag{10.50}$$

In a model of the mechanical behavior of a material, the internal energy can be expressed as the sum of two terms:

$$E_i = E_p + E_e \tag{10.51}$$

E_e represents the aliquot of energy that the material absorbs without developing permanent local deformations and is linked to the values of the internal damping of the material, occurring when the applied load is below the fatigue limit. The material can therefore absorb the quantity of energy E_e for an infinite number of cycles and, having passed the transitory, the increase in temperature is practically negligible.

E_p represents that amount of energy that generates permanent local deformations in the material. It assumes values other than zero only for the loads higher than the fatigue limit and is linked, according to the hypothesis of Miner, to cumulative damage. If the relation between these deformations is assumed to be constant (an acceptable hypothesis to limit the temperature range) for mass and heat per unit volume, then:

$$a\sigma^2 - \Delta T = b \tag{10.52}$$

where a and b are constants that depend on the mechanical and thermal characteristics of the material. Considering that for loads below the fatigue limit, the variations in temperature ΔT are negligible compared to the increases observed above this limit σ_0:

$$\sigma_0 = \sqrt{\frac{b}{a}} \tag{10.53}$$

and, therefore, the previous expression for $\sigma > \sigma_0$ becomes:

$$a\left(\sigma^2 - \sigma_0^2\right) = \Delta T \qquad (10.54)$$

For values of $\sigma > \sigma_0$, the temperature increments are due to local plastic deformations, the consequence of reaching the corresponding local plastic stress. These increments increase until they reach thermal equilibrium between the heat produced and that dissipated.

Fatigue failure can, therefore, only occur if the stress created in a point of the material (due to the presence of microcracks, dislocations, or microcavities) reaches local values above the elastic limit. Under static loading, local plasticization and the redistribution of the stress on the surrounding material does not generate a particularly critical condition and the material fails only under much higher loads. Under dynamic loading with repeated applications of the load, reaching the condition of local plasticization leads to an increase in the microcracks until the material breaks.

The hypothesis according to which the fatigue failure of a material occurs when the energy of plastic deformation reaches a cumulative constant limit value E_c, characteristic for each material, makes it possible to determine not only the fatigue limit but also the entire fatigue curve using thermographic investigation.

From the energy point of view, if E_c is the energy per unit volume and E_p is the energy of plasticization per unit volume and per cycle, it is possible to define the cumulative damage in the first N_0 cycles as:

$$\int_0^{N_0} E_p dN \qquad (10.55)$$

and, correspondingly, the residual life as:

$$E_c - \int_0^{N_0} E_p dN \qquad (10.56)$$

The expression for energy equilibrium written for an elementary volume indicates that, under operating conditions, the energy of plastic deformation E_c is for all practical purposes proportional to the temperature on the exposed surface of the element under consideration. Given that the part relative to the variation in internal energy is small compared to the energy dissipated as heat, the integral of the temperature measured on the surface of the elementary volume of the specimen under examination is equal to (less a constant, which is dependent on the coefficient of thermal transmission) that energy limit E_c, characterizing the material in terms of fatigue failure.

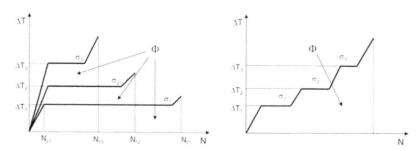

FIGURE 10.18   Evaluation of Φ_i and N_{fi} using different specimes or by the step method.

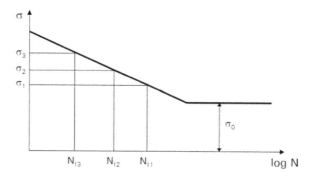

FIGURE 10.19   Determination of the whole Wöhler curve by N_{fi}.

The constancy of the energy limit makes it possible to obtain, with a limited number of trials, the entire fatigue curve and the behavior of the material or mechanical component over time. In fact, presuming that E_c is a constant that characterizes a material or a given mechanical component, it can be evaluated by subjecting the test specimen to a preset stress σ (greater than the fatigue limit), taking it to failure, and measuring the characteristic data of the thermal investigation. For the applied stress value, the function $\Delta T = f(N)$, whose integral Φ is proportional to the limiting energy E_c, can be determined. In practice, at the fatigue limit or lower stress values, thermal variations slightly above the theoretical value of zero are observed.

The temperature values to be introduced into the calculation of Φ are the increments in temperature evaluated with respect to the value, ΔT_0 corresponding to the fatigue limit. It can be, therefore, held that:

$$\Phi = \int_0^{N_f} E_p dN \tag{10.57}$$

With the procedure used to determine the fatigue limit it is possible, for each stress value, to determine the temperature of the second phase ΔT_2 and the number of cycles N_s, of stabilization.

Then, considering that the integral for the calculation of the area Φ can be approximated using the following expression, it is possible to determine the value N_f for any other stress value:

$$\Phi \cong \Delta T_2 \cdot \left(N_f - \frac{N_s}{2} \right) \cong \Delta T_2 \cdot N_f \qquad (10.58)$$

The whole curve can be determined, as shown in Figures 10.18 and 10.19, by associating the value of N_{fi}, derived from (10.58) after measuring N_{si} and ΔT_{2i}, to the i-th applied stress σ_i (Fargione et al., 2002).

10.5 Summary

This chapter briefly introduced certain phenomena having a direct influence on conventional mechanical design and also reviewed the corresponding analysis methods used for the characterization of materials. One of these, the Risitano method developed by the Department of Industrial and Mechanical Engineering at Catania University, Italy, not only allows the fatigue curve of materials to be determined very rapidly, but also affords a saving in test material.

It is clear that knowledge of phenomena relating design parameters to variations in the characteristics of materials over time is an important factor in a design approach directed at the environmental quality of product. This knowledge is fundamental in programming the behavior of components and systems during their useful life, and in planning possible scenarios of recovery on their retirement by evaluating their corresponding impacts.

10.6 References

ASTM E606–92, Standard Practice for Strain-Controlled Fatigue Testing, *Annual Book of ASTM Standards*, Vol. 03.01, American Society for Testing and Materials, Philadelphia, PA, 2004.

Barsom, J.M., Fatigue crack propagation in steels of various yield strengths, *Journal of Engineering for Industry*, B73(4), 1190–1196, 1971.

Basquin, O.H., The exponential law of endurance test, *Proceeding of the American Society for Testing and Materials*, 10, 625–630, 1910.

Coffin, L.F., A study of the effects of cyclic thermal stresses in a ductile metal, *Transactions of American Society of Mechanical Engineers*, 76(4), 931–950, 1954.

Corten, H.T. and Dolan, T.J., Cumulative fatigue damage, in *Proceedings of ASME International Conference on Fatigue of Metals*, Institution of Mechanical Engineering, London, 1956, 235–246.

Craig, D., Kujawski, D., and Ellyn, F., An experimental technique to study the behaviour of small corner cracks, *International Journal of Fatigue*, 17(4), 253–259, 1995.

Curti, G. et al., Analisi tramite infrarosso termico della temperatura limite in prove di fatica, in *Proceedings of 14th AIAS Conference*, Catania, Italy, 1986, 211–220.

Fargione, G. et al., Rapid determination of the fatigue curve by the thermographic method, *International Journal of Fatigue*, 24(1), 11–19, 2002.

Fatemi, A. and Yang, L., Cumulative fatigue damage and life prediction theories: A survey of the state of the art for homogeneous materials, *International Journal of Fatigue*, 20(1), 9–34, 1998.

Freudenthal, A.M. and Heller, R.A., On stress interaction in fatigue and a cumulative damage rule, *Journal of the Aerospace Sciences*, 26, 431–442, 1959.

Geraci, A.L. et al., Determination of the fatigue limit of an austempered ductile iron using thermal infrared imagery, in *Proceedings of Digital Photogrammetry and Remote Sensing '95*, St. Petersburg, Russia, 1995, SPIE 2646–38.

Geraci, A.L., La Rosa, G., and Risitano, A., Influence of frequency and cumulative damage on the determination of fatigue limit of materials using the thermal infrared methodology, in *Proceedings of the 15th Polish National Symposium on Experimental Mechanics of Solids*, Warsaw, Poland, 1992, 63–65.

Golos, K. and Ellyin, F., Generalization of cumulative damage criterion to multi-level cyclic loading, *Theoretical and Applied Fracture Mechanics*, 7, 169–176, 1987.

Ibrahim, M.F.E. and Miller, K.J., Determination of fatigue crack initiation life, *Fatigue and Fracture of Engineering Materials and Structures*, 2, 351–360, 1980.

La Rosa, G. and Risitano, A., Thermographic methodology for rapid determination of the fatigue limit of materials and mechanical components, *International Journal of Fatigue*, 22(1), 65–73, 2000.

Leis, B.N., An energy-based fatigue and creep-fatigue damage parameter, *Journal of Pressure Vessel Technology*, 99(4), 524–533, 1977.

Lemaitre, J., *A Course on Damage Mechanics*, 2nd ed., Springer-Verlag, Berlin, 1996.

Locati, L., Un ausilio per la determinazione del limite di fatica in ricerca e in produzione, *Metallurgia Italiana*, 27, 188–204, 1935.

Manson, S.S., Behaviour of Materials under Conditions of Thermal Stress, NACA Report 1170, National Advisory Commission on Aeronautics, Lewis Flight Propulsion Laboratory, Cleveland, OH, 1954.

Manson, S.S., Fatigue: A complex subject—Some simple approximations, *Experimental Mechanics*, 5(7), 193–226, 1965.

Marco, S.M. and Starkey, W.L., A concept of fatigue damage, *Transactions of American Society of Mechanical Engineers*, 76(4), 627–632, 1954.

McClintock, F.A., A criterion for ductile fracture by the growth of holes, *Journal of Applied Mechanics*, 35, 363–371, 1968.

Miller, K.J. and Zachariah, K.P., Cumulative damage laws for fatigue crack initiation and stage I propagation, *Journal of Strain Analysis*, 12, 262–270, 1977.

Miner, M.A., Cumulative damage in fatigue, *Journal of Applied Mechanics*, 12(3), A159-A164, 1945.

Morrow, J.D., Cyclic plastic strain energy and fatigue of metals, in *Internal Friction, Damping, and Cyclic Plasticity*, ASTM STP 378, American Society for Testing and Materials, Philadelphia, PA, 1965, 45–87.

Neuber, H., *Theory of Notch Stresses*, Springer, Berlin, 1958.

Palmgren, A., Die lebensdauer von kugellagern, *Zeitschrift des Vereines Deutscher Ingenieure*, 68(14), 339–341, 1924.

Peterson, R.E., Notch sensitivity, in *Metal Fatigue,* Sines, G. and Waisman, J.L., Eds., McGraw-Hill, New York, 1959, 293–306.

Prot, M., Resultats d'essais de fatigue sous charge progressive, *Revue de Metallurgie,* 48, 822–824, 1951.

Shigley, J.E. and Mischke, C.R., *Mechanical Engineering Design,* 5th edition, McGraw-Hill, New York, 1989.

Rice, J.R. and Tracey, D.M., On the ductile enlargement of voids in triaxial stress fields, *Journal of the Mechanics and Physics of Solids,* 17, 201–217, 1969.

Wheeler, O.E., Spectrum loading and crack growth, *Journal of Basic Engineering,* 94, 181–186, 1972.

Willenborg, J.D., Engle, R.M., and Wood, H.A., A Crack Growth Retardation Model Using an Effective Stress Concept, AFFDL TM-71-1-FBR, Air Force Flight Dynamics Lab, Wright-Patterson Air Force Base, Fairborn, OH, 1971.

Wöhler, A., Über die festigkeitsversuche mit eisen und stahl, *Zeitschrift Fuer Bauwesen,* 20, 73–106, 1870.

Youn, B.D., Choi, K.K, and Tang, J., Structural durability design optimisation and its reliability assessment, *International Journal of Product Development,* 1(3–4), 383–401, 2005.

Part III

Methods, Tools, and Case Studies

Chapter 11

Product Constructional System Definition Based on Optimal Life Cycle Strategies

The design of products with good environmental performance over their entire life cycle requires the development of methods and models that provide as complete a vision of the problem as possible, and allow the optimization of product architecture and components while respecting the constraints imposed by their main functional performances.

This chapter presents a method for Life Cycle Design, focusing on the analysis of strategies extending the useful life (maintenance, repair, upgrading, and adaptation of the product) and strategies of recovery at the end-of-life (direct reuse of components and recycling of materials). This analysis and design method is able to evaluate the most suitable strategies for each component and subassembly comprising the product, and to define the best redesign choices in terms of certain characteristics of product architecture and components. It is based on a process of analysis and decomposition of the conventional constructional system and its reinterpretation in terms of the life cycle strategies, by means of the modularity concept and the Design for Disassembly approach. To clarify this method, the development of a redesign proposal for a widely used household appliance, the refrigerator, is described.

The fundamental issues in this chapter were previously introduced in Chapter 9.

11.1 Aims and Approach

This chapter presents a method, complete with fundamental mathematical modeling, to aid the study of product constructional systems and investigate their environmental efficiency. The latter can be determined in various ways. The approach proposed here seeks to optimize the life cycle strategies that appear to be more effective for an environmentally efficient life cycle (Section 9.1, Chapter 9):

- Those strategies aimed at maintaining performance and functionality of product during the phase of use (maintenance, repair, upgrading,

297

and adaptation of the product), in that they can favor the extension of the product's useful life

- Those strategies oriented toward the planning of recovery processes at the end of the product's useful life (direct reuse of components and recycling materials in the primary production cycle or in external cycles), in that they are directed at reducing the environmental impact of disposal and at the recovery of resources

These types of interventions can be translated into clear environmental benefits: reduction of the volumes of the virgin materials required; extension of the product's working life; closing the cycles of the resource flows in play by recovery operations. The proposed tool is conceived to support two action typologies:

- Analysis of conventional constructional systems for a correct definition of the life cycle strategies most appropriate to preexisting products
- Redesign of architectures and components for the improvement of environmental performance and for the development of new, environmentally acceptable products

To translate the strategies of extension of useful life and recovery into requisites of product architecture and components, we propose the modularity concept (Chapter 9, Section 9.4) and the Design for Disassembly approach (Chapter 9, Section 9.3.3), which focus on harmonizing product layout, geometries, and joining systems in terms of the separability of the parts.

11.2   Method and Tools for Analysis and Design

An analysis and design intervention characterized by the premises outlined above is complex and requires a methodology that provides a procedure and supporting tools for the definition and correct interpretation of environmental requirements. Such a methodology must also identify the most effective elements of successful product redesign.

The method developed here is divided into certain successive moments, summarized in Figure 11.1. The first phase is the analysis and decomposition of the conventional constructional system, which identifies the determining characteristics of the product architecture, the unavoidable design constraints, and the primary functional components. Then follows an evaluation of the most appropriate strategies of extension of useful life and of recovery, according to the system characteristics. Finally, the conventional product architecture (reinterpreted using the evaluation tool) is mapped to

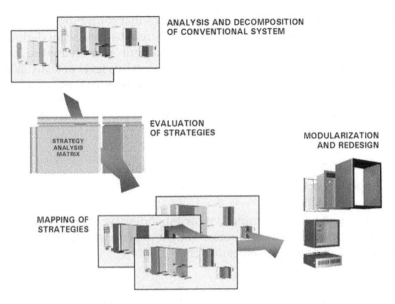

ANALYSIS AND DECOMPOSITION
OF CONVENTIONAL SYSTEM

EVALUATION
OF STRATEGIES

MODULARIZATION
AND REDESIGN

STRATEGY
ANALYSIS
MATRIX

MAPPING OF
STRATEGIES

FIGURE 11.1   Summary of analysis and design method.

evidence the distribution of the most appropriate strategic options in relation to the characteristic properties of the product's various parts. This offers
views of the system that can suggest the most effective design interventions
and recommend the structure of the architecture and modularization of the
system that best respect these options.

At the component and junction design level, this last phase requires the
Design for Disassembly approach (Chapter 9, Section 9.3.3), directed at ensuring the separability of the new constructional system (allowing the disassembly
of the main components) and making it possible to apply the optimal strategies.
A product that provides for relatively easy separation of its parts or components can facilitate product maintenance, repair, and updating, and the separation of components and materials for recovery at the end of its useful life.

This type of investigation can have dual goals: the definition of the most
suitable strategies to apply to predefined, conventional constructional systems,
and development of new architectures aligned with the most effective strategies for extending useful life and recovery of resources.

11.2.1   Product Constructional System and Design Choices

In general, "product architecture" refers to the arrangement and relationships of the physical blocks comprising the functional elements of a product
(Ulrich and Eppinger, 2000). "Functional elements" are those units that

perform single operations and transformations that contribute to the overall product function. Defining product architecture thus consists of defining its approximate geometric configuration (layout) and identifying the interactions between its main units or modules. A successive level of analysis refers to the definition of component characteristics (dimensions, shape, material) and of junction systems.

Product constructional system is defined in two successive levels of design choices:

- Modularity and layout (embodiment design)
- Properties of components (detail design)

These choices, in turn, determine two corresponding typologies of component characteristics: separability and accessibility, and performance (durability, reliability and other physical characteristics).

11.2.2   Analysis and Decomposition of Product Architecture

Among the different approaches to architecture decomposition, decomposition by modularity is considered more appropriate in this context because it analyzes the independence of functional and physical components. Unlike structural decomposition, which is restricted to a hierarchical model of the system, decomposition by modularity exploits the lack of dependency between physical components of the design (Kusiak and Larson, 1995). This choice is motivated by the strategic value that architecture modularity has in relation to the design of product life cycle, first described in Chapter 9, Section 9.4.

In the method proposed here, the analysis and decomposition of product architecture consists of three phases:

- Definition of the main functional units
- Analysis of the interaction between units (and definition of the consequent layout constraints)
- Analysis of the characteristic performances required of each unit

The functional units are those that collectively produce the overall functioning of the system, divided into physical blocks that perform the single operations. Therefore, the definition of functional units requires an initial, function-based decomposition (Kirschman and Fadel, 1998). First, the overall function is determined for the system. Then, depending on its complexity, it is broken down into subfunctions which, if necessary, may be decomposed again to produce a functional graph that approximates the subsystem boundaries and translates the functional units into physical blocks.

The results of the analysis of the interactions between the units may be expressed by a symmetrical interaction matrix:

$$IU = \left\lfloor iu_{ij} \right\rfloor_{nxn} \tag{11.1}$$

where iu_{ij} represents the interaction (value of 1 or 0) between the i-th and j-th units (Kusiak, 1999).

Analyzing the characteristic performance of the functional units consists of defining the performance constraints that, for each unit, can be expressed by one or more functions of the type:

$$Pf = Pf(Gf, Gv, Sh, MtPp) \tag{11.2}$$

where Pf represents the characteristic performance, Gf and Gv are the fixed and variable geometric parameters, respectively, Sh represents the form characteristics, and MtPp represents the properties of the material (Giudice et al., 2005).

11.2.3 Investigation Typologies

As mentioned above, the method proposed here supports two different investigation typologies:

• Analysis of conventional constructional systems for a correct definition of the most suitable interventions for preexisting products and an evaluation of environmental criticality
• Product redesign for the improvement of environmental performance in the life cycle

11.2.3.1 Analysis of Criticality and Potentiality of the Conventional System

At this level of intervention, the proposed method is directed at the best mapping of strategies for extending useful life and recovery at end-of-life, according to the properties of the preexisting construction units. This mapping is achieved using the matrix of strategy evaluation described below. The matrix translates certain determinant factors for the single strategies into component suitability to the strategy. The determinant factors, as shown below, are classified as dependent on, or independent from, the design choices.

In the case where a preexisting structure is analyzed, the design choices have already been made and therefore the entire set of these factors must be evaluated to define the optimal strategies. From the analysis of the conventional constructional system it is possible to:

- Define the main components and their constituent materials
- Identify the functional units
- Evaluate the modularization of the architecture by analyzing the correspondence between functional units and components
- Analyze the interactions between the components (which must respect the necessary interactions between functional units)

This provides a matrix of component interaction:

$$IC = \left\lfloor ic_{ij} \right\rfloor_{m \times m} \qquad (11.3)$$

Using the matrix of strategy evaluation, it is possible to quantify the relevance of each main component of the product in relation to each strategy of useful life extension and end-of-life recovery. Then, to ensure that the most suitable strategies are actually feasible, the architecture must allow the necessary separability of the components. To evaluate separability, which represents the main criticality of the architecture, the matrix defined by Equation (11.3) must be transformed into a matrix of the irreversible junctions (each interaction is translated into junction)

$$IC^* = \left\lfloor ic^*_{ij} \right\rfloor_{m \times m} \qquad (11.4)$$

where ic^*_{ij} is 1 if the junction between the i-th and j-th components is irreversible, and 0 if it is reversible or nonexistent, or if $i = j$. The separability of the components can then be expressed by the following vector:

$$SC = \left(sc_1 \ldots sc_i \ldots sc_m \right) \text{ where } sc_i = \prod_{j=1}^{m} (1 - ic^*_{ij}) \qquad (11.5)$$

The i-th component is separable if $sc_i = 1$, otherwise it is inseparable ($sc_i = 0$)."

11.2.3.2 Redesign of Product

From the viewpoint of Life Cycle Design, the modularization of the constructional system must achieve two main objectives regarding life cycle requirements: the independence of components belonging to different modules, and the affinity of components of the same module (Gershenson et al., 1999). This assumption is the foundation of redesign intervention.

The first phase of system redesign is the analysis of opportunities for architecture redesign based on the functionality and performance constraints imposed on the main units, introduced in Section 11.2.2 and expressed by the interaction matrix (11.1) and by a function set of type (11.2). In architecture redesign, the tool used for the evaluation of optimal strategies ignores the determinant factors directly dependent on design choices (which must subsequently be optimized), and takes into account only those dependent on factors external to the design choices (required characteristics and functionality, conditions of use).

The results of this first analysis, dependent on solely external factors, indicate which design choices would respect the predisposition of each component to useful life extension and end-of-life strategies. With these results it is also possible to identify any affinities that may exist between components. Components similar in terms of suitability for both the strategies and the required functional performance can be appropriately grouped and modularized, in order to facilitate, for each module identified, the most appropriate servicing or recovery operations (Marks et al., 1993). These indications are then implemented in the first level of design choices (layout, modularity) that falls within the domain of the embodiment phase of the design process (Chapter 7). Having redefined the main components, it is necessary to modify the interaction matrices of the functional units (11.1) and of the components (11.3).

The next level of design choices (that of components—typology of materials, durability, reliability) that falls within the domain of the detail design is approached in terms of:

- Required performance characteristics, expressed by (11.2)
- Indications obtained from the preliminary evaluation of the optimal strategies

The optimal choice is identified by varying the design parameters and evaluating the subsequent effects on the strategy distribution.

To complete redesign, the degree of appropriate separability can be identified for each module, in order to ensure a reduced impact (generally economic) of the disassembly phase. Therefore, the system of junctions must be defined so that it ensures:

- Functional interaction between components
- Separability, enabling the strategies identified as optimal for each component

Also in this case, separability also depends on the system of junctions through a matrix of type (11.4) and can be expressed using a vector as in (11.5). The required separability becomes the objective of the Design for Disassembly intervention, which guides the final phase of redesign at the component and junction system levels (refer to Chapter 9, Part Design and Joint Design, Section 9.3.3.1 and Table 9.2).

11.2.4   Verification Tools

The results of redesigning must be analyzed to evaluate their effectiveness in terms of reaching the desired goals (Section 3.2.3, Chapter 3). With respect to extending the product's useful life, these results can be evaluated using the tools for the analysis of product serviceability, which quantify its level of maintainability and reparability as a function of constructional system efficiency (Chapter 9, Section 9.2.2,). To evaluate performance in terms of environmental impact, it is possible to apply the tools of Life Cycle Assessment (LCA), which allow the evaluation of the environmental impact of the optimized product's life cycle (Chapter 4).

By evaluating the redesigned product in this way and comparing the results with those obtained on the conventional system, it is possible to determine the effectiveness and the success of the redesign, and its resulting benefits.

11.3   Optimal Life Cycle Strategy Evaluation Tool

The evaluation tool that enables useful life extension and recovery strategies to be related to the product parts and subsystems consists of a set of matrices that quantify the relevance of each main component in terms of each practicable strategy. This quantification is obtained by evaluating the potentiality of the components in relation to the determinant factors for each strategy (Chapter 9, Sections 9.2.3 and 9.3.5).

11.3.1   Determinant Factors for Strategies

The determinant factors are properties of components that render them appropriate for the application of one or more of the life cycle strategies under examination. For example, a component that requires frequent cleaning and is particularly susceptible to deterioration is a good candidate for regular maintenance; thus, the need for cleaning and the susceptibility to physical deterioration can be considered determinant factors for the maintenance strategy.

Determinant factors, as noted above, are distinguished by their dependence on, or independence from, the design choices. The former (durability, reliability, resistance) are directly dependent on choices made at the component level (materials, geometry). They are generally quantifiable by evaluating physical–mechanical properties and by applying tools and techniques for the analysis of component duration and life prediction (Chapter 10). The latter type depend on factors external to design choices (required characteristics and functionality, conditions of use). Generally, their quantification can only be based on qualitative evaluations.

The determinant factors that will be considered here are summarized in Tables 11.1 and 11.2 in relation to each strategy under examination. Those depending on design choices are displayed in italics.

11.3.2 Implementation of Matrices for Analysis of Strategies

Figure 11.2 shows the basic set of matrices for evaluation of strategies. To create a strategy analysis matrix, the main components must first be entered according to the indications obtained from the preliminary analysis and decomposition of product architecture.

Each component has a line of evaluation terms, one term for each determinant factor for the strategy for which the potential of the components is to be evaluated. In this way a matrix can be developed for each strategy, completed

TABLE 11.1 Extension of useful life strategies and determinant factors

MAINTENANCE	CLEANING NEED
	PHYSICAL DETERIORATION
	DURATION
REPAIR	DAMAGE RISK
	RELIABILITY
	DURATION
UPGRADING or ADAPTATION	OBSOLESCENCE
	USE MODE CHANGES
	USE ENVIRONMENT CHANGES

The determinant factors that depend on design choices are italicized.

TABLE 11.2   End-of-life strategies and determinant factors

REUSE OF PARTS	PHYSICAL DETERIORATION
	TECHNOLOGICAL OBSOLESCENCE
	DURATION
RECYCLING OF MATERIALS	PHYSICAL DETERIORATION
	TECHNOLOGICAL OBSOLESCENCE
	DURATION
	RECYCLABILITY

The determinant factors that depend on design choices are italicized.

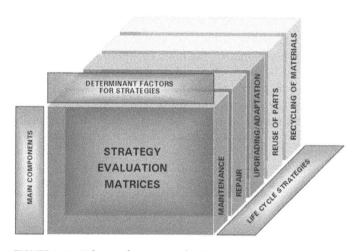

FIGURE 11.2   **Scheme of strategy evaluation matrices.**

by a final column consisting of global evaluation terms representing the overall evaluation of each component. These final terms are the sum of the terms in the corresponding matrix line, appropriately weighted according to the importance of each determinant factor.

For example, to evaluate a component's potential in relation to maintenance strategies, the corresponding matrix will consist of three columns, one for each determinant factor for maintenance (Table 11.1). The first column will consist of terms quantifying each component's need of cleaning operations. The second column will consist of terms quantifying each component's susceptibility to deterioration. The third column will consists of terms quantifying each component's lifespan. Further (conclusive) column will report the overall evaluations.

In a first analysis, the terms of intermediate and overall evaluation could be based on a qualitative evaluation of the determinant factors for each strategy. A quantitative evaluation approach may be formulated (Giudice et al., 2002), based on the mathematical model summarized as follows.

Have C_i indicate the i-th of the m components comprising the product, and have DF_j^X indicate the j-th of the n^X determinant factors for the practicable strategy X. The matrix M^X for the evaluation of strategy X can be expressed as:

$$M^X = \left[m_{ij}^x \right]_{i=1,2,\ldots,m}^{j=1,2,\ldots,n^x} \tag{11.6}$$

where the term m_{ij}^x quantifies the j-th determinant factor for the strategy X, relative to the i-th component.

Then have w_j^x indicate the weight of the j-th determinant factor for strategy X. The aptness A_i^X of the i-th component C_i to strategy X represents the Index of Strategy X for the i-th component:

$$A_i^x = \sum_{j=1}^{n^x} w_j^x \cdot m_{ij}^x \tag{11.7}$$

A correct use of the proposed model requires not only an appropriate quantification of the strategy determinant factors, but also their normalization to render them homogeneous in relation to the application of (11.7), and an evaluation of the weighting of each factor in relation to the strategy.

11.4 Case Study: System Analysis and Redesign of a Household Refrigerator

Large, domestic electrical appliances (refrigerators, washing machines, dishwashers) generally have an average life estimated at about 10 years. This period represents the potential life of the product since it corresponds to a prevision of usefulness (i.e., to the time it is anticipated the appliance can maintain its primary functions—physical life, see Section 9.1 of Chapter 9). This potential life can be significantly reduced by the design cycle (i.e., the interval of time between successive generations of the product), whereby the production of a model is discontinued four to five years after its first appearance on the market. This characteristic of brief effective useful life, together with their widespread use in all domestic environments, makes electrical appliances particularly sensitive to problems of retirement and recovery.

In the case of the refrigerator, this problem is amplified by the large variety of product typologies produced to meet varying consumer demands, which can make it vulnerable to a further reduction in its useful life. This is compounded by a problem of recovery resulting from the conventional product constructional system that, at present, combines a wide variety of different and incompatible materials inseparably (EC-VHK, 1999). The walls of refrigerators, in particular, are composed of metals, plastics, and PUR foam glued together; because they cannot be disassembled or dismantled, they are usually cut up, shredded, or ground up (Lambert and Stoop, 2001).

These problems have led to legislative pressures and incentives intended to limit the environmental impact of this specific manufacturing sector, operating on many phases of the life cycle. In the European Union, for example, use of refrigerating fluids and foaming agents responsible for increased deterioration of the ozone layer and global warming (chlorofluorocarbons—CFCs) has been discouraged or suppressed. The introduction of energy labels showing products' energy consumption has been complemented by certification of products' overall eco-compatibility, within the EC's eco-label award scheme (Chapter 1, Section 1.6.1). The first EC directive on criteria for awarding the eco-label to refrigerators was issued in 1996; in its 1999 revision (2000/40/EC), the previous criteria were integrated with explicit requirements regarding the extension of useful life (lifetime extension) and the facilitation of product disassembly for recovery and recycling at end-of-life. That revised directive required that joints must be easy to find and be accessible, electronic assemblies and the whole product must be easy to dismantle, and incompatible and hazardous materials must be separable. These key criteria were confirmed in the last revision (2004/669/EC).

11.4.1   Preliminary Analysis of System

Following this methodology, the first phase consists of a preliminary analysis and decomposition of the constructional system to define the main functional units, the interactions between units (and consequent layout constraints), and the characteristic performances required of each unit. In the case of the refrigerator, the six main functional units summarized in Table 11.3 were identified and associated with their main performance characteristics. The matrix of the interactions between the main units (11.1) is reported in Table 11.4.

11.4.2   Analysis of Criticality and Potentiality of the Conventional Architecture

From the analysis of the conventional system it is possible to define the main components and their materials and to identify the functional units, as shown

TABLE 11.3 Functional units and main performances requested

FUNCTIONAL UNITS	MAIN PERFORMANCES REQUESTED
U_1 External casing	Shock resistance, structural sturdiness, thermal insulation
U_2 Rear panel	Thermal insulation
U_3 Thermal insulator	High thermal insulation, structural sturdiness
U_4 Internal lining	Hygienic and washable, structural sturdiness, thermal insulation
U_5 Plant	Thermodynamic performance
U_6 Plant–cell interface	Thermal insulation

TABLE 11.4 Functional interaction between main units

	U_1	U_2	U_3	U_4	U_5	U_6
U_1	0	1	1	0	1	0
U_2	1	0	1	0	0	0
U_3	1	1	0	1	0	0
U_4	0	0	1	0	0	1
U_5	1	0	0	0	0	1
U_6	0	0	0	1	1	0

in Figure 11.3. It can be seen that in this case, unit 6 (which transfers the cooling action generated by the cooling plant 5 to the internal cell 4) coincides with part of the plant itself (evaporation plate); thus, units U_5 and U_6 are grouped together in a single component C_5 (cooling plant).

From the conventional architecture analysis it is also possible to determine the main criticality—the impossibility of separating the parts at the end of the working life because of the foam insulation element that joins all the cabinet components and part of the cooling system. This criticality is expressed by the matrix reported in Table 11.5, where the irreversible junctions (11.4) are reported in the upper part and the consequent vector of component separability (11.5) is given on the lower line.

This analysis was directed at producing the most correct mapping of the strategies of extension of useful life and recovery at end-of-life, in relation to the properties of the components. This mapping was achieved using the strategy evaluation matrix described above (Section 11.3). Use of the matrix quantifies the relevance of each essential component of the product in relation to each feasible strategy.

As an example, Figure 11.4 shows the application of the matrix corresponding to the reuse strategy. Using the matrix in an analogous manner, it is

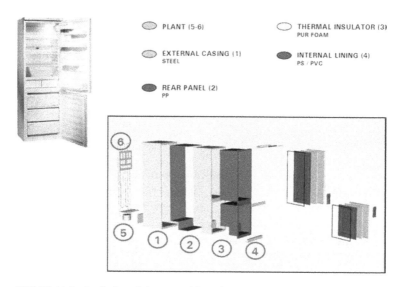

FIGURE 11.3   Analysis and decomposition of a conventional system.

possible to map the various strategies under examination, indicated by a specific color for each strategy, where the color intensity represents the suitability of the components in relation to the strategy. Figure 11.4 shows the mapping relative to reuse, highlighting the components with the greatest relevance in relation to this end-of-life strategy.

Figures 11.5 and 11.6 show the strategy evaluation matrices regarding each strategy under examination (maintenance, repair, reuse, and recycling). Of the determinant factors for each strategy, those dependent on design choices (material typology, reliability, and durability), which in this type of analysis are taken as preestablished parameters, are highlighted. Once quantified, the parameters are broken down into value ranges of four different levels (0, zero; 1, low; 2, medium; 3, high). The figures also show the corresponding strategy indices calculated according to the weighting method introduced above (Section 11.3.2).

If cleaning operations are excluded, the component requiring the most servicing (maintenance and repair) is the cooling plant. This component is not completely separable from the rest of the structure, as is confirmed by the information reported in Table 11.5 (the evaporators are embedded in the polyurethane foam). Therefore, only some parts of the cooling plant have a good level of serviceability.

Strategies for end-of-life involve the polyurethane insulation (for reuse) and the metal and polymer casings. However, once again the zero separability of these components does not permit optimal strategies to be applied.

TABLE 11.5 Irreversible junctions and separability of components

	U_1C_1	U_2C_2	U_3C_3	U_4C_4	$U_{5+6}C_5$
U_1	0	0	1	0	0
U_2	0	0	1	0	0
U_3	1	1	0	1	1
U_4	0	0	1	0	0
U_{5+6}	0	0	1	0	0
	0	0	0	0	0

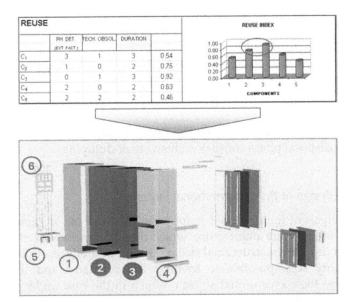

FIGURE 11.4 Application of a strategy evaluation matrix and mapping of the system: Reuse.

In conclusion, the desired environmental potentials cannot be realized in this case, highlighting the criticality of the conventionally manufactured product due to the poor separability of its components. Figure 11.7 shows the results of the LCA, performed with SimaPro 4.0® software (Pré Consultants BV, Amersfoort, The Netherlands), using the Eco-indicator 95 impact assessment method (Chapter 4, Section 4.2 and Table 4.3). The main processes making up the entire life cycle were incorporated in the analysis. As a result, the environmental impacts of the manufacturing, use (hypothesizing a life of eight years), and disposal phases are quantified (the Eco-indicator method

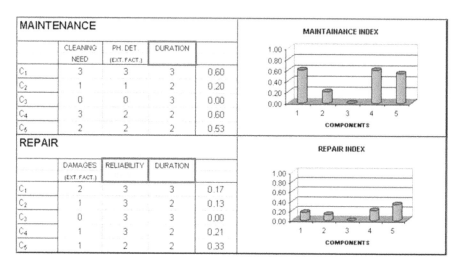

MAINTENANCE

	CLEANING NEED	PH. DET (EXT. FACT.)	DURATION	
C_1	3	3	3	0.60
C_2	1	1	2	0.20
C_3	0	0	3	0.00
C_4	3	2	2	0.60
C_5	2	2	2	0.53

MAINTAINANCE INDEX

REPAIR

	DAMAGES (EXT. FACT.)	RELIABILITY	DURATION	
C_1	2	3	3	0.17
C_2	1	3	2	0.13
C_3	0	3	3	0.00
C_4	1	3	2	0.21
C_5	1	2	2	0.33

REPAIR INDEX

FIGURE 11.5 Conventional system: Matrices for life cycle strategy evaluation and strategy indices (Maintenance and Repair).

expresses the impacts in Point–Pt). The first two phases lead to the greatest impact; the disposal phase consists exclusively of dumping.

11.4.3 Redesign of the Constructional System

Careful analysis of the strategies mapped in this way can suggest ways to reinterpret the product architecture, which then becomes the starting point for redesigning the constructional system. From the analysis of potentiality and criticality it is possible to identify the problems and limitations presented by the conventional system, which in the case under examination are principally:

- Dispersion of the thermodynamic plant in the entire system
- Heterogeneity of the materials used (metals and polymers)
- Impossibility of separating the parts at end of use due to the foam insulating element that binds together all the components of the refrigerator cabinet

Having defined the main problems resulting from the conventional system analysis, it is possible to identify the priority objectives of redesign:

- Separation of the refrigeration system from the structure—The thermodynamic system must be separated from the rest of the refrigerator

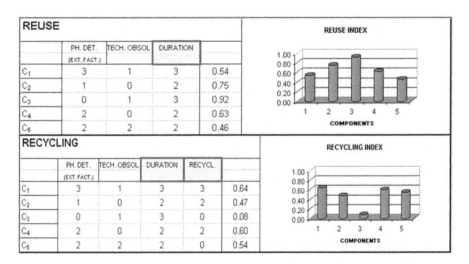

REUSE

	PH. DET. (EXT. FACT.)	TECH. OBSOL	DURATION	
C_1	3	1	3	0.54
C_2	1	0	2	0.75
C_3	0	1	3	0.92
C_4	2	0	2	0.63
C_5	2	2	2	0.46

RECYCLING

	PH. DET. (EXT. FACT.)	TECH. OBSOL	DURATION	RECYCL.	
C_1	3	1	3	3	0.64
C_2	1	0	2	2	0.47
C_3	0	1	3	0	0.08
C_4	2	0	2	2	0.60
C_5	2	2	2	0	0.54

FIGURE 11.6 Conventional system: Matrices for life cycle strategy evaluation and strategy indices (Reuse and Recycling).

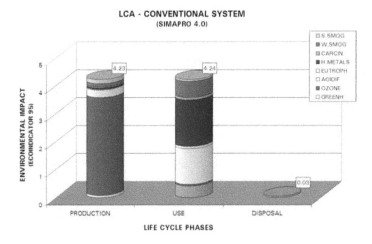

FIGURE 11.7 LCA for the conventional system.

and enclosed in a cooling module easily accessible for servicing operations and separable for substitution and reuse.

- Separability of the cabinet—The main components of the refrigerator cabinet (external covering, insulation, internal lining) must be designed in a way that allows simple, stable, and reversible disassembly to facilitate disassembly at end of use, while meeting structural and thermal requirements.

- Choice of materials—The materials must be chosen in a way that provides optimal stratification from the point of view of the specific properties and functions required of each component.

The first phase of redesign involves the use of tools to evaluate the optimal strategies, ignoring determinant factors directly dependent on design choices (which must be optimized subsequently) and considering only those dependent on factors external to design choices. The results of this first phase are reported in Figures 11.8 and 11.9. As shown by the first two matrices (Figure 11.8), if cleaning operations and damage on external casing due to accidents (unit 1) are excluded, the need for maintenance and repair is concentrated in the cooling plant (unit 5). This suggests making design choices that respect this predisposition so that servicing is concentrated on the single most sensitive unit, making all its components separable from the product and easily accessible.

The other two matrices (Figure 11.9) identify the units offering the best opportunities for reuse (units 2, 3, 6) and those most suitable for recycling (units 1, 4). Unit 5 offers broadly equivalent opportunities (the complexity of the cooling system requires, however, a deeper level of analysis). In this case, the results obtained provide indications for the most appropriate design choices. Furthermore, they reveal the close affinity between unit 2 (the rear panel) and unit 6 (the element transferring the cooling action from the cooling plant into the cell); these units have identical needs for servicing and requirements for characteristic performance (Table 11.3).

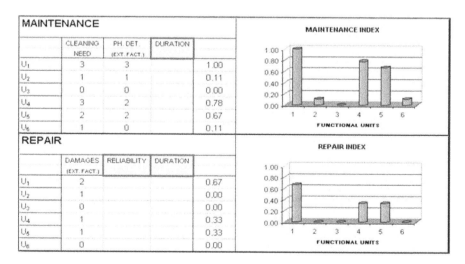

MAINTENANCE

	CLEANING NEED	PH. DET. (EXT. FACT.)	DURATION	
U_1	3	3		1.00
U_2	1	1		0.11
U_3	0	0		0.00
U_4	3	2		0.78
U_5	2	2		0.67
U_6	1	0		0.11

REPAIR

	DAMAGES (EXT. FACT.)	RELIABILITY	DURATION	
U_1	2			0.67
U_2	1			0.00
U_3	0			0.00
U_4	1			0.33
U_5	1			0.33
U_6	0			0.00

FIGURE 11.8 Redesign: Matrices for life cycle strategy evaluation and strategy indices (Maintenance and Repair).

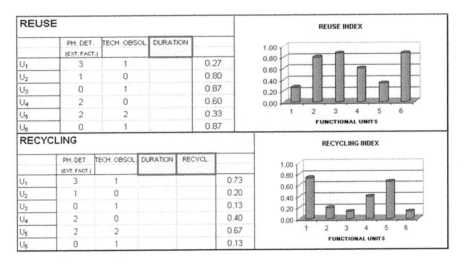

REUSE				
	PH. DET. (EXT. FACT.)	TECH. OBSOL.	DURATION	
U_1	3	1		0.27
U_2	1	0		0.80
U_3	0	1		0.87
U_4	2	0		0.60
U_5	2	2		0.33
U_6	0	1		0.87

RECYCLING					
	PH. DET. (EXT. FACT.)	TECH. OBSOL.	DURATION	RECYCL.	
U_1	3	1			0.73
U_2	1	0			0.20
U_3	0	1			0.13
U_4	2	0			0.40
U_5	2	2			0.67
U_6	0	1			0.13

FIGURE 11.9 Redesign: Matrices for life cycle strategy evaluation and strategy indices (Reuse and Recycling).

To interpret these indications in a first level of design choices (layout):

- Unit 5 (cooling plant) is subdivided into two main components, the cooling plant C_5 and an external case C_6 that houses the entire system and separates it from the rest of the manufactured product.
- Units 2 and 6 are combined in a single component C_2.

The next level of design choices (component definition) is evaluated with respect to:

- The required performance characteristics, reported in Table 11.3
- The indications obtained from preliminary evaluation of the optimal strategies (Figures 11.8 and 11.9)

The optimal choice is identified by varying the design parameters, quantifying the respective determinant factors (previously neglected), and evaluating the consequent effect on the distribution of strategies, which is quantified by the values assumed by the strategy indices. In the case under examination, the optimal choices are realized in the architecture shown in Figure 11.10, which summarizes the layout of the functional units, the general geometry, the materials chosen for each component, and the distribution of optimal strategies.

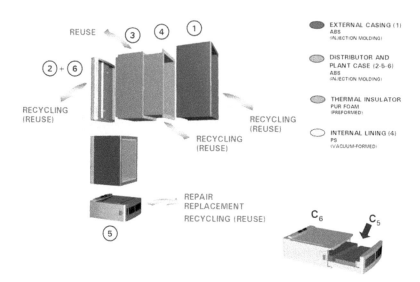

FIGURE 11.10 Redesigned system: Layout, materials, and distribution of optimal strategies.

To complete the redesign, it is necessary to define a junction system that:

- Ensures the redefined functional interactions
- Ensures separability, allowing the execution of the strategies identified as optimal

The junction system proposed, respecting the functional interactions, involves a single juncture between the external casing and the rear component that closes the cell and transmits the refrigerating action of the cooling plant into the cell itself. The overall junction system could present a single irreversibility in the connection between the cooling plant (C_5) and its casing (C_6), however these together make up the cooling unit. Table 11.6 summarizes the matrix of irreversible junctions [Equation (11.4)], the component separability vector [Equation (11.5)], and the separability vector of the functional units. It can be seen that the single irreversibility does not affect the complete separability of the main units.

Conducting an LCA on the redesigned architecture and comparing the results with that performed on the conventional architecture, it is possible to evaluate the environmental benefits conferred and, therefore, the effectiveness of the redesign method used. Figure 11.11 demonstrates the environmental impacts relevant to the main phases of the life cycle. With respect to

TABLE 11.6 Irreversible Junctions, Separability of Components, Separability of functional units

	U_1 C_1	U_{2+6} C_2	U_3 C_3	U_4 C_4	U_5 C_5	U_5 C_6
C_1	0	0	0	0	0	0
C_2	0	0	0	0	0	0
C_3	0	0	0	0	0	0
C_4	0	0	0	0	0	0
C_5	0	0	0	0	0	1
C_6	0	0	0	0	1	0
	1	1	1	1	0	0
	1	1	1	1	1	

the conventional system (Figure 11.7), the new solution is characterized by a marked increase in impact both during production (+11%) and during use (+19%), due to an increase in electricity consumption because the new system requires a more powerful cooling plant. In compensation, the complete separability of the system allows a disposal phase so efficient that these negative effects are balanced, resulting in an environmental impact over the entire life cycle that is better than that of the conventional system (−25%). This is shown in Figure 11.12, which directly compares the whole life cycles of the conventional and redesigned systems. This confirms the effectiveness and good outcome of the redesign.

11.4.4 Focus on the Results of the Modularity Concept and Ease of Disassembly Approach

After the objectives of redesign were defined and the proposed method was applied, it was possible to arrive at some preliminary considerations that supported product concept development and were centered on the fundamental design aspects shown in Figure 11.10. The structure of the constructional system, the component geometry, the choice of materials, and the junction systems of the final solution summarized in the figure are the result of the specific design approach chosen (decomposition of product architecture, modularization, ease of disassembly). The goal of this design approach was to enhance the optimal strategies identified for each component

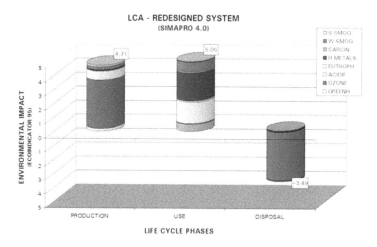

FIGURE 11.11   LCA for the redesigned system.

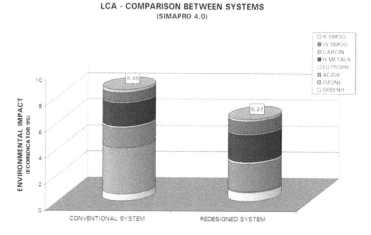

FIGURE 11.12   LCA: Comparison between the conventional and redesigned system.

in the decomposition and analysis phase (shown for each component in Figure 11.10).

In more detail, some characteristics of the final proposal are:

- Modular architecture of constructional system (Figure 11.13)—
  A refrigerating unit placed at the base is integrated with easily

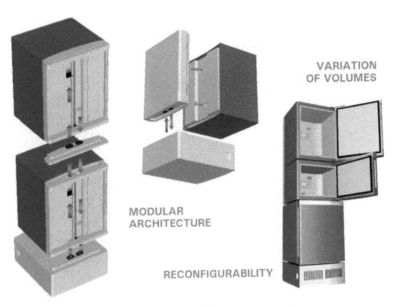

VARIATION
OF VOLUMES

MODULAR
ARCHITECTURE

RECONFIGURABILITY

FIGURE 11.13 Modular architecture of redesigned constructional system.

connectable refrigerator modules to give maximum flexibility in variation of volume, modes of their thermal use (volumes at different
temperatures), and the possibility of reconfiguring the entire system.
This type of architecture requires a frost-free thermodynamic system
that generates flows of cold air channeled into the refrigerator
modules using an appropriate distribution system.

- Separation of the thermodynamic module from the constructional
  system (Figure 11.14)—The refrigerating unit contains the entire
  frost-free system divided in two separate areas: the hot area (motor–
  compressor, condenser, cooling fan) and the cold area (evaporator,
  fan distributing refrigerating flows). The unit is also housed in a part
  of the module that is extractable to allow immediate access and facilitate maintenance, repair, temporary or permanent substitution,
  upgrading, and recovery.
- Constructive architecture of the refrigerator modules in separable
  monomaterial components (Figure 11.15)—The architecture of the
  refrigerator modules harmonizes the different functions of the parts
  with the requirements identified by the analysis of recovery strategies, through the separability of each monomaterial component and
  the choice of materials. The latter has strategic value in the context

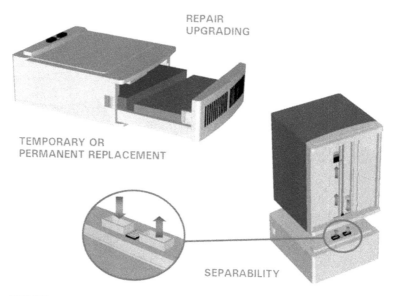

REPAIR
UPGRADING

TEMPORARY OR
PERMANENT REPLACEMENT

SEPARABILITY

FIGURE 11.14   Separability of thermodynamic unit.

of Design for Environment; material choice is important and complex, since materials must be chosen not only according to functional requirements but also in relation to the possibilities of recovery and recycling and the environmental impact of these materials. Chapter 12 addresses this issue in more detail. There are specific studies on the optimal choice of insulation for refrigerators (Weaver et al., 1996). Here, because the choice was directed at disassembly, two characteristics were considered: the separability of the parts requires valid alternatives to the conventional polyurethane foam insulation, possibly by premolding the polyurethane layer or using vacuum insulation panels; and the need to develop an easily separable system suggests the use of polymer materials, easily molded to provide shapes favoring disassembly and reversible integral junctions.

- Reversible junction system (Figure 11.16)—The monomaterial components of each refrigerator module are designed in such a way that they require only a single junction for assembly and disassembly. This junction must be efficient and easily reversible, and can be realized by exploiting the ease of working polymer materials, integrating interlocking shapes into the components.

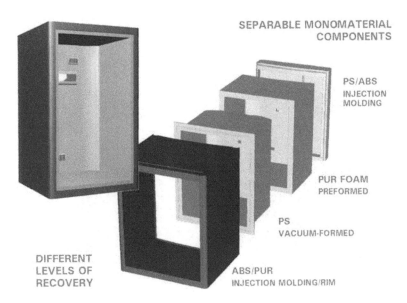

SEPARABLE MONOMATERIAL
COMPONENTS

PS/ABS
INJECTION
MOLDING

PUR FOAM
PREFORMED

PS
VACUUM-FORMED

DIFFERENT
LEVELS OF
RECOVERY

ABS/PUR
INJECTION MOLDING/RIM

FIGURE 11.15   Structure of modules.

11.4.5   Product-Service Integration

The characteristics of the solution developed here suggest the possibility of integrating the product with service that improves its potential for adaptation and evolution in line with user needs, and that ensures the recovery of the retired product, enabling closure of the life cycle. This is in line with recent acknowledgments of the strategic value of a product-service system in the field of environmental protection (Roy, 2000).

On the basis of the specific particulars of the product concept developed, the product-service system can be associated with several mileposts in the intermediate and final phases of the product life cycle (use and end-of-life):

- Acquisition—Support in the choice of configuration according to user needs
- Use—Replacement of modules for upgrading, installation of new modules, and reconfiguration of product
- Service—Maintenance and repair operations, possible substitution of refrigeration system
- Retirement—Recalling modules or entire products for recovery and disposal

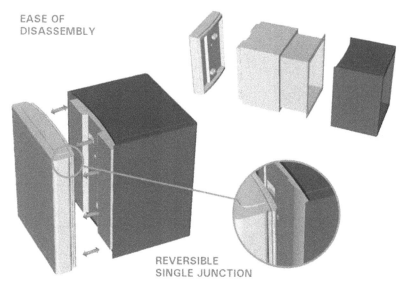

EASE OF
DISASSEMBLY

REVERSIBLE
SINGLE JUNCTION

FIGURE 11.16   Reversible junction system.

11.5   Final Remarks

The most significant conclusions concern the two main contents of this chapter, the definition of the method and design tools and the development of the metadesign proposed as a possible solution for the case study presented.

The method and tools of design analysis, decomposition, and redesign have interesting potential as aids in the design of eco-compatible products in terms of their maintenance, repair, upgrading, and recovery at end of useful life— that is, the most effective strategies improving exploitation of resources used in production (Chpater 9, Section 9.1). Although oriented toward the particular product typology under examination, the method described here appears sufficiently versatile to allow its application in other contexts.

The design proposal synthesizes the main concepts identified as fundamental (product architecture modularization, ease of disassembly) and outlines a product whose life cycle responds to the guidelines traced by the reference frame. The strategies of extension of useful life and recovery at end of life are manifested in innovative product architecture and in the integration of the product and how it is serviced, which exploits the potential of the constructional system.

Finally, the experience described here provides an insight into the definition of environmentally sustainable products—the development of products oriented toward environmental requirements, where these requirements are transformed from limiting constraints into a positive stimulus for innovation.

11.6 Summary

The chapter describes the development of a methodological tool, complete with the fundamental mathematical modeling, for the study of product constructional systems with the aim of determining their environmental efficiency. Environmental efficiency is pursued through two intervention typologies on the life cycle: those directed at maintaining performance and functionality during use (strategies for the extension of useful life) and those oriented toward the recovery of resources (strategies of recovery at the end-of-life).

The method is based on certain key phases: the preliminary analysis and decomposition of product architecture; the evaluation of optimal strategies for each main component; and the definition and implementation of modularity and the separability that allows the optimal strategies to be applied. Furthermore, the method supports design intervention at two different levels: definition of layout and modularity (embodiment design), and choice of the main characteristics of components (detail design).

The refrigerator case study examined here highlights the versatility of this method used as a tool for analysis of environmental criticality and potentiality of conventional constructional systems, for a correct definition of the most suitable intervention strategies on preexisting products; and for product redesign, taking account of unavoidable requisites and integrating them with new requirements for the environmental efficiency of the product's life cycle.

11.7 References

2000/40/EC, Commission Decision establishing the ecological criteria for the award of the Community eco-label to refrigerators, Official Journal of the European Communities, L 13, 19/1/2000, 22–26, 2000.

2004/669/EC, Commission Decision establishing revised ecological criteria for the award of the Community eco-label to refrigerators, Official Journal of the European Communities, L 306, 2/10/2004, 16–21, 2004.

EC-VHK, Revision European Eco-label Criteria for Refrigerators, Draft final report VHK 1999, European Commission, Brussels, 1999.

Gershenson, J.K., Prasad, G.J., and Allamneni, S., Modular product design: A life-cycle view, *Journal of Integrated Design and Process Science*, 3(4), 13–26, 1999.

Giudice, F., La Rosa, G., and Risitano, A., An ecodesign method for product architecture definition based on optimal life-cycle strategies, in *Proceedings of Design 2002—7th International Design Conference*, Dubrovnik, Croatia, 2002, 1311–1322.

Giudice, F., La Rosa, G., and Risitano, A., Materials selection in the life-cycle design process: A method to integrate mechanical and environmental performances in optimal choice, *Materials and Design*, 26(1), 9–20, 2005.

Kirschman, C.F. and Fadel, G.M., Classifying functions for mechanical design, *Journal of Mechanical Design*, 120, 475–482, 1998.

Kusiak, A. and Larson, N., Decomposition and representation methods in mechanical design, *Transactions of the ASME—Special 50th Anniversary Design Issue*, 117, 17–24, 1995.

Kusiak, A., *Engineering Design: Products, Processes, and Systems*, Academic Press, San Diego, CA, 1999.

Lambert, A.J.D. and Stoop, M.L.M., Processing of discarded household refrigerators: Lessons from the Dutch example, *Journal of Cleaner Production*, 9, 243–252, 2001.

Marks, M., Eubanks, C.F., and Ishii, K., Life-cycle clumping of product designs for ownership and retirement, in *Proceedings of ASME Design Theory and Methodology Conference*, Albuquerque, NM, 1993, DE-Vol. 53, 83–90.

Roy, R., Sustainable product-service systems, *Futures*, 32, 289–299, 2000.

Ulrich, K.T. and Eppinger, S.D., *Product Design and Development*, 2nd ed., McGraw-Hill, New York, 2000.

Weaver, P.M. et al., Selection of materials to reduce environmental impact: A Case study on refrigerator insulation, *Materials and Design*, 17(1), 11–17, 1996.

Chapter 12

Environmental Characterization of Materials and Optimal Choice

A product's environmental impact is directly influenced by the environmental properties of the materials used, such as energy costs, emissions involved in production and manufacturing phases, and recyclability. The choice of materials, therefore, assumes strategic importance and requires an extension of the characterization of materials, integrating conventional characterization (aimed at defining physical–mechanical properties) with a complete characterization of environmental behavior. To enable the designer to make an optimal choice of materials that harmonizes performance characteristics and properties of eco-compatibility, the selection process must take account of a wide range of factors: constraints of shape and dimension, required performance, technological and economic constraints associated with the manufacturability of materials, and environmental impacts of all the phases of the life cycle.

In accordance with the Life Cycle Design approach, this chapter proposes a definition of the environmental characterization of materials and processes, and a systematic method that introduces environmental considerations in the selection of the materials used in components. This definition and method are directed at meeting functional and performance requirements while minimizing the environmental impact associated with the product's entire life cycle. The proposed selection procedure elaborates data on the conventional and environmental properties of materials and processes, relates this data to the required performance of product components, and calculates the values assumed by functions that quantify the environmental impact over the whole life cycle and the cost resulting from the choice of materials. As shown in the case study presented, the results can then be evaluated using multiobjective analysis techniques.

12.1 Materials Selection and Environmental Properties

"New materials inspire designers; but even more, design drives material development" (Ashby, 2001). This statement highlights the close connection between materials and the design activity, confirmed by the significance of the issues related to the efficient integration of materials selection in the product development process (Edwards, 2003; Lu and Deng, 2004).

The enormous variety of materials available for engineering applications and the complexity of the requirements conditioning the choice of the most appropriate materials and processes lead to a taxing problem of multiple-criterion optimization (Brechet et al., 2001). In recent years, several systematic methods have been proposed to help the designer in the selection of materials and processes (Charles et al., 1997; Farag, 1997; Asbhy et al., 2004). Of the more commonly used quantitative selection methods, that developed by Ashby is based on the definition of material indices consisting of sets of physical–mechanical properties which, when optimized, maximize certain performance aspects of the component under examination (Ashby and Cebon, 1995). Defining these indices makes it possible to compile selection charts summarizing the relations between properties of materials and engineering requirements (Ashby, 1999).

Usually taking into consideration the physical-mechanical properties of materials, these selection charts can be extended to introduce some environmental properties (Navin-Chandra, 1991). From this standpoint, several important studies have been based on the development of indices able to express the environmental performance of materials by introducing the energy consumption and emissions (into the atmosphere or water) associated with the materials (Holloway, 1998), or eco-indicators developed on the basis of Life Cycle Assessment methods (Wegst and Ashby, 1998). An alternative approach is that of translating environmental impact in terms of economic cost of production, introducing functions of environmental cost such as energy consumption and toxicity that depend on the properties of the materials (Chen et al., 1994).

All the methods proposed are limited to quantifying the environmental impact of the choice of materials on the basis of their environmental properties associated with the production phase. Only a few studies have considered the influence of the choice of materials on the impact associated with the working life of the component (Kampe, 2001). To date, the problem of choice of materials from the viewpoint of Life Cycle Design (taking into account the environmental impacts involved in all phases of the life cycle, from production to retirement) has been considered only in general terms, with the aim of defining guidelines for choices that integrate properties of materials, manufacturing demands, and end-of-life impacts, and suggesting a distinction of selection criteria between component design and assembled product design (Stuart, 1998).

12.2 Environmental Characterization of Materials and Processes

The influence that the materials used to manufacture a product have on its environmental impact is manifested in the energy costs and emissions associated with the production and end-of-life processes of the material, and in the intrinsic properties of the material and production process that constrain its level of recyclability. Complete environmental characterization of a material should, therefore, consist of defining the environmental impact linked to its production and disposal, and of evaluating the margins of recyclability in terms of decline in performance of the recycled material and recovery costs. Therefore, the optimal choice of materials, in relation to environmental demands, requires this complete environmental characterization, with particular regard to the following aspects:

- Environmental impact associated with production processes (energy costs and overall impact)
- Environmental impact associated with phases of end-of-life (recycling or disposal)
- Suitability for recycling (expressed by the recyclable fraction)

Information on the energy costs and recyclable fractions of more common materials can be obtained from commercially available databases, such as that of the CES® (Cambridge Engineering Selector, Granta Design Ltd., Cambridge, UK) materials selection software. Overall environmental impact can be evaluated using the techniques of Life Cycle Assessment (LCA), the analysis method used to quantify the environmental effects associated with a process or product through the identification and quantification of the resources used and the waste generated. As was discussed in Chapter 4, LCA evaluates the impact of using these resources and of the emissions produced. Quantification of the impacts is based on inventory data that is subsequently translated into eco-indicators such as those used here. These are evaluated according to the Eco-indicator 99 method (Chapter 4, Section 4.2 and Table 4.3) and calculated using SimaPro 5.0® software (Pré Consultants BV, Amersfoort, The Netherlands).

Environmental characterization is also extended to common primary (forming) and secondary (machining) manufacturing processes, evaluating the indicators that quantify the impacts of standard processes per unit of process parameter or of the volume or weight of material processed.

12.2.1 Data on Materials and Processes

For each material it is necessary to integrate the information used in conventional design with that regarding environmental properties to obtain:

- General properties (density, cost)
- Mechanical properties (e.g., modulus of elasticity, hardness, fatigue limit)
- Thermal and electrical properties (e.g., conductivity and thermal expansion, operating temperature, electrical resistance)

PP
Polypropylene

Density	890 – 900	kg/m³
Cost	1.00 – 1.20	EURO/kg

Mechanical Properties

Young's Modulus	896 – 1240	MPa
Shear Modulus	315 – 435	MPa
Poisson's Ratio	0.41 – 0.42	
Elastic Limit	20 – 29	MPa
Tensile Strength	27 – 38	MPa
Compressive Strength	25 – 55	MPa
Elongation	200 – 500	%
Hardness	62 – 90	MPa
Fatigue Limit	11 – 15	MPa

Thermal and Electrical Properties

Thermal Conductivity	0.14 – 0.16	W/m.K
Thermal Expansion	122 – 170	10⁻⁶/K
Glass Temperature	248 – 258	K
Max Service Temp	350 – 370	K
Specific Heat	1880 – 1950	J/kg.K
Resistivity	3.3×10^{22} – 3×10^{23}	μ ohm.cm
Dielectric Constant	2.2 – 2.3	

Environmental Properties

EI Material Production	306	mPt/kg
EI Disposal (Incineration)	–6.71	mPt/kg
EI Disposal (Landfill)	3.36	mPt/kg
EI Recycling	–195	mPt/kg
Energetic Cost	100 – 120	MJ/kg
Recyclable Fraction	0.45 – 0.55	

Other Properties

Flammability	Low
Resistance to Sea Water	High
Resistance to Acid	High
Resistance to Alkalis	High
Resistance to UV	Low
Resistance to Wear	Average

FIGURE 12.1   Material datasheet: Polypropylene.

- Environmental properties (energy cost, environmental impact, recyclability)

As an example, the datasheet in Figure 12.1 relates to a widely used plastic material (polypropylene) and shows the data on its environmental properties. Eco-indicators were evaluated with SimaPro 5.0 software, using the Eco-indicator 99 method and expressing impacts in mPt (milliPoint). With this software it is possible to select the inventory data to be used for impact evaluation, in this specific case Buwal 250 data (Pré, 2003).

Likewise, the following information must be obtained for the primary and secondary manufacturing processes:

- Physical attributes of the final product
- Economic cost of standard process (fixed and variable costs)
- Environmental properties (energy consumption, environmental impact of standard process)

12.3   Summary of Selection Method

The reference method depicted in Figure 12.2 is based on calculation models that quantify and interrelate the various performances required of the material

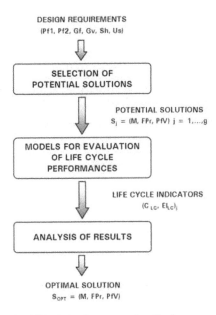

FIGURE 12.2   Summary of method.

in order to identify potential solutions, and a successive, multiobjective analysis aimed at harmonizing the conventional performance, costs, and environmental performance of the product.

The first phase consists of defining the set of design requirements and parameters:

- Primary performance (Pf1), in relation to the specific functionality of the component
- Secondary performance (Pf2), which can impose further restrictions to guide the selection
- Geometric parameters, distinguishing between fixed (Gf) and variable (Gv) geometric parameters
- Typology of shape and relative level of complexity (Sh), which greatly affects the choice of forming processes
- Use of component (Us), which can influence an initial selection of materials

The set of design requirements constitutes the input for the procedure of selecting potential solutions. This procedure is based on two different types of analysis, shown in Figure 12.3. In the first stage, the production feasibility of each hypothetical solution is evaluated by analyzing some of the information given in the set of design requirements (in particular, the typology of

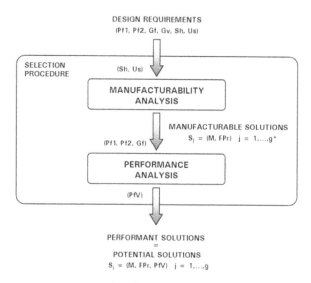

FIGURE 12.3 Procedure for selection of potential solutions.

shape required and the intended use). The solutions identified in the analysis of production feasibility must then be evaluated in terms of the required performances (Pf1, Pf2). The potential solutions obtained are then analyzed in subsequent phases of the selection method.

Each potential solution S is defined by pairs of material–primary forming process (M, FPr), and by the performance volume (PfV), representing the minimum volume needed to meet the requirements of primary performance. If appropriate, the definition of the generic solution S can also include any processes of secondary machining required after the initial forming.

In the following phase, the calculation models are applied to each potential solution in order to evaluate the indicators of environmental impact and cost over the entire life cycle. The final phase of the method involves analyzing the results and identifying the optimal choice.

12.4 Analysis of Production Feasibility

The first stage of the selection procedure must correlate material, process, shape, and function. The problem of the interaction between these factors is considered central to the selection of materials and has been thoroughly investigated (Ashby, 1999).

In the method proposed here, this problem is addressed by considering shape (Sh) and use (Us) to be design requirements, expressed using binary vectors V^{Sh} and V^{Us}, and introducing binary matrices correlating shape–process, material–use, and material–process:

$$\Phi^{S-P} = \left[\Phi^{S-P}_{sp} \right]^{p=1,\dots,np}_{s=1,\dots,ns} \quad \Phi^{U-M} = \left[\Phi^{U-M}_{um} \right]^{m=1,\dots,nm}_{u=1,\dots,nu} \quad \Phi^{P-M} = \left[\Phi^{P-M}_{pm} \right]^{m=1,\dots,nm}_{p=1,\dots,np} \quad (12.1)$$

where nm, np, ns, and nu are the numbers of, respectively, possible materials, processes, shape typologies, and uses. Considering processes of primary manufacture only, on the basis of the correlation matrices (12.1) and vectors V^{Sh} and V^{Us}, and following the calculation scheme summarized in Figure 12.4, it is possible to obtain the vectors V^{Pr} and V^{Mt}, indicating, respectively, the primary processes able to produce the required typology of shape, and the materials suitable for the intended use. The subsequent application of the material–process correlation matrix gives a matrix of producible solutions:

$$\Omega = \left[\omega_{pm} \right]^{m=1,\dots,nm}_{p=1,\dots,np} \quad \text{where } \omega_{pm} = \omega_{pm}\left(V^{Sh}, V^{Us}, \Phi^{S-P}, \Phi^{U-M}, \Phi^{P-M} \right) \quad (12.2)$$

This matrix indicates all the pairs of material–primary process that constitute the set of producible solutions.

The material–use correlation matrix constitutes a filter in the preselection of possible solutions in that it limits the choice to those materials conventionally employed for the intended use. For a broader preselection, it is possible to bypass this filter. In this case, the terms of matrix (12.2) would depend solely on V^{Sh}, Φ^{S-P}, and Φ^{P-M}.

Using the above approach in the analysis of production feasibility, it is possible to:

- Produce an analytical and exhaustive selection of all the possible solutions that can satisfy the intended form and use.

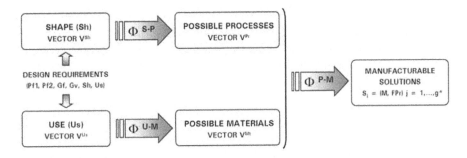

FIGURE 12.4 Summary of production feasibility analysis.

- Separate the selection conditioned by production feasibility from that conditioned by performance requirements, thereby evidencing the relationships between choice of material and effect on life cycle impacts; such relationships, as shown below, depend on the different performance capacities of the materials.

This approach requires the prior compilation of the correlation matrices (12.1). Given the ever-greater variety of engineering materials and related manufacturing processes, it is reasonable to consider compiling these matrices by typology of material. Alternatively, for a first selection of material–process pairs, it is possible to use existing software tools such as CES, which implements Ashby's methodology. It must be remembered, however, that tools of this type allow a selection that already takes account of the performances required.

12.5   Analysis of Performance

The second stage of the selection procedure identifies producible solutions that respect the required performance characteristics. In this way a set of potential solutions is obtained, which are then analyzed by applying the calculation models to evaluate their environmental and economic impacts over the entire life cycle.

In general, the analysis of performance can be simplified by considering three different typologies of mathematical relations:

- Function of performance volume (PfV)—Expresses the minimum volume necessary to meet the primary performance requirements. Generally, it is a function of the primary performance (Pf1), the geometric parameters (Gf, Gv), and the properties of the material (MtPp):

$$PfV = PfV(Pf1, Gf, Gv, MtPp) \qquad (12.3)$$

- Geometric conditions of performance—If the variable geometric parameters Gv are directly correlated with primary performance Pf1, the geometric conditions of performance can be expressed by functions constrained by a range of values (defined by the design requirements):

$$Gv = Gv\ (Pf1, Gf, MtPp) \qquad Gv \in (Gv_1, Gv_2) \qquad (12.4)$$

- Secondary conditions of performance—Conditions of this type can be generally expressed using functions dependent on the properties of the materials and the performance volume, to be compared with assumable limit values:

$$Pf2 = Pf2 \left(PfV, MtPp\right) \qquad Pf2 \leq \geq Pf2^{LIM} \qquad (12.5)$$

In conclusion, if a producible solution meets all the performance constraints and requirements, it then becomes a performing solution and can be selected for final evaluation. As shown in Figure 12.3, the set of potential solutions consists of all the performing material–primary process pairs, integrated by the corresponding performance volume. The latter parameter acquires particular relevance in the proposed method because it directly conditions the values assumed by the life cycle indicators which, defined below, guide the optimal choice. Using this approach, it is possible to correlate the search for environmentally and economically convenient solutions with the performance characteristics of the materials.

Only in the case of particularly simple design problems can the functions of type (12.3) be defined in analytical form (Giudice et al., 2001). More generally, the performance volume cannot be explicitly ascribed to the factors affecting it; it is the result of design procedures employing modern methods of engineering design, implemented in commonly used tools based on parametric CAD and FEM software for structural performance analyses.

12.6 Life Cycle Indicators

The final phases of the selection method consist of applying the calculation models to the set of potential solutions, evaluating the indicators of environmental impact and cost relative to the entire life cycle (Life Cycle Indicators), and then analyzing the results and identifying the optimal choice. The indicators are functions of the quantities of material necessary to produce the component, expressed by the performance volume.

12.6.1 Environmental Impact Functions

The Environmental Impact of the Life Cycle (EI_{LC}) is expressed by:

$$EI_{LC} = EI_{Mat} + EI_{Mfct} + EI_{Use} + EI_{EoL} \qquad (12.6)$$

where EI_{Mat} is the environmental impact of the material needed to produce the component; EI_{Mfct} is the impact associated with its manufacture; EI_{Use} is the impact related to the entire phase of use (which can depend on the choice of material); and EI_{EoL} is the impact of the end-of-life (recycling, disposal).

The first two terms of Equation (12.6) constitute the Environmental Impact of Production (EI_{Prod}), which can be expressed by:

$$EI_{Prod} = EI_{Mat} + EI_{Mfct} = ei_{Mat} \cdot W + ei_{Prss} \cdot \mu \left(+ ei_{Mchg} \cdot \eta \right) \qquad (12.7)$$

where ei_{Mat} is the eco-indicator per unit weight of material (expressed by W); ei_{Pcss} is the eco-indicator of the primary forming process per unit of μ, which can represent the characteristic parameter of the process or the quantity of material processed; and ei_{Mchg} is the eco-indicator of the secondary machining process per unit of characteristic parameter of process η. As mentioned above, these eco-indicators can be evaluated using the Eco-indicator 99 method.

The Environmental Impact of End-of-Life (EI_{EoL}) can be expressed by:

$$EI_{EoL} = ei_{Dsp} \cdot \left(1 - \xi \right) \cdot W + ei_{Rcl} \cdot \xi \cdot W \qquad (12.8)$$

where ei_{Dsp} and ei_{Rcl} are, respectively, the environmental impact of disposal and of recycling processes per unit of weight of material (ei_{Rcl} generally includes a quota of environmental impact recovered), and ξ is the recyclable fraction. So defined, Equation (12.8) refers to the optimal condition where, at the end-of-life, all of the recyclable fraction of material is recovered. Considering a more realistic scenario, it is possible to introduce an appropriate coefficient of reduced recyclability to obtain the fraction actually recycled.

Finally, the Environmental Impact of Use (EI_{Use}) cannot be expressed in general terms and must be defined each time, according to the specific case under examination. In this chapter, it will be defined in relation to the particular case study discussed below.

12.6.2   Cost Functions

Similar to the first life cycle indicator, which quantifies the environmental impact, the second life cycle indicator quantifies the economic cost related to the entire life cycle. Hypothesizing that both production and disposal costs are paid by a single entity (the manufacturer), the Cost of the Life Cycle (C_{LC}) can be expressed as:

$$C_{LC} = C_{Prod} + C_{EoL} \qquad (12.9)$$

The Cost of Production (C_{Prod}) can be expressed in a form analogous to Equation (12.7), as a function of the quantity of material to be employed and of the more significant process parameters. Alternatively, it is possible to use a conventional evaluation of the production costs of a component, distinguishing between variable and fixed costs and dividing the latter by the size of the production batch (Ulrich and Eppinger, 2000).

The Cost of End-of-Life (C_{EoL}) can be expressed as:

$$C_{EoL} = c_{Dsp} \cdot (1 - \xi) \cdot W + (c_{Rcl} - r_{Rcl}) \cdot \xi \cdot W \qquad (12.10)$$

where c_{Dsp}, c_{Rcl}, and r_{Rcl} are, respectively, the cost of disposal, the cost of recycling processes, and the proceeds from the sale of recycled material per unit weight of the material; ξ is the recyclable fraction.

12.7 Analysis of Results and Optimal Choice

By applying these models, the life cycle indicators (EI_{LC}, C_{LC}) are calculated for each potential solution. Various tools can be used to evaluate the fitness of each solution in order to identify the optimal choice. Two tools that are particularly simple but significant in terms of the proposed method are described below. More sophisticated tools are discussed in references to multiobjective optimization in general (Sawaragi et al., 1985), and in relation to the specific case of materials selection (Ashby, 2000).

12.7.1 Graphic Tools

Graphs of C_{LC}–EI_{LC} can clearly visualize the different fitness of the potential solutions. Graphic tools are particularly useful when a large number of solutions must be compared; an example is shown in Figure 12.5.

12.7.2 Multiobjective Analysis

In its simple form, multiobjective analysis is the analysis of a multiobjective function γ, which includes the more significant product properties, suitably normalized and weighted:

$$\gamma = \sum_{q=1}^{nq} \alpha_q \cdot B_q \qquad (12.11)$$

As already suggested for the comparison of alternative solutions in the problem of choice of materials (Farag, 2002), the following expression can be used

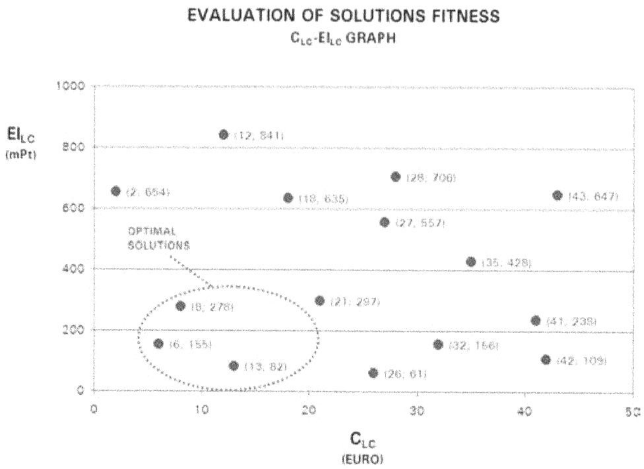

EVALUATION OF SOLUTIONS FITNESS
C_{LC}-El_{LC} GRAPH

FIGURE 12.5 Evaluation of solution fitness: C_{LC}–El_{LC} graph.

to calculate the normalized values B_q of the properties (in cases where the multiobjective function and all properties are to be minimized):

$$B_q = \frac{V_q}{V\max_q} \tag{12.12}$$

where V_q is the value assumed by the q-th property for the solution under examination and $V\max_q$ is the maximum value assumed by the q-th property among all the solutions to be compared. A set of B_q coefficients is obtained for each of the potential solutions to be evaluated. The optimal solution is that with the minimum value of the function γ.

12.8   Case Study: Selection of Material for an Automobile Brake Disk

The following case study illustrates the application of this method of selection and choice of materials and of the supporting calculation models. The design problem consists of the optimum choice for the material of an automobile brake disk, depicted in Figure 12.6.

A preliminary meaningful case study was conducted on a simpler design problem, the optimal polymeric material selection for a piping component (Giudice et al., 2001).

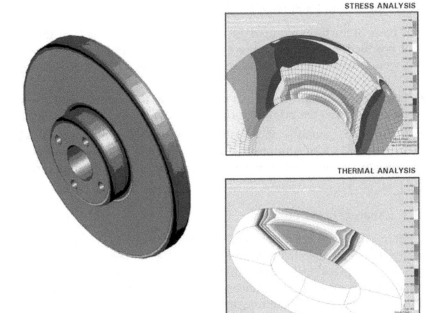

STRESS ANALYSIS

THERMAL ANALYSIS

FIGURE 12.6 Case study: Automobile brake disk.

12.8.1 Definition of Design Requirements

The first phase of this method is the definition of the set of design requirements:

- Primary performance required (Pf1) is that of ensuring, in relation to a reference condition of vehicle movement, efficient braking within a given distance. In physical–mechanical terms, this translates into the dissipation of energy through friction and structural performance correlated with the mechanical and thermal loading conditions (stress–strain analysis).
- Secondary performance required (Pf2) is that of limiting the weight W.
- Fixed geometric constraint (Gf) is the external radius of the disk R_e.
- Variable geometric parameters (Gv) are the thickness s and internal radius of the disk R_i.
- Shape required (Sh) is a three-dimensional rotation solid.

12.8.2   Analysis of Production Feasibility

On the basis of the form required (Sh) and of the expected use (Us), the analysis of production feasibility suggests some hypothetical solutions, two of which were considered (one conventional and one of recent introduction):

- Solution S_1 consists of grey cast iron BS 350 as the material, and green sand casting as the primary forming process.
- Solution S_2 consists of an aluminum matrix compound (F3K20S Duralcan®, Alcan Aluminum Ltd., San Diego, CA) as the material, and squeeze casting (liquid metal forging) as the primary forming process.

12.8.3   Analysis of Performance

By defining the weight of the automobile and imposing the required braking capacity, it was possible to determine the braking moment required on each wheel and the pressures at the disk–pad contact necessary to produce this moment. The primary performance was thus translated into the following conditions of correct functioning that must be ensured by the thermal–mechanical characteristics of the material:

- Thermal peaks below the maximum operating temperature of the materials
- Global stress state (due to superimposition of mechanical and thermal loading) below the mechanic resistance limits of the materials
- Global strain state (due to the superimposition of mechanical and thermal loading) within the elastic limit of the materials

Given the complexity of the problem, the performance analysis was conducted using the finite element software MSC Patran/Nastran® (MSC Software Corporation, Santa Ana, CA), which allowed the correlation of performance properties of the materials, variable geometric parameters, and the corresponding structural and thermal loading. As an example, Figure 12.6 shows some results of the stress and thermal analyses on the disk. These FEM analyses were calibrated on the basis of experimental data available in the literature (Bassignana et al., 1984; Brembo, 1998). Both of the producible solutions under examination were found to function. Table 12.1 shows the values that the performance volume PfV and the variable geometric parameters (thickness s and internal radius R_i) must assume in order to ensure the performance, together with the corresponding weights.

Comparing the two solutions under examination, the Duralcan option requires greater performance volume PfV (and therefore larger overall dimensions) to ensure primary performance. The conventional solution in

TABLE 12.1 Performance volume, weight, and variable geometric parameters

	PfV (dm³)	W (kg)	s (mm)	R_i (mm)
BS 350	0.82	6.00	25	110
F3K20S	1.36	3.83	30	90

cast iron reduces the overall dimensions but results in a greater weight (+56% compared to Duralcan).

12.8.4 Evaluation of Life Cycle Indicators and Analysis of Results

Equations (12.6) and (12.9) were used to calculate the indicators of environmental impact and cost for each performing solution. The results of the calculation models are reported in Table 12.2. The general models were simplified as follows:

- In the calculation of production impacts and costs, only the primary manufacturing processes were considered; secondary processes were ignored.
- In the evaluation of Equation (12.9), the end-of-life costs expressed by Equation (12.10) were ignored because of the difficulty of obtaining the relevant data. Thus, only the cost of production C_{Prod} was considered as the cost indicator.
- In this first phase, Equation (12.6) was evaluated ignoring the environmental impact related to use of the product.

From the values in Table 12.2, it is clear that the Duralcan solution leads to an impact (2272.2 mPt) two orders of magnitude greater than that of the solution in cast iron (43.5 mPt). The graph shown in Figure 12.7 describes the composition of the environmental indicator EI_{LC}; it is particularly interesting in that it demonstrates the different distributions of the environmental impact over the life cycle for each potential solution.

Comparing the two solutions, it is evident that:

- The overall impact of the Duralcan solution is essentially due to the impact of producing the material itself, which also offers a negligible recycling fraction (low recovery of impact).
- The solution in cast iron has a much lower production impact and, furthermore, its high recyclability allows a substantial recovery of impact at end-of-life.

TABLE 12.2   Results of the evaluation of Life Cycle Indicators

	LIFE CYCLE INDICATORS					
	EI_{PROD} (mPt)	EI_{EOL} (mPt)	C_{MAT} (EURO)	C_{PRSS} (EURO)	EI_{LC} (mPt)	C_{PROD} (EURO)
BS 350	208.9	−165.4	6.59	14.70	43.5	21.29
F3K20S	2293.3	−21.1	18.94	27.52	2272.2	46.46

COMPOSITION OF LIFE CYCLE INDICATORS
ENVIRONMENTAL IMPACT INDICATOR EI_{LC}

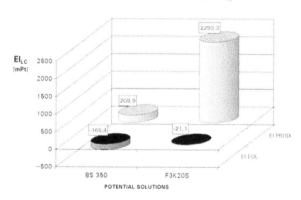

FIGURE 12.7   Composition of indicator EI_{LC} in relation to phases of life cycle.

The values reported in Table 12.2 clearly indicate which of the two solutions is more favorable—the conventional solution in cast iron—since it results in the lowest values of both C_{Prod} and EI_{LC}.

In conclusion, it is evident that when the properties considered most important for the final product are those of reduced cost and environmental impact of the life cycle, the best solution is that in cast iron. The alternative solution in Duralcan is favorable only when light weight is chosen as the primary property.

This is confirmed by applying the multiobjective analysis method introduced in Section 12.7. Considering EI_{LC}, C_{Prod}, weight W, and performance volume PfV as objective functions, different values of the function to minimize γ are obtained for the two alternative solutions according to how the set of weight coefficients α_i is defined. Figure 12.8 shows the results for four different orientations of investigation, corresponding to the different emphases given to the objective functions in the evaluation of γ:

- 1—Maximum importance given to environmental impact, medium to cost, low to W and PfV reductions

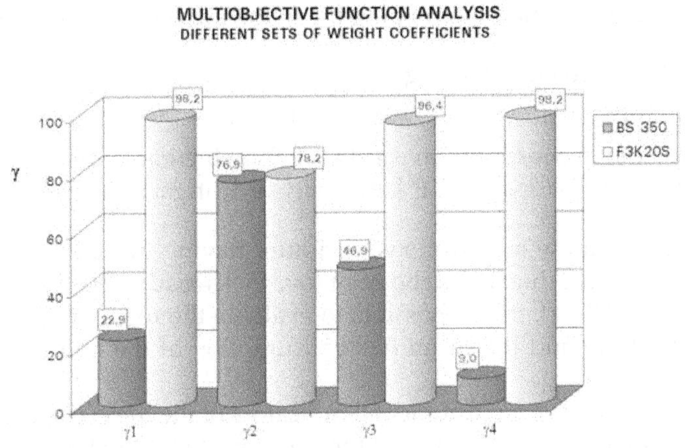

MULTIOBJECTIVE FUNCTION ANALYSIS
DIFFERENT SETS OF WEIGHT COEFFICIENTS

FIGURE 12.8   Study of multiobjective function γ.

- 2—Maximum importance given to W reduction, medium to cost, low to environmental impact and PfV reductions
- 3—Primary reduction of cost
- 4—Primary reduction of environmental impact

It can be seen that, compared to the solution in cast iron, the solution in Duralcan is interesting only in the second case.

12.8.5   Introduction of Environmental Impact of Use: Evaluation of Life Cycle Indicators and Analysis of Results

The Duralcan solution has the primary advantage of reducing the weight of the disk. The consequent lightening of the vehicle can result in a sufficient reduction in the environmental impact of use to recover the increased impact in production. To evaluate whether (and under what conditions) this is true, it is necessary to evaluate the term EI_{Use} in Equation (12.6), which was ignored previously. Apart from this, all the other simplifications introduced in Section 12.8.4 remain the same. Having established an overall reference distance traveled (mission), the environmental impact of use EI_{Use} can be expressed as:

$$EI_{Use} = EI_{Fuel} + EI_{Mission} = ei_{Fuel} \cdot q_{Fuel} + ei_{Mission} \cdot q_{Mission} \qquad (12.13)$$

where ei_{Fuel} is the eco-indicator per unit weight of fuel; $ei_{Mission}$ is the eco-indicator associated with the use of the vehicle powered with this kind of fuel per unit of distance traveled; q_{Fuel} is the quantity of fuel needed for the entire distance covered; and $q_{Mission}$ is the total expected distance.

To evaluate all the quantities in play, the following assumptions were made:

- Weight of vehicle = 1000 kg
- Mean fuel consumption = 0.085 L/km
- Reduction in consumption due to a 10% reduction in total weight of vehicle = 4.5% (*Source:* IKP, University of Stüttgart, Germany)

On the basis of these assumptions, and after having evaluated the overall reduction in weight due to the choice of four disks made of Duralcan instead of cast iron (−8.7 kg), it was possible to evaluate the reduced weight of the vehicle (991.3 kg) and the mean fuel consumption of the lightened vehicle (0.0847 L/kg).

Table 12.3 shows the environmental indicators of the life cycle, considering the environmental impact of use for an expected traveling distance of 150,000 km. It is clear that the solution in Duralcan results in a lower environmental impact than that in cast iron, in terms of both the phase of use alone (−0.4%) and the entire life cycle (−0.3%).

The percentage reduction in EI_{LC} also depends on the expected distance traveled. The graph in Figure 12.9 shows, for the two solutions, the break-even point of EI_{LC}. This represents the minimum distance that must be traveled for the EI_{LC} corresponding to the solution in Duralcan to be less than the EI_{LC} of the solution in cast iron (about 31,300 km).

The graph in Figure 12.10 describes the new composition of the environmental indicator EI_{LC} in relation to the different phases of the life cycle for each potential solution (distance traveled = 150,000 km). Because of the different orders of magnitude, the components regarding the phase of use are shown in Pt rather than in mPt.

In conclusion, when the environmental impact relating to the phase of use (influenced by the vehicle weight) is also taken into account, the solution in Duralcan is advantageous not only when lightness is chosen as the primary property, but also when the environmental impact of the entire life cycle is considered; this advantage becomes apparent after a minimum distance traveled of approximately 31,000 km.

TABLE 12.3 Results of the evaluation of Life Cycle Indicators (including phase of use)

	LIFE CYCLE INDICATORS			
	EI_{PROD} (mPt)	EI_{USE} (mPt)	EI_{EOL} (mPt)	EI_{LC} (mPt)
BS 350	208.9	2729884	−165.4	2729927
F3K20S	2293.3	2719201	−21.1	2721479

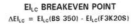

EI$_{LC}$ BREAKEVEN POINT

ΔEI$_{LC}$ = EI$_{LC}$(BS 350) - EI$_{LC}$(F3K20S)

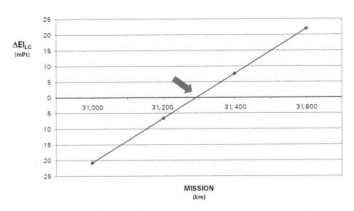

FIGURE 12.9   Breakeven point of EI$_{LC}$ for the two solutions.

COMPOSITION OF LIFE CYCLE INDICATORS
ENVIRONMENTAL IMPACT INDICATOR EI$_{LC}$

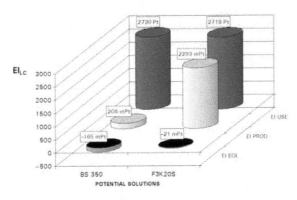

FIGURE 12.10   Composition of indicator EI$_{LC}$ in relation to phases of life cycle (including use).

This consideration is again confirmed by applying the multiobjective analysis method. Considering EI$_{LC}$ (which now also includes EI$_{Use}$, calculated for the reference distance of 150,000 km), C$_{Prod}$, the weight W, and the performance volume PfV as objective functions the results shown in Figure 12.11 are obtained for the four different investigation orientations described in Section 12.8.4. It can be seen that, again, the solution in cast iron is better for the first (maximum importance to environmental impact, medium to cost) and third (primary reduction of cost) investigation typologies, while that in Duralcan is better for the second one (maximum importance to weight reduction).

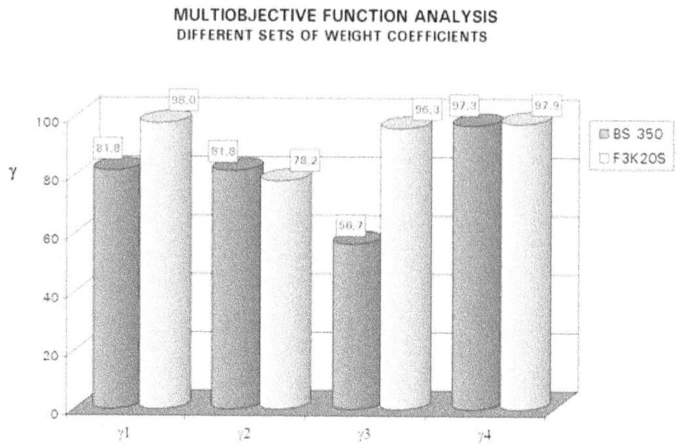

MULTIOBJECTIVE FUNCTION ANALYSIS
DIFFERENT SETS OF WEIGHT COEFFICIENTS

FIGURE 12.11   Study of multiobjective function γ (including use).

However, for the last investigation typology (directed at reducing primarily the environmental impact) the two solutions are essentially equivalent, while for distances over 150,000 km the solution in Duralcan tends to be more advantageous than that in cast iron (since the lower value of EI_{LC} due to the reduced weight tends to increase with the distance traveled).

12.9   Acknowledgments

The main contents of this chapter were previously published (Giudice, F., La Rosa, G., and Risitano, A., Materials selection in the life-cycle design process: A method to integrate mechanical and environmental performances in optimal choice, *Materials and Design*, 26[1], 9–20, 2005), and are reproduced with permission from Elsevier.

12.10   Summary

The proposed selection procedure elaborates data (both conventional and environmental) regarding the properties of materials and processes. It relates this data to the performance requirements demanded of the product and calculates the values assumed by functions that quantify the environmental impact over the entire life cycle, including the phases of use and retirement, and the costs resulting from the choice of materials.

A complete application of this method in the design of an automobile component allowed a direct comparison between the optimal choice made after a multiobjective analysis and that obtained in a conventional design approach. This experience demonstrated the need to use new tools in order to ensure, in the design phase, environmental safeguards in the development of industrial products, and the possibility of fully integrating such new tools with conventional design tools.

12.11 References

Ashby, M.F., *Materials Selection in Mechanical Design*, 2nd ed., Butterworth-Heinemann, Oxford, UK, 1999.

Ashby, M.F., Multi-objective optimization in material design and selection, *Acta Materialia*, 48(1), 359–369, 2000.

Ashby, M.F., Drivers for material development in the 21st century, *Progress in Materials Science*, 46, 191–199, 2001.

Ashby, M.F. and Cebon, D.A., Compilation of material indices, Granta Design Ltd., Cambridge, UK, 1995.

Ashby, M.F. et al., Selection strategies for materials and processes, *Materials and Design*, 25, 51–67, 2004.

Bassignana, P., Gavello, E., and Leonti, S., Analisi termica, termoelastica e meccanica di freni automobilistici, *ATA Ingegneria Automotoristica*, 37(11), 727–740, 1984.

Brechet, Y. et al., Challenges in materials and process selection, *Progress in Materials Science*, 46, 407–428, 2001.

Brembo, *Il Manuale del Disco Freno*, Giorgio Nada Editore, Milano, Italy, 1998.

Charles, J.A., Crane, F.A.A., and Furness, J.A.G., *Selection and Use of Engineering Materials*, 3rd ed., Butterworth-Heinemann, Oxford, UK, 1997.

Chen, R.W. et al., A systematic methodology of material selection with environmental considerations, in *Proceedings of IEEE International Symposium on Electronics and the Environment*, San Francisco, 1994, 252–260.

Edwards, K.L., Designing of engineering components for optimal materials and manufacturing process utilisation, *Materials and Design*, 24, 355–366, 2003.

Farag, M.M., *Materials Selection for Engineering Design*, Prentice Hall Europe, Hemel Hempstead, UK, 1997.

Farag, M.M., Quantitative methods of materials selection, in *Handbook of Materials Selection*, Kutz, M., Ed., John Wiley & Sons, New York, 2002, chap. 1.

Giudice, F., La Rosa, G., and Risitano, A., Optimal material selection in the design for environment process: Environmental characterisation of polymeric materials and a methodology of selection, in *Proceedings of ENTREE 2001—Environmental Training in Engineering Education*, Florence, Italy, 2001, 215–234.

Holloway, L., Materials selection for optimal environmental impact in mechanical design, *Materials & Design*, 19(4), 133–143, 1998.

Kampe, S.L., Incorporating green engineering in materials selection and design, in *Proceedings of the 2001 Green Engineering Conference: Sustainable and Environmentally Conscious Engineering*, Roanoke, VA, 2001, 7/1–7/6.

Lu, W.F. and Deng, Y.-M., A system modeling methodology for materials and engineering systems design integration, *Materials and Design*, 25, 459–469, 2004.

Navin-Chandra, D., Design for environmentability, in *Proceedings of ASME Design Theory and Methodology Conference*, Miami, FL, 1991, DE-31, 119–125.

Pré, SimaPro 5 Database Manual: The BUWAL 250 Library, Pré Consultants BV, Amersfoort, The Netherlands, 2003.

Sawaragi, Y., Nakayama, H., and Tanino, T., *Theory of Multiobjective Optimisation*, Academic Press, Orlando, FL, 1985.

Stuart, J.A., Materials selection for life cycle design, in *Proceedings of IEEE International Symposium on Electronics and the Environment*, Oak Brook, IL, 1998, 151–158.

Ulrich, K.T. and Eppinger, S.D., *Product Design and Development*, 2nd ed., McGraw-Hill, New York, 2000.

Wegst, U.G.K. and Ashby, M.F., The development and use of a methodology for the environmentally conscious selection of materials, in *Proceedings of 3rd Biennial World Conference on Integrated Design and Process Technology (IDPT)*, Berlin, 1998, 5, 88–93.

Chapter 13

Design for Disassembly and Distribution of Disassembly Depth

The disassembly of products is necessary whenever it is opportune to conduct servicing operations or to recover parts and materials at end-of-life. Design for Disassembly, therefore, has strategic value in the context of design activities oriented toward the environmental performance of a product. The objective of this design approach is to optimize the architecture of products and the characteristics of components in relation to the following aims: recover parts, components, and subassemblies reusable in new products; recover recyclable materials; access parts or components that may be subject to servicing operations (repair, maintenance, diagnostics); and limit disassembly time and costs.

After an overview of the main issues related to the definition of optimal disassembly level and the design approach to ease of disassembly, this chapter presents a method to analyze the distribution of the disassembly depth of the components comprising a constructional system (i.e., the difficulty of their disassembly based on the geometric and junction constraints conditioning their extraction and removal). The application of this method provides information that aids redesign actions directed at facilitating product disassembly. If suitably integrated into the design process, the method proposed can be used to optimize the main design choices (layout, geometries, junctions). This optimization consists of a redistribution of the disassembly depth, guided by the component properties themselves, in a way that improves the ease of disassembly of those parts of the system that may require frequent servicing interventions or that present heightened opportunities for recovery. The fundamental issues in this chapter were previously introduced in Chapter 9.

13.1 Design for Disassembly and Disassembly Level

A product must be disassembled whenever it is necessary to remove subassemblies or single components from that product. As previously discussed

(Chapter 9, Section 9.3.3.1), "disassembly" can be defined as the systematic removal of the required parts from an assembly, with the condition that the disassembly process does not cause any damage to the parts (Brennan et al., 1994).

Of the various themes treated in the field of Design for Environment, the optimal disassembly of constructional systems plays a strategic role, since it can affect both the phase of use (facilitating servicing operations—maintenance and repair) and the end-of-life phase (favoring operations to recycle materials and reuse components).

In general, improving the disassembly process of a product can be achieved by intervening at two different levels (Jovane et al., 1993, Gungor and Gupta, 1999):

- In the design phase, making choices that can favor the disassembly of the constructional system (in this case, this is Design for Disassembly—DFD).
- Attempting to best plan and optimize the disassembly process (in this case, this is Disassembly Process Planning—DPP).

Design for Disassembly can, therefore, be defined as a design approach wherein the objective is to optimize the architecture and all other constructional characteristics of a product in relation to the following main requirements:

- The simple and rapid separability of parts to be serviced or recovered
- Limiting the time and costs of disassembly

13.1.1 Design Approaches to Ease of Disassembly

In general terms, Design for Disassembly (DFD) is defined as an aid to product design such that high percentages of the parts making up the product can be reused or recycled (Zang et al., 1997). Entering more into the specifics of design practice, DFD arose as the natural extension and integration of Design for Assembly (DFA) (Boothroyd and Alting, 1992; Scheuring et al., 1994); consequently, it is understood as a design practice that uses assembly methods and configurations that allow the separation and recovery of components and materials in an economically efficient manner (Billatos and Basaly, 1997).

The fact that the concept of DFD itself arose from the need to limit the environmental impact of a product at the end of its life, has ensured that the more significant studies in this area were conducted in relation to the opportunities for recovery in products destined for disposal. In this context, complete methods (as well as computer-based methods) analyze the various typologies of subassemblies and junctions and their accessibility, and evaluate how far to go in the process of disassembly, what to recover, and what to dispose of as waste (Navin-Chandra, 1993; Johnson and Wang, 1995). Some of these methods attempt to exploit the full potential of the design action, giving particular emphasis to the variety of choices that can influence the recovery phase at end-of-life (Ishii et al., 1994), beginning with the choice of materials compatible with the aims of recycling (Di Marco et al., 1994). Many of these methods also developed as extensions to the useful potential of techniques for planning the disassembly process (Johnson and Wang, 1995; Lambert, 1997; Srinivasan et al., 1997), demonstrating the close connection between the two levels of approach to the problem of disassembly—DFD and DPP.

The recently heightened need to integrate environmental requirements into design practice has resulted in the tools of DFD developing in accordance with new approaches to product design and development. A significant study in this context was based on the application of Axiomatic Design for the development of integrated guidelines targeted at DFD. Axiomatic design allows, first, the generation of acceptable design solutions and, subsequently, the selection of the best of these (Chen, 2001).

13.1.1.1 Metrics for Design

The primacy of the economic aspect of disassembly and recovery is evidenced by most of the studies performed, and sometimes this factor becomes the main metric underlying the development of a design tool (Chen et al., 1994). Other alternative metrics that can be used to quantify the ease of disassembly of products, and as indicators of the excellence of the design intervention, are based on the estimated disassembly time (Kroll and Carver, 1999) or on more complex indicators taking into account the various aspects of disassembly operations (time required, use of tools, accessibility, force required, and others) (Das et al., 2000). This demonstrates the wide variety of factors affecting the process of disassembly.

Less common, instead, are studies into the optimization of disassembly, attempting to harmonize the indications of the metrics mentioned above with specific indicators of environmental impact. This is the case, for example, of tools that investigate the design alternatives with the aim of finding a compromise between the costs of disassembly and the environmental benefits of the recovery process at end-of-life, quantified by a single indicator that takes into account the impoverishment of resources, energy consumption and toxic emissions (Harjula et al., 1996). Studies taking such a complete approach to the environmental problem underscore how a reduction in the environmental impact can only be obtained through the implementation of additional design expedients.

13.1.1.2   *Orientations of the Design Intervention*

Any tool for aiding the designer in an efficient intervention of Design for Disassembly must be directed at harmonizing the layout, component geometries, materials, and junction systems, in relation to the ease of product disassembly. In the majority of cases, this translates into limiting the costs of disassembly. The goals of such a design intervention must not, however, be limited to only considering the requirements of the end-of-life phase. From the standpoint of designing for the entire life cycle, a statement of a DFD problem that is exclusively oriented toward the product's end-of-life must be extended to acknowledge the importance that disassembly has in relation to servicing operations (Dewhurst, 1992; Vujosevic et al., 1995; Sodhi et al., 2004). This suggests that in the design approach a substantial distinction must be established between DFD oriented toward the requirements of servicing and maintenance and DFD oriented toward recovery at the product's end-of-life:

- In the context of servicing and maintenance, the objective is to make design choices that most efficiently simplify the accessibility and ease of disassembly of certain predetermined components that require servicing interventions.

- In the context of recovery at the product's end-of-life, the design choices must be made in such a way that they allow the most efficient recovery at end of life that is possible without knowing a priori which components it will be appropriate to remove and which to eliminate as waste.

This distinction leads to a discrimination commonly made in differentiating the various typologies of disassembly:

- Selective disassembly, consisting of the disassembly of certain preset components

- Complete or partial disassembly, where the components to be disassembled are not predetermined but must be evaluated on the basis of (generally economic) expediency; this often limits the disassembly to a partial level rather than encouraging the complete disassembly of the constructional system

This last distinction introduces another aspect, specific to disassembly aimed at recovery at end-of-life, that deserves attention: the importance of determining the most appropriate disassembly level (Disassembly Leveling)—how far to go in the disassembly process on a constructional system in order to recover components and materials (Gungor and Gupta, 1999).

13.1.2   Optimal Disassembly Level

The costs of disassembly are proportional to the time required by the disassembly operations and to the percentage of components removed (Zhang et al., 1997). Therefore, even if complete disassembly can represent the ideal condition offering the greatest environmental benefit, it may not be economically sustainable since it generally involves disassembly costs much greater than the financial revenues generated. This is conditioned by market laws as well as by technological limitations in the recovery and recycling processes (Gungor and Gupta, 1999).

The problem of defining the optimal disassembly level (Disassembly Leveling Problem) thus consists of seeking a condition of equilibrium between the resources invested in the disassembly process and the resulting benefits. The equilibrium is generally determined using instruments based on cost analysis. This is the case in approaches that, for example, identify the optimal disassembly level and sequences by investigating alternative disassembly strategies for each subassembly on the basis of economic evaluations (Penev and de Ron, 1996).

Other authors have studied the disassembly level problem in relation to the possible advantages of product redesign, by analyzing the different options at end-of-life in order to minimize the waste for disposal and to maximize the economic benefits of recycling. Such analyses can identify the product's weak points that can be corrected with appropriate modifications to the design (Navin-Chandra, 1994; Zussman et al., 1994; Pnueli and Zussman, 1997).

In summary, the problem of product disassembly at end-of-life can be related to the problems of recovery that were introduced in Chapter 9, Section 9.3.2: Given a product, determine the recovery plan that can efficiently balance its costs of disassembly and recovery processes with the resulting profits, stated in terms of resources employed and recovered (generally economic resources). The literature contains approximate evaluations on curves of the costs, revenues, and profits of disassembly and recovery interventions. For example, Figure 9.5 in Chapter 9 combines the costs of disassembly with those relating to testing, to any possible remanufacturing, and to design expedients aimed at facilitating recovery. According to such evaluations, the recovery costs become prohibitive with increasing recovery and disassembly levels. On the contrary, beyond a certain disassembly level the revenues tend to stabilize and, consequently, the profit curve has a maximum (optimal disassembly level); after that, it tends to decrease as the disassembly level increases.

Theoretically, therefore, the optimization of the product's architecture and components functional to disassembly consists of making design choices that move the peak of the profit curve toward a greater disassembly level and increase the profit value at the peak.

13.2 Distribution of Disassembly Depth

The recovery profit function in Figure 9.5 is sensitive to the efficiency of the disassembly sequence, that, if optimized, would allow costs and times to be reduced. This is thus the first intervention factor necessary to obtain an advantageous disassembly, and is indirectly conditioned by the product's architecture, which can favor the efficiency of the disassembly sequence. The theme of optimal disassembly planning will be treated in the following chapter.

For the same efficiency of disassembly sequence, the profit can be greatly improved if the product architecture is such that the first disassembly actions free the parts of greatest value (Simon and Dowie, 1993). This consideration highlights the importance that product configuration can have for the goals of disassembly, since it is precisely this configuration that determines which parts of a constructional system can be removed at each stage of the disassembly process (Viswanathan and Allada, 2001).

Extending this vision to a design intervention taking into account not only the requirements of end-of-life but also those of servicing and maintenance, it is possible to say, more generally, that disassembly will be more efficient (i.e., reducing the cost of disassembly and, possibly, improving the profit of recovery) if the product's architecture is such that the first disassembly actions free:

- The parts that most frequently or most probably will need to be removed.
- The parts that must be removed, or that it may be most appropriate to remove, at end-of-life.

Again referring to the theoretical graph of Figure 9.5, the benefit of this criterion is evident:

- The components that must be removed, repaired, or substituted are readily accessible, with consequent reduction in disassembly costs and, therefore, in the overall economic cost associated with the corresponding servicing operations.
- The components characterized by environmental criticality, high economic value, or greater potential for recovery can be removed easily, with a consequent increase in the profit of recovery at end-of-life and the possibility of increasing the disassembly level.

These considerations evidence the direct dependence of the disassembly efficiency on the product architecture and, in particular, on how the difficulty of disassembling single components is distributed in the context of the entire constructional system. This difficulty, which can be called "disassembly depth," must be appropriate for each component in relation to the need or appropriateness of its disassembly. As will be shown below, disassembly need and appropriateness depend on the properties of duration, reliability, environmental impact, and economic value of each component.

13.3 Objectives and Approach to the Problem

With these premises, the approach proposed here is intended to define a method to analyze the disassembly depth of the components making up a constructional system (i.e., the difficulty of their disassembly on the basis of the geometric and junction constraints conditioning their extraction and removal). This method must elaborate data regarding the spatial configuration of the constructional system, the geometric characteristics of its constituent parts, and the relative junction systems. On the basis of this information, it must characterize the components in terms of disassembly difficulty. The final aim is to be able to use this method to:

- Evaluate the distribution of the disassembly difficulty of components or subsystems comprising a constructional system
- Aid product design directed at improving the ease of disassembly through a reasoned redistribution of disassembly depth, guided by certain characteristics of single components that indicate the necessity or opportuneness for disassembly (e.g., reliability, environmental impact, economic value)

13.4 Method Statement

Although the importance that the product configuration can have on the objectives of disassembly was emphasized in Section 13.2, it must be noted that an effective Design for Disassembly intervention requires an approach structured on several levels, each of which can be associated with diverse design requirements (Yamagiwa et al., 1999). These include study of the spatial and functional distribution of components study of the geometries and materials of parts, and study of the junction systems. A tool to aid a Design for Disassembly intervention must then harmonize the product layout, the geometries of the components, the materials, and the junction systems in relation to the ease of disassembly of the constructional system in question.

Starting from this presupposition and from the above considerations, to achieve the set target the method must be structured according to a procedure that interrelates:

- The characteristics defining each design solution to be analyzed (layout, geometries, materials, junctions)
- The objective properties of the components (i.e., those properties determining the necessity or opportuneness of their disassembly— duration and reliability, environmental impact, economic value)
- The disassembly depth of the components (quantified by the number of other components and junctions that must be removed in order to disassemble each component)

This analysis procedure is summarized in Figure 13.1. Having set the characteristics of the design solution to be analyzed, it is possible to calculate the disassembly depth and the objective properties of each component. These produce the distribution of the values assumed by appropriate functions that can quantify the efficiency of the design solution in relation to the need for disassembly. This analysis procedure can be used to

- Compare different design alternatives and identify the optimal
- Aid the improvement of a design solution

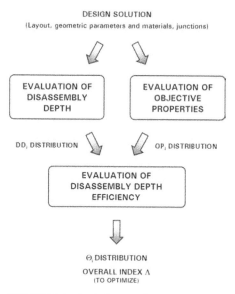

FIGURE 13.1   Analysis procedure.

13.5 Evaluation of Disassembly Depth

The distribution of the disassembly depth of components is evaluated using the procedure outlined in Figure 13.2. Given a constructional system (defined by layout, geometries of parts, and junctions) translated into a corresponding mathematical model through matrices describing spatial and junction constraints (possibly derived from a CAD model), the proposed method characterizes components or subassemblies comprising the constructional system on the basis of their disassembly difficulty, according to the direction of disassembly. From the results of the characterization, which indicates the components to be removed and the junctions to be eliminated, it is then possible to calculate a numerical index quantifying the disassembly depth. This index provides a theoretical evaluation of the disassembly depth, since it is based on an analysis of the components considered individually, as if each of them was, in turn, the single object of disassembly.

The accuracy of this evaluation further depends on the precision of the stage wherein the components are characterized on the basis of the difficulty of disassembly. The approach described below limits the problem to two dimensions, but can be extended to cover three-dimensional cases. With this approach it is possible to obtain an approximate evaluation of the maximum disassembly depth, since it is based on a simplified characterization that takes into consideration only one disassembly direction at a time, excluding the possibility of changing direction during disassembly. The conceptual limitations of this approach will be pointed out where appropriate, providing suggestions on how the precision of the evaluation given by the method may be improved.

13.5.1 Preliminary Modeling of the Constructional System

To represent the system in order to allow the analysis of disassembly depth, the modeling proposed here is a matrix type and is based on the geometric

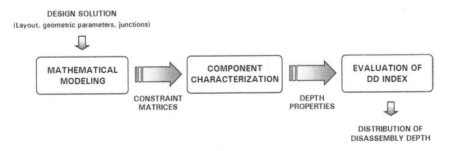

FIGURE 13.2 Calculation of disassembly depth.

characteristics of the system and on the consequent precedences of disassembly between its elements. Having translated the system compo- nents into objects with forms representing each of their characteristic overall dimensions, the problem of disassembling a component can be reduced to the problem of moving an object O in a spatial region R, avoiding any colli- sion with the other objects O_j present in the same region. This is part of a well-known class of geometric problems known as Spatial Planning Problems (Lozano-Perez, 1983). While the resolution of such problems is generally highly complex, the modeling proposed here is instead very simple, primarily because the analysis of component movement is limited to one direction at a time. This is inspired by geometric-type planning for selec- tive disassembly (Srinivasan and Gadh, 2002) and leads back, in part, to existing matrix models (Zhang and Kuo, 1997; Gungor and Gupta, 1998; Dini et al., 2001).

Given a constructional system composed of n components C_i, it is possible to compile a matrix of spatial constraints for each direction of disassembly based on the geometric parameters characterizing each component and on the resulting overall dimensions in the various directions:

$$A = \left[a_{ij} \right]_{i=1,2,\dots,n}^{j=1,2,\dots,n} \tag{13.1}$$

where the term a_{ij} is 1 if the disassembly of the i-th component in the direc- tion under examination requires the removal of the j-th component; otherwise it is 0.

Considering the constraints of the junctions that connect the components together, it is appropriate to make a distinction between a functional connec- tion (a liaison) and a true junction (a fastener) (Simon and Dowie, 1993). In general, a functional connection can be said to be produced using one or more fasteners. Consequently, for an approximate evaluation of the disas- sembly depth it is possible to limit the analysis to quantifying the functional connections between components. A more accurate evaluation would require a quantification of the fasteners and also distinguishing between them according to their typology (Das and Naik, 2002). With this premise, the matrix of junction constraints between components can be defined as:

$$B = \left[b_{ij} \right]_{i=1,2,\dots,n}^{j=1,2,\dots,n} \tag{13.2}$$

where the term b_{ij} is 1 if the i-th component is connected to the j-th compo- nent; otherwise it is 0.

For the analysis to consider true fasteners, it is necessary to introduce an additional matrix indicating which junctions connect each component. By indicating the total number of J_v junctions present in the constructional system with f, it is possible to define the matrix of junction distribution:

$$D = [d_{iv}]_{i=1,2,\dots,n}^{v=1,2,\dots,f} \qquad (13.3)$$

where the term d_{iv} is 1 if the v-th junction connects the i-th component; otherwise it is 0.

For an even more accurate analysis, it is necessary to consider the different typologies of junctions, and this is possible by compiling the matrix appropriately. By indicating the number of different junction typologies with h, the number of junctions of k-th typology present in the constructional complex with f_k, and the generic junction of k-th typology with J_v^k, the matrix (13.3) must be compiled by ordering the columns for junction typology, as shown in Figure 13.3.

13.5.2 Characterization of Components on the Basis of Disassembly Difficulty

The characterization of the generic component C_i on the basis of the difficulty of disassembly consists of determining the set {Cd} of components and the set {Jd} of fasteners to be removed in order to disassemble C_i. As shown above, the characterization method proposed here is directed at a theoretical evaluation of the disassembly depth. This characterization is based on two simplified presuppositions:

- The components are considered singly, as if each in turn was the only objective of the disassembly.

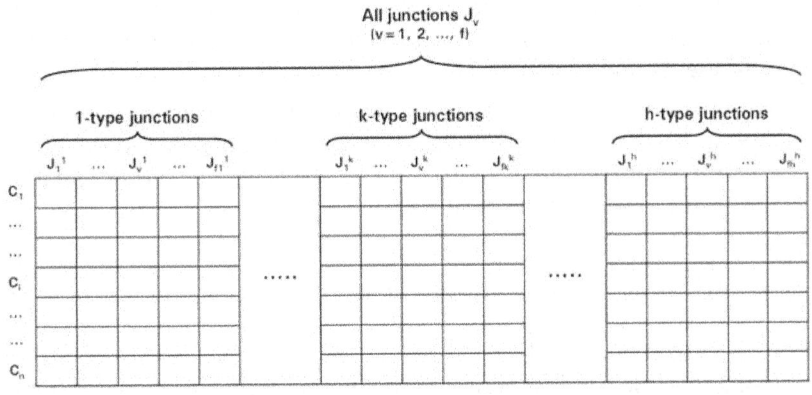

FIGURE 13.3 Junction distribution matrix: Columns ordered by junction typology.

- The analysis does not include the possibility of changing direction during disassembly; thus, each characterization is specific for a single direction of disassembly.

13.5.2.1 Procedure and Rules of Characterization

Having translated the constructional system into a mathematical model using matrices (13.1), (13.2), and (13.3), characterization is performed following the successive steps:

- Define the component to be characterized.
- Select the direction of disassembly.
- Apply the rules of characterization.

The rules of characterization must be applied in the order in which they are presented here. Divided into rules for spatial constraints and those for junction constraints, they allow the identification of, respectively, the sets {Cd} and {Jd} for the component under examination. Considering the generic component C_i as the component to be characterized, the rules of characterization, complete with the relevant analytical expressions, are presented below.

Characterization rules for spatial constraints:

- Rule 1—Remove the set {Cd}$_1$ of components directly obstructing the disassembly of C_i in the direction under examination.

$$R1: \{Cd\}_1 = \left\{ C_j, \forall j \text{ such that } a_{ij} = 1 \right\}$$

  This rule relates to the first level of spatial constraint to which component C_i is subjected (i.e., that constituted by components directly blocking its removal).

- Rule 2—Remove the set {Cd}$_2$ of components obstructing the disassembly of components belonging to the set {Cd}$_1$, in the direction under examination, and so on for all the successive levels of spatial constraint to which C_i is subjected.

$$R2: \{Cd\}_r = \left\{ C_q, \forall q \text{ such that } \begin{matrix} a_{iq} = 0 \\ a_{pq} = 1 \end{matrix} \text{ with } C_p \in \{Cd\}_{r-1} \right\}$$

  Using this rule, applied iteratively on the entire constructional system, it is possible to identify all the components indirectly obstructing the disassembly of C_i in the direction under examination, differentiating these components according to successive levels of spatial constraint. With this differentiation, the set $\{Cd\}_r$ of components indirectly obstructing the disassembly of C_i at the generic r-th level of spatial constraint consists of all the components impeding the disassembly of components belonging to the set $\{Cd\}_{r-1}$, relative to the preceding level of spatial constraint.

As can be seen from the analytical expressions, these two rules operate only on matrix A (13.1). By applying them, it is possible to determine the set $\{Cd\}$ of all the components to be removed in order to disassemble C_i following a preset direction. In fact, if n_r is the total number of levels of spatial constraint to which C_i is subjected, then:

$$\{Cd\} = \bigcup_{r=1}^{n_r} \{Cd\}_r \qquad (13.4)$$

Figure 13.4 shows a simple example illustrating the application of these first two rules of characterization. Having numbered the components making up the system (n = 5), having compiled matrix A, and having defined the component to be characterized (C_1) and the direction of disassembly, rules R1 and

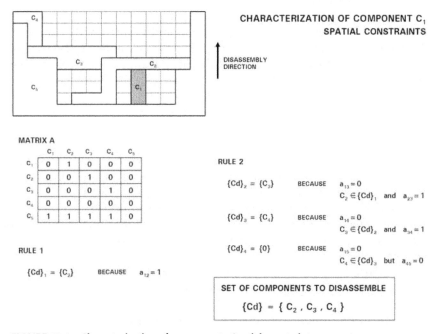

FIGURE 13.4 Characterization of components: Spatial constraints.

R2 were applied to determine the set {Cd}. It can be seen that, in this case, C_1 is subject to three levels of spatial constraint.

Characterization rules for junction constraints:

- Rule 3—Remove the set {Jd}$_1$ of junctions directly connecting C_i.

$$R3: \{Jd\}_1 = \{J_v, \forall v \text{ such that } d_{iv} = 1\}$$

This rule expresses the need to remove all the junctions constraining the component C_i, an unavoidable condition for its disassembly.

- Rule 4—Remove the set {Jd}$_2$ of junctions connecting each component that must be removed (i.e., those belonging to the set {Cd}) to components whose removal is not necessary for the disassembly of C_i.

$$R4: \{Jd\}_2 = \left\{J_v, \forall v \text{ such that } \begin{smallmatrix} b_{pq} = 1 \text{ with } C_p \in \{Cd\}, Cq \notin \{Cd\} \\ d_{pv} = d_{qv} = 1 \end{smallmatrix}\right\}$$

This rule expresses the necessity to free all the components to be disassembled from those that do not need to be removed (ignoring, however, those junction constraints connecting together the components to be disassembled).

These two rules operate on matrices B (13.2) and D (13.3) and also use the information obtained through the first two rules, R1 and R2. By applying them it is possible to determine the set {Jd} of all the fasteners to be removed in order to disassemble C_i. In fact:

$$\{Jd\} = \{Jd\}_1 \bigcup \{Jd\}_2 \tag{13.5}$$

Figure 13.5 shows the application of the second two rules on this system. Having also identified and numbered the fasteners present in the system (f = 6, not differentiated for typology), and having compiled matrices B and D, rules R3 and R4 were applied to determine the set {Jd}.

This example demonstrates that with this approach the junction J_4 does not need to be removed, since the components it connects (C_2 and C_3) can be removed as a single subassembly. This would not have been true, however, if C_2 was connected to C_5 by the junction J_3', instead of by J_3. In that case, the removal of J_3' would require the separation of C_2 from C_3 through the removal of junction J_4.

This observation highlights one of the typical problems of disassembly, the accessibility of fasteners; to eliminate it would require additional

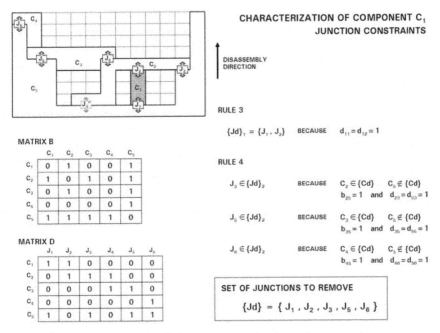

FIGURE 13.5   Characterization of components: Junction constraints.

characterization rules. For example, to eliminate the specific problem of the accessibility of junction J_3' would require the introduction of:

- Rule 5—Remove the set of junctions connecting each component to be disassembled to all other components to be disassembled, when the disassembly process makes this necessary.

The preliminary modeling proposed here, however, does not allow a rule of this kind to be expressed in analytical form. The characterization method proposed is based on only the first four rules defined above. In this way, additional aspects such as that just discussed are ignored and the accuracy of the evaluation of disassembly depth is limited.

13.5.2.2   Improving the Characterization

The problem of the accessibility of fasteners described above leads to other aspects that should not be disregarded in an accurate analysis of the disassembly process. Referring to rule R5, for example, in some cases it may be necessary to extend the rule to include junctions that connect components whose spatial constraints do not require them to be removed. This would be the case where access to a junction to be removed according to rules R3 and R4 necessitates the removal of a junction connecting components not requiring removal according to rules R1 and R2. Again, because of problems of junction accessibility it can happen that the removal of a junction identified by rules R3 and R4 necessitates the disassembly of a component that, according to rules R1 and R2, did not need to be removed.

An accurate characterization of the difficulty of disassembling a component ultimately requires the complete resolution of the problem of planning the selective disassembly of the component in question; this problem will be treated in detail in the following chapter. Chapter 14 will also propose a tool able to generate the optimal sequences of selective disassembly, taking into account questions raised here and allowing the determination of the complete sets {Cd} and {Jd}. This tool also considers the possibility of changing the direction of disassembly during the sequence, and thus constitutes a valid aid for accurately characterizing components with the aim of analyzing disassembly depth.

13.5.3   Index of Disassembly Depth

After the characterization phase, it is possible to determine the distribution of disassembly depth of components, initially quantified by the number of other components to be removed for the disassembly of each component (n_D), given by the set {Cd}, and by the number of junctions (f_{Dk}) to be removed for each junction of k-th typology, given by the set {Jd}. This initial data is further elaborated in order to obtain, for each component, a single index of disassembly depth dd, which takes account both of the other components to be removed because of spatial constraints (using the term dd_{SC}), and of the junctions and the disassembly difficulty of the junctions (using the term dd_{JC}):

$$dd = dd_{SC} + \beta \cdot dd_{JC} = \frac{1+n_D}{n} + \beta \cdot \frac{\sum_{k=1}^{h} \alpha_k \cdot f_{D_k}}{f} \qquad (13.6)$$

where $1+n_D$ is the total number of components to be removed (including the component whose disassembly depth is being evaluated), n is the total number of components, f_{Dk} is the number of junctions of the k-th type to be removed, f is the total number of junctions present in the system, α_k is the disassembly difficulty of junctions of the k-th type, h is the number of junction types, and β is a coefficient taking into account the greater weight of dd_{JC} compared to dd_{SC} ($\beta > 1$). Allowing for values of the coefficients α_k in the interval [0,1], $\alpha_k = 1$ indicates the maximum difficulty of disassembly. The index dd can then assume values belonging to the interval [1/n, 1+β].

The maximum value $1+\beta$, expressing the maximum disassembly depth, is found when, to remove a component, it is necessary to disassemble:

- All the other components present in the constructional system ($dd_{SC} = 1$)
- All the fasteners present in the constructional system, and these are all of typology characterized by the maximum disassembly difficulty ($dd_{JC} = 1$)

So defined, finally, the index dd can be compared to the maximum value present in the constructional system under analysis, obtaining for each component the normalized value:

$$DD_i = \frac{dd_i}{dd_{MAX}} \qquad (13.7)$$

The set of values assumed by DD_i for each of the n components comprising the constructional system describes the distribution of the normalized disassembly depth.

13.5.3.1 Disassembly Difficulty of Junction Systems

As seen in Equation (13.6), the term dd_{JC} differentiates the junction typologies in terms of disassembly difficulty. The suitability of this expedient has already been described by other authors, both in relation to the diversity of difficulty of access to different types of junction (Das and Naik, 2002), and in relation to other attributes of junction types (e.g., dimensions, form, and operational stratagems that affect the ease of their removal) (Sodhi et al., 2004).

   In the evaluation of disassembly depth using Equation (13.6), it is, therefore, important to make an appropriate choice of the values to attribute to the coefficients α_k that express the disassembly difficulty of each k-th type junction. To quantify these difficulties, it is possible to turn to times of disassembly, which has already been estimated by other authors applying general methods for the evaluation of work times (Vujosevic et al., 1995; Kroll and Carver, 1999), or to indications provided by approaches based on specific metrics of the disassembly difficulty of junctions (Sodhi et al., 2004).

13.6 Efficiency of Ease of Disassembly Distribution

Having evaluated the distribution of the values assumed by Equations (13.6) or (13.7) for all the components of the system under examination, the first objective of the proposed method has been achieved: to analyze the distribution of the disassembly difficulty of components or subassemblies making up a constructional system. This first type of analysis allows the identification of the main criticalities of the system and can provide the first indications for design modifications aimed at improving the product in the context of DFD:

- Improve the reliability and duration of components at great disassembly depth
- Improve the potential for reuse and recycling of components at low disassembly depth
- Limit the economic value of components at great disassembly depth

The second objective of the proposed method has a more radical approach, that of supporting product redesign aimed at improving its ease of disassembly through a reasoned redistribution of the disassembly depth. To be accurately reasoned, this redistribution must be guided by certain characteristics of the single components that quantify the necessity or opportuneness of their disassembly (e.g., reliability, environmental impact, economic value).

To achieve this second objective, it is necessary to introduce additional analytical tools, described below, which make it possible to quantify:

- The appropriateness of the disassembly depth of each component with respect to the corresponding necessity or opportuneness of disassembly
- The efficiency of the distribution of the disassembly depth of all components of the system (i.e., the overall efficiency of the design solution under examination)

This information constitutes a valid aid in the redesign of the system in light of the chosen aims. In particular, the mathematical functions expressing the system properties listed above become the objective functions of the redesign problem, which is thus reduced to evaluating the benefits that possible design modifications might have on the values assumed by these functions.

13.6.1   Evaluation of the Objective Properties of Components

Beginning with the main characteristics of each component (geometries, forms, materials) it is possible to evaluate those that may be the objective properties of disassembly (i.e., the properties that determine the necessity or opportuneness of disassembly). If this generic property is indicated by op_i, it must be defined in such a way that the greater its value, the greater the necessity or opportuneness of disassembly. So defined, it can be compared to the maximum value present in the assembly under analysis to give the normalized form:

$$OP_i = \frac{op_i}{op_{MAX}} \tag{13.8}$$

Some suggestions on how to define the objective property are given below.

- Reliability—Components with poor reliability, and therefore high frequency of failure, require lower disassembly depths. For the objective property of poor reliability, use of a function appropriately derived from reliability parameters is recommended. An example of such a parameter is Mean Time Between Failure (MTBF), the mean value of the length of time of component use between the occurrence of two successive malfunctions.

- Frequency of maintenance—Components with a high frequency of maintenance require lower disassembly depth. For the objective property of high frequency maintenance, use of a function appropriately derived from maintenance analysis parameters is recommended. An example of such a parameter is Mean Time Between Maintenance (MTBM), the mean length of time of component use between the commencement of two successive maintenance interventions.

- Environmental impact—Components with a high environmental impact require lower disassembly depths to facilitate their recovery. For the objective property of high environmental impact, use of functions expressing the environmental impacts associated with the main phases of the component's life cycle are recommended. These can be evaluated using the Life Cycle Assessment techniques (Chapter 4) and the now widely available supporting tools for impact quantification (such as those used in Chapters 11 and 12).

- Economic value—Components of high economic value require lower disassembly depths to facilitate their recovery. For the objective property of high economic value, use of the component's production cost, or the economic value attributed to it in the market for used or recycled, is recommended.

On a final note, it should be pointed out that the concept of "objective properties of disassembly" (i.e., the properties that determine component necessity or opportuneness of disassembly) is closely linked to the concept of "determinant factors for the life cycle strategies" that was extensively treated in Chapters 9 and 11.

13.6.2 Evaluation of the Efficiency of Disassembly Depth

As presented in Figure 13.1, after the disassembly depth and objective properties of each component have been evaluated it is necessary to define suitable functions able to quantify the efficiency of the design solution in relation to the requirements of disassembly. With this aim, an index of disassembly depth efficiency Θ_i, which is expressed as a function of Equations (13.7) and (13.8) for each i-th component, is introduced:

$$\Theta_i = \left| OP_i - (1 - DD_i) \right| \tag{13.9}$$

Minimizing the indices Θ_i, which can assume values in the interval [0,1], corresponds to seeking a configuration characterized by lower disassembly depths for components with a greater need for maintenance or recovery, and vice versa by greater disassembly depths for components with lower need for maintenance or recovery.

Finally, to have a single objective function able to quantify the overall efficiency of the design solution, it is possible to introduce:

$$\Lambda = \sum_{i=1}^{n} \gamma_i \cdot \Theta_i = \sum_{i=1}^{n} \gamma_i \cdot \left| OP_i - (1 - DD_i) \right| \tag{13.10}$$

where γ_i are coefficients introduced to vary the weight of single Θ_i, terms relative to the importance of each component. With the objective function Λ so defined, its minimization involves identifying the distribution of optimal disassembly depths through the whole constructional system, giving greater importance to components characterized by greater necessity or opportuneness for disassembly.

13.7 Case Study: Electromechanical System

A case study developed on an electromechanical assembly is presented as an example of applying this approach. The assembly consists of six main components, as shown in Figure 13.6, together with the abstraction representing a

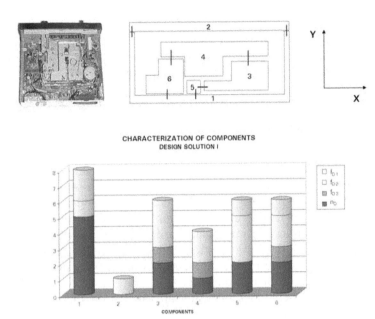

FIGURE 13.6 Case study: Electromechanical system, abstraction for disassembly analysis and component characterization.

first construction solution, with the approximate overall dimensions of the components and the distribution of the junctions.

13.7.1 Evaluation of Disassembly Depth

Given the layout typology, the analysis of disassembly depth was performed for the direction Y only. The results of the characterization of the components based on the disassembly difficulty, applied on the basis of the indications reported in Section 13.5, are shown in the graph in Figure 13.6, indicating for each component:

- How many other components obstruct its disassembly. Following the notation used in Equation (13.6), this number is n_D.
- How many junctions, differentiated according to three different typologies, must be removed to allow the disassembly. Following the notation used in Equation (13.6), this number is given by the sum of f_{D1}, f_{D2}, and f_{D3}.

This graph offers a first purely quantitative visualization of the distribution of disassembly difficulty among the system components, without making qualitative distinctions between spatial and junction constraints. Instead, the first series of histograms in Figure 13.7 shows the distribution of normalized disassembly depth DD, evaluated for each component using Equations (13.6) and (13.7), distinguishing the three junction typologies by different coefficients expressing the difficulty of their removal (Section 13.5.3.1).

13.7.2   Analysis of the Design Solution

With regard to evaluating the objective property of disassembly OP_i, in this case the economic value of the components was considered. Figure 13.7 shows, together with the values of distribution of disassembly depth DD, those of the objective property OP and of the resulting index of disassembly depth efficiency Θ, evaluated using Equation (13.9), for all the components of the first design solution. In particular, the function Θ provides indications regarding the criticality of the solution under examination, aiding a design intervention aimed at improving the ease of disassembling of the system, in relation to the value of the components.

Comparing the three series of histograms in Figure 13.7, the following observations can be made:

- Component 1 is characterized by an optimal efficiency (very low value of Θ), since it has great disassembly depth (for the direction Y) associated with very low economic value, and therefore poor suitability for disassembly.

- Components 3 and 4 are characterized by a good compromise between disassembly depth and economic value, verified by the low value of Θ.

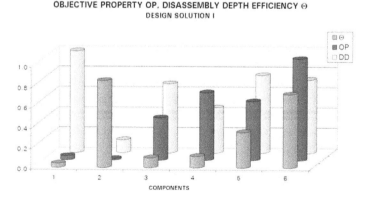

FIGURE 13.7   Analysis of design solution I.

- Component 5 is characterized by an intermediate efficiency, since it has good economic value associated with a rather high disassembly depth.

- Component 2 is characterized by an elevated inefficiency, since it has a very low disassembly depth but negligible economic value. In this case, therefore, such a low disassembly depth is completely inappropriate, and in a redesign intervention can be sacrificed in favor of other components.

- Component 6 constitutes the principal criticality of the system, as evidenced by the high value assumed by Θ. It is, in fact, the most valuable component of the entire system and, conversely, is characterized by one of the greatest disassembly depths.

13.7.3 Redesign and Optimization

Having identified the characteristics and criticalities of the design solution under examination, it is possible to undertake an efficient redesign of the system. To this end, it is possible to develop design alternatives such as those depicted in Figure 13.8, together with their analysis summarized in graphs entirely analogous to those of Figure 13.7.

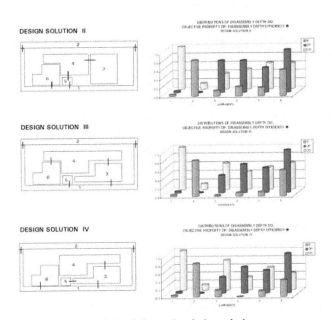

FIGURE 13.8 Analysis of alternative design solutions.

DISASSEMBLY DEPTH DD
DISTRIBUTIONS COMPARISON

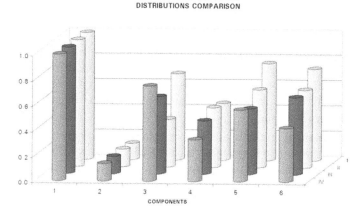

FIGURE 13.9   Distribution of disassembly depth DD (comparing all solutions under examination).

The proposed alternative solutions basically maintain the original layout, but differ in the geometry of the components and in the junction systems. These modifications produce variations in:

- The distribution of disassembly depth, as can be seen in the graph in Figure 13.9 comparing the different distributions corresponding to the four solutions considered

- The production costs of the components, and therefore in the values of the objective property OP (only solution III makes use of the same components constituting solution I).

Comparing the distributions of the Θ functions (to be minimized) for all four design solutions (Figure 13.10), it can be seen how the redesign process tends to gradually better harmonize the disassembly depths with the economic values of the system's components potentially recoverable at end-of-life. This conclusion is confirmed by the graph in Figure 13.11, showing the values of the overall objective function Λ. This value, calculated using Equation (13.10) for each of the design solutions, is inversely proportional to the efficiency of each solution.

13.8   Summary

With the method proposed in this chapter it is possible not only to analyze the distribution of disassembly depth in a product, but also to aid its redesign with the aim of facilitating its disassembly. If integrated into the design process, it can allow the optimization of a vast range of design choices (layout,

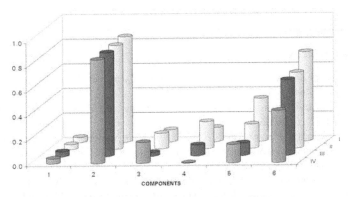

FIGURE 13.10 Distribution of disassembly depth efficiency Θ (comparing all solutions under examination).

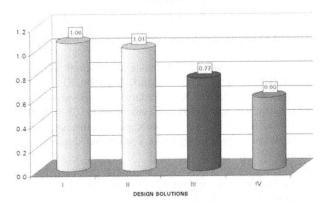

FIGURE 13.11 Comparison of overall objective function Λ between all solutions under examination.

geometries, junctions) to improve the ease of disassembly at different levels of the design process (particularly embodiment and detail design), in agreement with the basic principles of Design for Disassembly.

Furthermore, this optimization is guided by the component properties themselves, in a way that facilitates the disassembly of those parts of the system that are most likely to require frequent servicing or that present an elevated suitability for recovery. Based on these results, it is possible to introduce cost functions allowing the feasibility of a disassembly to be evaluated in terms of cost and added revenue from the extraction and removal of the objective component. This last aspect can have a determining influence on the practicability of a servicing plan and the improvement of the level of disassembly and recovery at end-of-life.

13.9 References

Billatos, S.B. and Basaly, N.A., *Green Technology and Design for the Environment*, Taylor & Francis, Washington, DC, 1997.

Boothroyd, G. and Alting, L., Design for assembly and disassembly, *Annals of the CIRP*, 41(2), 625–636, 1992.

Brennan, L., Gupta, S.M., and Taleb, K.N., Operations planning issues in an assembly/disassembly environment, *International Journal of Operations and Production Management*, 14(9), 57–67, 1994.

Chen, K.Z., Development of integrated design for disassembly and recycling in concurrent engineering, *Integrated Manufacturing Systems*, 12(1), 67–79, 2001.

Chen, R.W., Navin-Chandra, D., and Prinz, F.B., A cost–benefit analysis model of product design for recyclability and its applications, *IEEE Transactions on Components, Packaging, and Manufacturing Technology*, 17(4), 502–507, 1994.

Das, S.K. and Naik, S., Process planning for product disassembly, *International Journal of Production Research*, 40(6), 1335–1355, 2002.

Das, S.K., Yedlarajiah, P., and Narendra, R., An approach for estimating the end-of-life product disassembly effort and cost, *International Journal of Production Research*, 38(3), 657–673, 2000.

Dewhurst, P., Design for Disassembly: The Basis for Efficient Service and Recycling, Report 63, The University of Rhode Island, Kingston, RI, 1992.

Di Marco, P., Eubanks, C.F., and Ishii, K., Compatibility analysis of product design for recyclability and reuse, *Computers in Engineering*, 1, 105–112, 1994.

Dini, G., Failli, F., and Santochi, M., A disassembly planning software system for the optimization of recycling processes, *Production Planning and Control*, 12(1), 2–12, 2001.

Gungor, A. and Gupta, S.M., Disassembly sequence planning for complete disassembly in product recovery, in *Proceedings of Northeast Decision Sciences Institute Conference*, Boston, 1998, 250–252.

Gungor, A. and Gupta, S. M., Issues in environmentally conscious manufacturing and product recovery: A survey, *Computers and Industrial Engineering*, 36, 811–853, 1999.

Harjula, T. et al., Design for disassembly and the environment, *Annals of the CIRP*, 45(1), 109–114, 1996.

Ishii, K., Eubanks, C.F., and Di Marco, P., Design for product retirement and material life-cycle, *Materials and Design*, 15(4), 225–233, 1994.

Johnson, M.R. and Wang, M.H., Planning product disassembly for material recovery opportunities, *International Journal of Production Research*, 33(11), 3119–3142, 1995.

Jovane, F. et al., A key issue in product life cycle: Disassembly, *Annals of the CIRP*, 42(2), 651–658, 1993.

Kroll, E. and Carver, B.S., Disassembly analysis through time estimation and other metrics, *Robotics and Computer Integrated Manufacturing*, 15(3), 191–200, 1999.

Lambert, A.J.D., Optimal disassembly of complex products, *International Journal of Production Research*, 35(9), 2509–2523, 1997.

Lozano-Perez, T., Spatial planning: A configuration space approach, *IEEE Transactions on Computers*, C-32(2), 108–120, 1983.

Navin-Chandra, D., ReStar: A design tool for environmental recovery analysis, in *Proceedings of ICED International Conference on Engineering Design*, The Hague, The Netherlands, 1993, 780–787.

Navin-Chandra, D., The recovery problem in product design, *Journal of Engineering Design*, 5(1), 67–87, 1994.

Penev, K.D. and de Ron, A.J., Determination of a disassembly strategy, *International Journal of Production Research*, 34(2), 495–506, 1996.

Pnueli, Y. and Zussman, E., Evaluating the end-of-life value of a product and improving it by redesign, *International Journal of Production Research*, 35(4), 921–942, 1997.

Scheuring, J., Bras, B., and Lee, K.M., Significance of design for disassembly on integrated disassembly and assembly processes, *International Journal of Environmentally Conscious Design and Manufacturing*, 3(2), 21–33, 1994.

Simon, M. and Dowie, T., Quantitative Assessment of Design Recyclability, Report DDR/TR8, Manchester Metropolitan University, Manchester, UK, 1993.

Sodhi, R., Sonnenberg, M., and Das, S.K., Evaluating the unfastening effort in design for disassembly and serviceability, *Journal of Engineering Design*, 15(1), 69–90, 2004.

Srinivasan, H. and Gadh, R., A non-interfering selective disassembly sequence for components with geometric constraints, *IIE Transactions*, 34, 349–361, 2002.

Srinivasan, H., Shyamsundar, N., and Gadh, R., A framework for virtual disassembly analysis, *Journal of Intelligent Manufacturing*, 8, 277–295, 1997.

Viswanathan, S. and Allada, V., Configuration analysis to support product redesign for end-of-life disassembly, *International Journal of Production Research*, 39(8), 1733–1753, 2001.

Vujosevic, R. et al., Simulation, animation, and analysis of design disassembly for maintainability analysis, *International Journal of Production Research*, 33(11), 2999–3022, 1995.

Yamagiwa, Y., Negishi, T., and Takeda, K., Life cycle design achieving a balance between economic considerations and environmental impact with assembly–disassembly evaluation design, in *Proceedings of EcoDesign '99: 1st International Symposium on Environmentally Conscious Design and Inverse Manufacturing*, Tokyo, 1999, 760–765.

Zhang, H.C. and Kuo, T.C., A graph-based disassembly sequence planning for end of life product recycling, in *Proceedings of IEEE/CPMT International Electronics Manufacturing Technology Symposium*, Austin, TX, 1997, 140–151.

Zhang, H.C. et al., Environmentally conscious design and manufacturing: A state of the art survey, *Journal of Manufacturing Systems*, 16(5), 352–371, 1997.

Zussman, E., Kriwet, A., and Seliger, G., Disassembly-oriented assessment methodology to support design for recycling, *Annals of the CIRP*, 43(1), 9–14, 1994.

Chapter 14

Optimal Disassembly Planning

From the viewpoint of environmental protection, an efficient planning of disassembly operations takes on strategic importance, since it can improve both the useful life of the product (facilitating service interventions) and the end-of-life phase (favoring the recycling of materials and the reuse of components). After an overview of the main issues and current approaches to the problem of Disassembly Planning, this chapter proposes a new approach aiding the identification of optimal disassembly in relation to service operations and the recovery of resources at the end-of-life. Despite their diverse aims, the two tools proposed here have in common both the modeling typology on which they operate and the logic underlying the search algorithms used. The choice of genetic algorithms is dictated by the complexity inherent in the complete mathematical solution of the problem of generating the disassembly sequences, which suggests the use of a nonexhaustive approach.

The results of a wide-ranging series of simulations indicate that both tools can be used not only for the characteristic purposes of disassembly planning, but also for aiding design. This is particularly true for the second tool, which offers a complete approach to the problem of disassembly aimed at recovery; it combines evaluations of economic and environmental impacts and extends these evaluations over the product's entire life cycle. Its structure provides the designer with an autonomous capacity to decide on both the level of disassembly to be achieved and the definition of the optimal recovery plan (i.e., the best destination for the disassembled components on the basis of their properties).

The fundamental issues in this chapter were previously introduced in Chapters 9 and 13.

14.1 Disassembly Planning

The disassembly of products is necessary every time it is appropriate to remove subsystems or single components comprising the product itself. From the viewpoint of environmental protection, disassembly can have several goals (Lambert, 1997): recovering parts, components, and subassemblies that can be reused in new products; recovering recyclable materials; and accessing parts or components that may be subject to service operations (repair, maintenance, diagnostics).

As highlighted in the previous chapter, interventions to improve the process of product disassembly can be made at two different levels (Jovane et al., 1993; Gungor and Gupta, 1999):

- At the design phase, adopting choices that can favor the disassembly of the constructional system (Design for Disassembly—DFD)
- Seeking to best plan and optimize the process of disassembly (Disassembly Process Planning—DPP)

Disassembly Process Planning, which includes all the problems relating to disassembly of constructional systems, is considered in terms of two different levels of analysis (Lambert 2003):

- Sequence Level—Starting from a mathematical representation of the assembled system, the analysis considers the problem of generating and optimizing the sequence of disassembly (Disassembly Sequencing).
- Detailed Level—Starting from the physical and geometric properties of components and fasteners, disassembly analysis takes into consideration the handling of components, directions of disassembly, conditions of obstruction, and choice of equipment, as far as determining the trajectories of any possible automatic manipulators (Disassembly Path Planning).

In a more complete view, these aspects should also be complemented by an analysis of the optimal disassembly level (Disassembly Leveling), which is the study of the level of disassembly that bests reconciles the requirements and the advantages of disassembling a system's components with the costs entailed by this disassembly. This concept was introduced in Chapter 13, Section 13.1. The literature contains general evaluations of cost–revenue curves for disassembly interventions aimed at repair and recovery at end-of-life (refer to Chapter 9, Figure 9.5). According to these evaluations, the costs of disassembly become prohibitive as the level of disassembly increases. On the contrary, beyond a certain level the revenues tend to stabilize. Consequently, the profit curve has a maximum, after which it tends to decrease as the level of disassembly increases.

The profit of recovery can be improved if the product architecture is such that the first disassembly results in freeing the most critical or most valuable parts. This consideration highlights the direct dependence of the efficiency of disassembly on the product architecture, and particularly on the depth of disassembly of single components in relation to certain characteristics that can make their removal opportune or necessary during use or at end-of-life. Depth of disassembly was thoroughly discussed in Chapter 13.

The profit function of disassembly is also strongly sensitive to the efficiency of the disassembly sequence. With a judiciously optimized disassembly sequence it is possible to:

- Reduce the times and costs of disassembly operations
- Perform the most appropriate disassemblies, avoiding those that are superfluous or less profitable

This is, therefore, the first intervention factor for profitable disassembly. It is indirectly conditioned by the characteristics of the product's architecture, which, depending on the typology, can favor the efficiency of the disassembly sequence.

In conclusion, the problem of disassembly planning includes the main aspects:

- Analysis of the system characteristics (component geometries, relations between components, junctions)
- Generation of possible disassembly sequences
- Identification of the most efficient and economical disassembly sequences
- Determination of optimal level of disassembly

14.1.1 General View of the State of the Art

Early approaches to the problem of disassembly planning were developed on the basis of understanding previously acquired in relation to the problems associated with assembly. These approaches came from the assumption that disassembly sequences could be assimilated to assembly sequences in reverse (Homem De Mello and Sanderson, 1991). However, various authors have subsequently described the profound differences existing between the two problems (principally, the fact that assembly processes are often not completely reversible, and the frequent need for selective or partial disassembly). These two problems demonstrate the need to treat disassembly planning in a specific and more appropriate manner (Srinivasan et al., 1997; Dini et al., 2001; Lambert, 2003). With this new perspective, a large number of approaches to disassembly have been proposed over the last decade, suggesting diverse attempts at classification

(O'Shea et al., 1998; Tang et al., 2000; Lambert, 2003). In particular, authors have emphasized:

- The different levels of approach (oriented toward components, or oriented toward products)
- The differences in the aims of disassembly (maintenance and servicing, or removal and recovery at end-of-life)
- The differences in disassembly modeling, in generating the disassembly sequences, and in methods to identify the optimum solution

14.1.2   Extension to Design of the Life Cycle

In environmental terms and in relation to the phases of a product's life cycle, the great importance of disassembly is its functionality in the phases of use (supporting maintenance and servicing operations) and end-of-life (supporting operations of recovery and disposal). With regard to servicing operations, several different approaches are described in the literature (Subramani and Dewhurst, 1991; Yokota and Brough, 1992; Vujosevic et al., 1995). While these have in common the criterion of determining the optimum disassembly sequence based on the minimization of costs, they differ in the typologies of models used to represent the disassembly process (connection diagrams or component–junction diagrams, "AND/OR" graphs). However, the problem is generally rooted in planning the selective disassembly of components requiring maintenance.

The subsequent consideration of the operations of recovery and disposal at end-of-life required an extension of the problem of disassembly planning. Here, disassembly sequences must be optimal not only in strictly economic terms, but also in terms of environmental impact of the end-of-life phase (Gungor and Gupta, 1999). This led to the introduction of the concept of Recovery Planning, based on a quantitative evaluation of a product's value at end-of-life in terms of its potential for reuse, remanufacturing, and the recycling of materials (Navin-Chandra, 1993; Navin-Chandra, 1994; Zussman et al., 1994; Pnueli and Zussman, 1997). These approaches all use graphic models to represent disassembly processes.

In contrast, other approaches tend toward analytical mathematical modeling. Beginning with component–junction diagrams, these introduce matrix representations to express the precedences of component disassembly (Disassembly Precedence Matrices) (Zhang and Kuo, 1997). This allows the elaboration of more data and makes it possible to study the problem at a greater level of detail, going so far as to consider even the times and costs of changing a tool or the direction of disassembly. Nearly optimal disassembly sequences can then be developed on the basis of a much more detailed analysis than those possible with graphic models (Gungor and Gupta, 1998). The introduction of matrix modeling also allows direct integration with CAD modeling of the assembly under examination (Dini et al., 2001).

14.1.3 Application of Artificial Intelligence

Extending the problem of disassembly planning and placing it in the context of the product's life cycle results in a significant increase in the amount of data to be elaborated and in the complexity of the performances to be optimized. To deal with the resulting increased complexity of the problem, researchers have turned to certain instruments of artificial intelligence (Lambert, 2003): neural networks, fuzzy logic, and genetic algorithms. The latter, in particular, have been applied in various cases, by virtue of their characteristic of working efficiently in open research domains and of identifying solutions close to optimal. In this context, some studies have described the use of genetic algorithms as aids in:

- The optimization of disassembling components or subassemblies for the purpose of maintenance, based on a component–contact constraints diagram model (Li et al., 2002)
- The economic and environmental analysis of disassembly processes, based on an AND/OR graph model and converting environmental factors into economic costs (Seo et al., 2001)
- The definition of the most efficient disassembly strategy in relation to the requirements of recovery at end-of-life, based on criteria of maximum revenue and minimum number of components to be disposed of as waste (Caccia and Pozzetti, 2000)

14.1.4 Concluding Considerations

The great variety of methods proposed in the literature, summarized above, is directly correlated to the specific problems the various authors intended to resolve. At present, there is little effort directed at the definition of a systematic methodology that is able to integrate the different approaches and to guide in solving different problems using the most effective approach. Apart from the specificity of the various methods found over the entire spectrum of studies on disassembly planning, it is also worth noting the limited vision of the environmental problem, which in this context is usually treated by translating environmental aspects into economic costs and considering only the end-of-life phase in the analysis.

14.2   Objectives and Approach to the Problem

With clear reference to these observations on the state of the art, this chapter proposes an approach to disassembly planning characterized by:

- Matrix-type modeling, based on the analysis of component geometries and on the relations of junctions and movement, in a way that facilitates direct interfacing with conventional CAD modeling of assemblies

- A calculation program with a structure allowing the identification of the disassembly sequence that is optimal in terms of the aspects considered most significant—servicing operations and planning recovery at end-of-life

The two different aspects treated here, service and the recovery of resources at end-of-life, require that a distinction be made between the two different typologies of disassembly (Srinivasan et al., 1997):

- Selective disassembly, where the objective is the disassembly of one or more preselected components (an approach oriented toward servicing operations)

- Partial or complete disassembly, where the components to be disassembled are not chosen a priori, but are defined by the research algorithm itself on the basis of certain important properties characterizing the components (an approach oriented toward recovery operations)

With these objectives, it is possible to develop two algorithms to solve the problem of optimizing disassembly in the two distinct cases. Despite their different aims, the two proposed tools operate on the same typology of modeling and share the logic followed in developing the two algorithms. It is interesting to note, however, that while the first is limited to a conventional approach to the problem (selective disassembly optimized by minimizing disassembly times), the second tool deals with the environmental problem of recovery through an innovative approach, in that it is:

- Based on a complete analysis (bringing together functions of cost and environmental impact) and extends the evaluations to cover the entire life cycle

- Characterized by an autonomous capacity to determine both the level of disassembly to be achieved and the optimal recovery plan (i.e., the best destination for the disassembled components, on the basis of their properties)

14.3 Common Structure of the Proposed Tools

In general, the complete mathematical solution to the problem of generating disassembly sequences requires such a large number of calculations that it becomes extremely complex (Lambert, 1997; Moyer and Gupta, 1997). While heuristic methods can be used to manage this problem, they do not ensure the determination of optimal solutions (Gungor and Gupta, 1997; Kuo et al., 2000).

To reach the objectives that have been set, the tools proposed here are directed at the definition of optimal (or near optimal) disassembly sequences, using algorithms of the genetic type (GA—Genetic Algorithms) (Holland, 1975; Goldberg, 1989). This choice is principally motivated by two characteristics of the research space to be explored that make the use of GA advisable (Mitchell, 1998): the space is vast and is not unimodal. Starting from this core choice of GA, the two problems discussed above can be treated by developing two tools that are distinct but nevertheless share the same calculation code structure. Furthermore, in both cases the code elaborates the same mathematical model for the description of the geometries of, and relations between, the various components (expressed through constraint matrices). The code is thus able to take into account the changes occurring in the system's structure during the progressive disassembly of the single parts. Figure 14.1 shows the scheme common to both the tools proposed. It postulates a preliminary

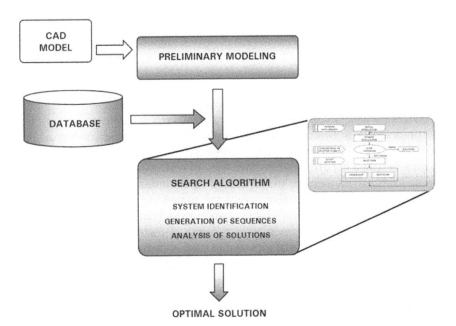

FIGURE 14.1   Common schematization of tools.

modeling, which has the aim of interpreting the assemblies under analysis in mathematical terms, allowing the subsequent elaboration of this information by the resolving algorithm. The latter, drawing on a set of data (included in a database) with varying typologies depending on the tool, identifies the optimal solution through three main phases:

- Identification of the system, requiring the formalization of the solution type to investigate and the definition of the objective function
- Generation of possible disassembly sequences
- Identification of the optimal solution

The main elements common to both tools for disassembly planning are described below. In later sections, they will be described in greater detail, delineating their more specific characteristics.

14.3.1   Common Preliminary Modeling

The system to be disassembled is represented here as consisting of a finite number of independent elements that can be removed individually. System elements are taken to mean:

- Single components linked to the system by reversible fasteners
- Any possible subgroups of components linked together by irreversible fasteners
- All reversible fasteners and junction systems (screws, clips, snap-fits, etc.)

With this type of approach, it is not possible to consider subgroups of elements to be treated as a single entity during the disassembly sequence, unless they are predefined on the basis of their homogeneity, the compatibility of their materials, or other criteria of affinity between the elements (see Chapter 11).

Considering a generic system consisting of n elements, the properties relative to each i-th element E_i are expressed by an index defining the typology of the element. This index allows one component, whose removal does not require particular operations other than simple translation in space, to be distinguished from a fastener, whose removal instead requires a specific intervention with a resulting increase in cost and time. Indicating the total number of different element typologies by n_e, each index of element typology e_j (which can assume integer values in the range $[0, n_e-1]$) is represented by data quantifying the difficulty of removing each element of that type. This property can be quantified using disassembly times (already assessed by other authors applying general methods for the evaluation of work times

[Vujosevic et al., 1995; Kroll and Carver, 1999]), or developing specific new approaches (Sodhi et al., 2004). In this way, each element type can be described by a mean disassembly time and the corresponding term can be normalized with respect to the simplest intervention, that of horizontal translation. Table 14.1 shows the indexing of element typologies together with the characterization adopted in the present study, based on the approximate disassembly times reported in the literature for the most common types of unfastening operations (Dowie and Kelly, 1994).

The elements defined in this way can be subjected to different elementary operations; these, too, are described by an index expressing the typology of disassembly operation. Indicating the total number of operation typologies by n_o, it is simple to distinguish between operations given that, once the diversification of the fasteners and other junction systems has been included in the analysis of the elements, the operations are reduced exclusively to movements of linear translation. In the three-dimensional case, such translations occur along the directions X, –X, Y, –Y, Z, and –Z; therefore, $n_o = 6$. Each generic index of operation typology o_k, which can assume integer values in the range $[0, n_o-1]$, is associated with an execution time for the operation, ultimately consisting of translation along the k-th direction. Also in this case, execution times are normalized with respect to the simplest operation of horizontal linear translation and can be compiled in a table analogous to Table 14.1. In this way, if it is considered opportune, it is possible to take into account the potentially greater difficulty that may be ascribed to operations of vertical translation, planning for a longer execution time than that for horizontal translation.

In representing the system in function of disassembly planning, the modeling proposed here is of a matrix type (such as that proposed in Chapter 13 for disassembly depth analysis), based on the geometric characteristics of the system and on the consequent precedences for disassembly among the elements constituting the assembly. Although this modeling is based on a geometric-type approach for the analysis of movements (Woo and Dutta,

TABLE 14.1 Indices of element typologies and characterization

ELEMENT TYPOLOGY INDEX (e_j)	DESCRIPTION	MEAN TIME (sec)	NORMALIZED TIME (tne_j)
0	Component (to remove)	1.25	1
1	Screw (to remove)	0.6 (per turn)	0.48 (per turn)
2	Snap–fit (to open)	1.5	1.2
3	Clip (to remove)	1	0.8
4	Connection (to break)	2	1.6
5	Wires (to disconnect)	1.5	1.2

1991; Srinivasan and Gadh, 2002), and makes use of preexisting matrix models (Zhang and Kuo, 1997; Gungor and Gupta, 1998; Dini et al., 2001), this modeling is particularly simple and comes down to the compilation of binary constraint matrices, one for each possible direction of disassembly. In the case of a system consisting of n elements, for the disassembly direction X this matrix is expressed by:

$$V^X = \left[v_{ij} \right]_{i=1,2,\ldots}^{j=1,2,\ldots} \qquad (14.1)$$

where the term v^X_{ij} has a value of 1 if the j-th element obstructs the removal of the i-th element in the X direction; otherwise it is 0. Analogously, the spatial constraint matrices can be defined in the other directions (in the case of three-dimensional analysis there will be six matrices, one for each direction of disassembly X, –X, Y, –Y, Z, and –Z).

Concerning the compilation of these matrices, it should be specified that:

- If the i-th element is a component, it is necessary to consider as obstacles all the other components impeding its movement in the direction under examination, and all the fasteners and junction systems acting directly on it, constraining it to other components (the latter may also not appear as direct obstacles to the movement of the component in question).

- If the element is a fastener, it is necessary to consider as obstacles all the components impeding its accessibility and movement in the direction under examination.

14.3.2 Disassembly Sequence and Operation Time

Using the preliminary modeling described above, the resolving algorithm will generate the possible disassembly sequences in a random manner, as explained in detail below. It will then evaluate the real practicability of each sequence based on a simple rule: To disassemble any system component, it is necessary to begin by disassembling the more external components whose removal is unobstructed, until the final objective is reached. A sequence will, therefore, consist of a series of element–operation couplets, ordered in such a way that all the constraint matrices of the type defined by Equation (14.1) are respected. Beginning with the removal of the most external elements, and updating the constraint matrix step-by-step as the elements are removed, a correctly ordered (i.e., truly practicable) disassembly sequence is obtained.

On the basis of this statement for the generation of the sequences, it is possible to define a function that quantifies the excellence of the sequence. With this aim, many of the studies present in the literature propose mathematical models to evaluate the cost of disassembly, which usually becomes the objective function to be optimized (or part of it). To quantify the efficiency of the disassembly sequences, it is preferable to examine a function that expresses the total time necessary for the disassembly process, a factor upon which the cost of disassembly is, in any case, directly dependent (Kroll and Carver, 1999).

Considering the generic i-th element E_i, its disassembly time TS_i is defined by the expression:

$$TS_i = \sum_{p=1}^{q_i} \left(tne_p \cdot tno_p + \delta \cdot cd_p \right) \qquad (14.2)$$

where the summation is extended to all the q_i elements that must be removed before disassembly E_i (the total number q_i includes the element E_i itself); tne_p is the normalized disassembly time corresponding to the typology of the p-th element (this correspondence depends on the index of element typology, according to Table 14.1); tno_p is the normalized execution time corresponding to the operation required for the removal of the p-th element (this correspondence depends on the index of operation typology); δ is a penalization coefficient taking into account a possible change in the direction of disassembly when passing from the removal of element E_{p-1} to that of element E_p ($\delta > 1$ expresses the time of changing direction, again normalized with respect to the operation of horizontal translation); and cd_p is a binary coefficient that assumes a value of 1 if this change in direction occurs; otherwise it is 0. For a more thorough analysis, it is possible to introduce additional penalization coefficients that take into account other delaying factors, such as changing tools or the removal of dangerous or high-value elements requiring greater care in handling and removal operations (Gungor and Gupta, 1998).

In the context of both tools proposed here, the choice of the disassembly sequence is based on checking the value assumed by the function (14.2). In the simplest case of selective disassembly directed at servicing interventions, function (14.2) becomes the actual objective function of the optimization. In fact, the problem comes down to the disassembly of a predetermined component in the shortest time possible (and thus with the lowest cost). In the case of disassembly for recovery, function (14.2) will be part of the objective function, which includes other terms as described below.

14.3.3 Structure and General Characteristics of the Resolving Algorithm

The following description of the resolving algorithm refers to specific issues related to Genetic Algorithm theory. For further details, refer to the specialized resources in the literature (Holland, 1975; Goldberg, 1989; Mitchell, 1998; Gen and Cheng, 2000).

As stated above, the search algorithm proposed here is of a genetic type and, therefore, elaborates a set of points (individuals) in the objective function domain, called the population. The individuals, which represent the possible solutions, are codified in structures that resemble the configuration of a chromosome (Holland, 1975). The structure of the algorithm used in the present study is entirely conventional and, for clarity, is shown in Figure 14.2, revealing certain properties that are explained below.

The algorithm initially generates a random population. From this population, each execution cycle (generation) selects individuals on the basis of their fitness and then applies genetic operators to them, with the most common being cross-over and mutation. These operators are applied according to parameters (called "probabilities of execution") that remain constant throughout the whole evolutionary process. The fitness of the individuals is quantified by the value assumed by a function of fitness (the fitness of an individual is an expression of the objective function to be optimized). In a conventional structure, the greater the fitness of an individual, the higher the probability that it is selected for reproduction (Goldberg, 1989). Having generated a new population, the fitness of the new generation is evaluated, on the basis of which either a new cycle is executed or the iteration is halted and the solution to the problem is identified. This depends on whether or not the condition defined by a criterion halting the algorithm (stop criterion) is achieved.

On the other hand, beginning with a random population (justified by the opportunity to initially explore as wide a domain of potential solutions as possible) can limit the effectiveness of the evolutionary process and its capacity to converge toward valid solutions (i.e., those representing disassembly sequences that are really feasible). This choice is, however, corroborated by the studies of authors applying genetic algorithms to the optimization of

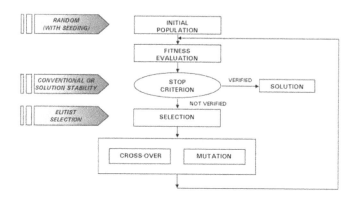

FIGURE 14.2   Structure of genetic algorithm implemented.

assembly sequences. In this respect, it should be noted how research has moved from evolutionary algorithms using already practicable initial sequences (Bonneville et al., 1995) to those adopting random initial populations while still obtaining convergence on optimal solutions (Dini et al., 1999).

In addition to its conventional structure, other particular characteristics of the genetic algorithm used here are:

- The fitness of the population is evaluated at each step of the evolutionary process. If the best individual is less fit than the champion from the preceding generation, the latter is reinserted in place of the least fit individual of the new generation, thereby applying a technique of elitism in the selection phase (elitist selection) (De Jong, 1975).

- The initial population is generated in a completely random manner. In the case where there are no valid individuals in this first generation (i.e., no feasible disassemblies and no fitness of an acceptable level), the search is guided by inserting into the population an individual that is already known to have a good level of excellence (i.e., a "seed" to facilitate the evolution of the population). This operation of seeding is equivalent to pointing the algorithm toward an area of the solutions domain that shows itself to be promising in terms of the search for the optimal (Oman and Cunningham, 2001).

Finally, regarding the stop criterion of the algorithm, it should be noted that genetic algorithms do not usually show formal properties of convergence toward the objective function assuming a stable value (and, therefore, toward the identification of the corresponding optimal solution). For this reason, in some cases it is necessary to define a heuristic stop criterion. Here, two alternatives are considered:

- Following a conventional approach, the investigation halts when a preset maximum number of generations or level of fitness is reached (Hulin, 1997). In this case, the algorithm stops when one of these two conditions is achieved.

- Following an approach based on requiring the solutions to be stable over the course of the evolution of successive generations, two parameters with preset values are introduced, f_1 and f_2. The algorithm halts when the fitness of the best individual in the generation under examination does not differ by more than f_1% from the mean of the maximum fitness values found for the last f_2 generations.

14.4   Development of the First Tool: Goals of Servicing

To resolve the first type of problem, that of selective disassembly aimed at interventions of maintenance and servicing during the product's useful life, the tool to be applied here is based on the genetic algorithm previously described, used to identify the best disassembly sequence for a given target element in an assembly.

This tool is confined to identifying a sequence of disassembly operations that minimize a simple objective function, given that it quantifies only the disassembly times. Despite this, its use is not limited exclusively to the planning and optimization of servicing operations on preexisting products. If employed as a tool for simulations in the phase of product design and development, it can constitute a valid support for a design intervention aimed at optimizing the system by facilitating the servicing operations on critical components, in accord with the fundamental principles of Design for Serviceability (Chapter 9, Section 9.2.1). This tool could also be particularly useful for characterizing the difficulty of disassembling a component, which is a basic issue in the method of disassembly depth analysis proposed in Chapter 13 (See Section 13.5.2).

14.4.1   Preliminary Modeling

With reference to the general schematization in Figure 14.1, preliminary modeling consists of defining the n elements making up the system; their numeration; characterizing them on the basis of their typology; and compiling the constraint matrices of type (14.1), which describe the system mathematically. The elements are characterized through the compilation of a vector expressing the properties:

$$P_{elem} = [p_i]_{i=1,2,\ldots,n} \tag{14.3}$$

Given that the typology is the only property of the elements of interest in this case, p_i coincides with the typology index corresponding to the i-th element. The database of the system is limited to collecting the data relative to the element and operation typologies, as defined in Section 14.3.1 (i.e., ultimately, the values of the removal and handling times relative, respectively, to the typologies of the elements and of the disassembly operations).

14.4.2   Identification of the System

Proceeding with identifying the system by the search algorithm, the formalization of the solution type to be investigated becomes a vector of integer

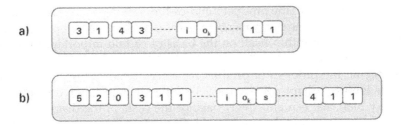

FIGURE 14.3 Formalization of solution type: (a) selective disassembly; (b) partial or complete disassembly.

numbers, consisting of couplets of indices (Figure 14.3a). The first corresponds to the index of the element i, following the preliminary numeration. The second corresponds to the typology index o_k of the operation required to remove that element.

The objective function coincides with the function of disassembly time expressed by function (14.2), calculated for the final element of the disassembly sequence (i.e., the target element).

14.4.3 Generation of Disassembly Sequences and Identification of Optimal Solution

The generation of the disassembly sequences is intrinsic to the functioning of the genetic algorithm. Although generated in a completely random manner, each individual of each generation (represented by the vector in Figure 14.3a) actually represents a potential disassembly sequence. If this respects the constraint matrices of type (14.1), and is therefore actually practicable in the sense discussed in Section 14.3.2, its fitness is evaluated through the calculation of the value assumed by the objective function. Otherwise, the sequence is not feasible and will be treated appropriately in the evolution of subsequent generations.

When the algorithm stops, it has found the optimal (or near optimal) solution, again expressed in the form given in Figure 14.3a, consisting of a series of element–operation couplets ordered in such a way that the constraint matrices are respected. These represent the feasible disassembly sequence that allows the target element to be removed while minimizing the objective function.

14.5 Development of the Second Tool: Goals of Recovery

The second tool is similar to the first in its general scheme, in its base modeling, and in its execution procedure. Unlike the first, however, it is directed at recovery at the product's end-of-life, seeking a partial or complete disassembly where the components to be disassembled are not preset but are defined by the algorithm itself on the basis of certain important properties characterizing them.

In general, the problem of product recovery at end-of-life can be formulated as follows (Chapter 9, Section 9.3): For a given product, determine the recovery plan that can effectively balance the costs of the disassembly and recovery processes and the resulting profits, in terms of resources employed and recovered. In this second case, therefore, the problem is extended to include not only the optimal disassembly sequence but also the evaluation of which elements are worth disassembly and what final destination to assign to them. This requires the introduction of a multiobjective function that takes into account the times of the disassembly operations together with other factors, such as the costs of production, recovery, and disposal of components, and the indicators that quantify the environmental impact of every phase of the product's entire life cycle. The terms of the objective function thus depend on new factors:

- The properties of the materials and elements constituting the system
- The level of disassembly
- The final destination of disassembled components (which can be reused, recycled, or disposed of as waste)

Furthermore, it will be defined in such a way that it is possible to treat the environmental problem of recovery through a complete analysis, combining functions of cost and environmental impact and extending the evaluation over the entire life cycle.

By virtue of these characteristics, the proposed new tool can be effectively employed in planning the end-of-life phase of preexisting products and as an aid in the product development phase. In fact, it can be used as a tool for the simulation and optimization of a system, for design specifically aimed at favoring the end-of-life phase of a new product, or in a more attentive design intervention that concentrates on the product's overall environmental performance over the entire span of its life cycle. Ultimately, this second tool, while fulfilling its role as an aid in disassembly planning like the first, is also an excellent support tool in Design for Disassembly.

14.5.1 Preliminary Modeling

Again referring to the general scheme in Figure 14.1, preliminary modeling includes defining the elements making up the system, numbering them, characterizing them in terms of their properties, and compiling the constraint matrices of type (14.1), describing the system mathematically. In this case, however, the elements are not characterized through the compilation of a vector, but of a matrix. This is because different properties must be expressed for each element. Indicating the number of properties to be taken into account with n_p, this matrix can be represented by:

$$P_{elem} = \left[p_{ij} \right]_{i=1,2,\ldots,n}^{j=1,2,\ldots,n_p} \qquad (14.4)$$

where, generally, the term p_{ij} expresses the j-th property of the i-th element. In this specific case, we consider six different properties ($n_p = 6$), defined as:

- p_{i1}, typology index corresponding to the i-th element (Table 14.1)
- p_{i2}, index of the material constituting the i-th element
- p_{i3}, weight W of the i-th element
- p_{i4}, reusability ϵ of the i-th element (assumes a value of 1 if the element can be reused; otherwise it is 0)
- p_{i5}, production cost C_{prod} of the i-th element
- p_{i6}, environmental impact of producing the i-th element

The compilation of the matrix (14.4) requires certain considerations. Above all, it is necessary to evaluate the weight of each element, the cost of production, and the environmental impact associated with its production (the latter evaluation will be treated below). The reusability of an element depends on the designed duration of the element's useful life, as a ratio of the anticipated working life of the entire product. If the duration is at least twice the working life of the product, the element is reusable in the same product. In the case where an element consists of several components characterized by different durations, the shortest one must be considered. A more detailed discussion of the reusability of product components and their possible recovery cycles will be treated in Chapter 15.

Finally, it is necessary to introduce an index expressing the material making up the i-th element. This requires the creation of a table analogous to Table 14.1 for element typology. In this case, however, each material index is associated with a set of data, as shown in Table 14.2, which summarizes the complete characterization.

Having chosen a congruous number of materials, each with an identifying index, it is possible to associate each material with the following terms that characterize it in relation to the requirements of the specific case:

- The recyclable fraction ξ (quantifying the recyclability of the material)

TABLE 14.2 Characterization of materials

MATERIAL INDEX	DESCRIPTION	RECYCLABILITY	ECONOMIC TERMS			ENVIRONMENTAL TERMS	
m	Material name	ξ	c_{Dsp}	c_{Rcl}	r_{Rcl}	ei_{Dsp}	ei_{Rcl}

- The costs of disposal c_{Dsp} and recycling c_{Rcl}, and the revenues from recycling r_{Rcl} per unit weight of material (terms necessary for an economic evaluation)
- The environmental impacts of disposal ei_{Dsp} and recycling ei_{Rcl} per unit weight of material (terms necessary for a strictly environmental evaluation)

Consequently, the database of the system in this case contains different types of data:

- Those concerning the typologies of elements and operations, as defined in Section 14.3.1 (i.e., the values of removal and handling times relative, respectively, to the typologies of elements and disassembly operations)
- Those concerning the complete characterization of the materials, as defined in Table 14.2

As was suggested in Chapter 12, information on the recyclable fractions of the more common materials may be obtained from commercially available databases, such as that of the material selection software CES® (Cambridge Engineering Selector, Granta Design Ltd., Cambridge, UK). The properties of environmental impact are discussed in detail below. To complete the considerations on preliminary modeling, it is necessary to complement the operation typologies o_k considered in Section 14.3.1 with an additional operation typology, the null operation (with an execution time of zero). The reasons for this will be explained below.

14.5.2   Advanced Modeling

The studies on disassembly planning found in the literature are generally characterized by a limited view of the environmental problem, which is treated by translating environmental aspects into economic costs, with the analysis confined to the end-of-life phase alone. The tool proposed here extends this viewpoint, integrating the cost functions with specific functions of environmental impact and including the potential to extend the analysis over the product's entire life cycle. This latter aspect, in particular, deserves more detailed discussion. In the case where it is necessary to disassemble and recover components from a preexisting product, it is sufficient that the economic and environmental analyses are limited to only the end-of-life phase in order to identify the most efficient recovery plan.

In the case where the problem of recovery is already being considered in the product design and development phase, rather than simply predicting the product's behavior only in the context of the end-of-life, it is more appropriate to have a complete vision of the product's behavior over the entire life cycle. It is, therefore, necessary to quantify the economic costs and environmental impacts beginning with the production phases, because it can happen that a design intervention aimed at improving the possibility of recovering the product at end-of-life can also lead to an increase in the environmental impact of production, annulling the environmental improvement obtained in the recovery phase. With this premise, the tool proposed requires further modeling described below, which is similar to the modeling introduced in Chapter 12.

14.5.2.1 Functions of the Environmental Impact of the Life Cycle

The properties of a material's environmental impact, expressed by the terms previously introduced as ei_{Dsp} and ei_{Rcl}, can be evaluated using the techniques of Life Cycle Assessment (Chapter 4) and the Eco-indicator 99 method, and calculated using SimaPro 5.0® software (Pré Consultants BV, Amersfoort, The Netherlands). With the same instruments, it is possible to evaluate the environmental impact associated with the production of each i-th element $EI_{Prod\,i}$, introduced in matrix (14.4). This can be expressed as:

$$EI_{Prod_i} = ei_{Mat_i} \cdot W_i + ei_{Prss_i} \cdot \mu_i \left(+ ei_{Mchg_i} \cdot \eta_i \right) \qquad (14.5)$$

where $ei_{Mat\,i}$ is the eco-indicator per unit weight of the material constituting the i-th element (weight is expressed by W_i) and $ei_{Pcss\,i}$ is the eco-indicator of the primary forming process per unit μ_i, representing the characteristic parameter of the process or the quantity of material processed. For a more accurate evaluation, it is also possible to consider the impact quota due to secondary machining processes (the term in parentheses, $ei_{Mchg\,i}$ is the eco-indicator of the machining process per unit of the characteristic process parameter μ_i). These eco-indicators can also be evaluated using the Eco-indicator 99 method and SimaPro software. If the i-th element consists of more than one component, its production impact will be the sum of the impacts of the single components.

Again using the same instruments to quantify the environmental impacts, given a system composed of n elements, the environmental impact of the life cycle of the i-th element $EI_{LC\,i}$ can be expressed as:

$$EI_{LC} = EI_{Prod_i} + EI_{EoL_i} \tag{14.6}$$

where the first term indicates the environmental impact of production, expressed by Equation (14.5). The second term indicates the environmental impact at the end-of-life of element $EI_{EoL\,i}$ and can, in turn, be defined as:

$$EI_{EoL_i} = \alpha_i \cdot \left(-EI_{Prod_i}\right) + \beta_i \cdot \left[ei_{Dsp_i} \cdot \left(1 - \xi_i\right) \cdot W_i + ei_{Rcl_i} \cdot \xi_i \cdot W_i\right] + \gamma_i \cdot \left(ei_{Dsp_i} \cdot W_i\right) \tag{14.7}$$

All the terms in this expression are already defined in Section 14.5.1, except the binary coefficients α_i, β_i, and γ_i. These depend on the final destination of the element. In fact, for each element all but one of these coefficients is zero, so that, depending on the case, Equation (14.7) quantifies the impact of reuse (first term), of recycling (second term), or of disposal (third term).

It should be noted that:

- The first term, in reality, expresses a recovery of impact, since reusing an element allows the recovery of the environmental impact associated with its production.

- The second term quantifies the impact of recycling the recyclable fraction of material (this impact can also assume a negative value, that is, it can also be a recovery of impact), and the impact of disposing of the remaining nonrecyclable fraction.

- The third term quantifies the impact of disposing of the material of the whole element.

- In the case where the i-th element consists of more than one component (e.g., subgroups of components linked by irreversible fasteners), the three terms must be calculated separately for each component. The same is true when evaluating the environmental impact of production $EI_{Prod\,i}$.

For the whole system of n elements, it is ultimately possible to express the environmental impact functions of the entire life cycle EI_{LC}, or of only the end-of-life phase EI_{EoL}:

$$EI_{LC} = \sum_{i=1}^{n} EI_{LC_i}, \quad EI_{EoL} = \sum_{i=1}^{n} EI_{EoL\,i} \tag{14.8}$$

14.5.2.2 Recovery Planning

From the analytical point of view, the coefficients α_i, β_i, and γ_i may be considered coefficients of recovery planning and can be expressed as functions of the reusability of element ϵ_i, and of a term that, as will become clear below, constitutes a variable of the optimization process. This latter variable indicates whether it is preferable that the element under examination is recovered or disposed of as waste. Labeling this term with the binary index s_i, called the destination index, it will assume a value of 1 or 0 depending on whether the corresponding i-th element is destined for disposal or recovery. The recovery planning coefficients can then be defined as:

$$\begin{cases} \alpha_i = \epsilon_i \cdot (1 - s_i) \\ \beta_i = (1 - \epsilon_i) \cdot (1 - s_i) \\ \gamma_i = s_i \end{cases} \tag{14.9}$$

It should be noted that if an element is to be recovered ($s_i = 0$), preference is given to reusing material rather than recycling it; recycling is chosen only when ϵ_i is 0 (i.e., when the element is not reusable).

Again considering the index s_i, it is appropriate to set the condition that this can assume a value of 0 (recovery of the element) only for disassembled elements. This equates to forcing any nondisassembled elements to be disposed of as waste in all cases. In fact, they could be recovered only in the particular case where they constitute a recoverable subgroup (i.e., composed of a single recyclable material, of materials compatible with the aims of recycling, or of elements that are all reusable, a condition that is not examined here).

The index s_i can, instead, assume a value of 1 (disposal of the element) independent of whether or not the corresponding element is disassembled. It would, therefore, seem superfluous (if not incorrect) to examine the possibility of having to dispose of a disassembled element. However, this is necessary since it may be advantageous to recover an element even when its removal requires the disassembly of another element that can only be destined for disposal, or that (although already disassembled) it is more advantageous to consider waste.

14.5.2.3 Functions of the Costs of the Life Cycle

With regard to the cost functions, analogously to Equation (14.7), it is possible to define the end-of-life cost of element $C_{EoL\,i}$:

$$\begin{aligned} C_{EoL_i} = {} & \alpha_i \cdot \left(- C_{Prod_i} \right) + \beta_i \cdot \left[c_{Dsp_i} \cdot (1 - \xi_i) \cdot W_i \right. \\ & \left. + \left(c_{Rcl_i} - r_{Rcl_i} \right) \cdot \xi_i \cdot W_i \right] + \gamma_i \cdot \left(c_{Dsp_i} \cdot W_i \right) \end{aligned} \tag{14.10}$$

The qualitative considerations made previously for Equation (14.7) again apply here. The only formal difference between Equations (14.10) and (14.7) is found in the second term, which in Equation (14.10) includes the revenues obtained from recycling the material, balancing the cost of recycling. In Equation (14.7), there is no term of impact recovery since, as noted, it is already included in the calculation of ei_{Rcl}. For the whole system of n elements, it is possible to express the cost functions of the entire life cycle C_{LC}, or of only the end-of-life phase C_{EoL} as:

$$C_{LC} = \sum_{i=1}^{n} \left(C_{Prod_i} + C_{EoL_i} \right) + C_{Dis} \quad C_{EoL} = \sum_{i=1}^{n} C_{EoL_i} \qquad (14.11)$$

From the first term of Equation (14.11), it is seen that the cost of the entire life cycle includes the production costs for each element $C_{Prod\,i}$ as well as the end-of-life costs $C_{EoL\,i}$ already defined, and the total cost of the entire process of disassembly C_{Dis}. The production costs can be expressed in a form analogous to Equation (14.5), as a function of the quantity of material to be used and of the more significant process parameters. Alternatively, it is possible to turn to a conventional evaluation of a component's production costs (Ulrich and Eppinger, 2000).

Finally, the cost of disassembly C_{Dis} can be evaluated using:

$$C_{Dis} = c_{Dis} \cdot \left(TS_{i*} \cdot t_{OR} \right) \qquad (14.12)$$

where c_{Dis} is the disassembly cost per unit time and TS_{i*} is the disassembly time defined in function (14.2), calculated for the final element i* of the disassembly sequence. Since this last term expresses a time normalized with respect to the operation of horizontal translation, it must be multiplied by the real time t_{OR} for this reference operation. In the case where it is necessary to take into account the variable difficulty of handling an element according to its weight or bulk, t_{OR} can be expressed as a function of the element's weight or volume rather than a constant value.

14.5.3   Identification of the System

With regard to the identification of the system by the search algorithm, also in this second case, the formalization of the solution type to be investigated comes down to a vector of integer numbers. Here, however, the vector is composed of triplets of indices (Figure 14.3b). Again, the first index corresponds to the element index i, following the preliminary numeration, and the second index corresponds to the operation typology index o_k, relating to the operation that must be performed to remove the i-th element. It should be noted that in this case it is also necessary to consider a "null" operation typology.

The first two indices are complemented by a third index, the destination index (s) introduced in Section 14.5.2.2. Depending on whether this assumes a value of 0 or 1, the corresponding element will be destined for recovery or for disposal. In this case, the objective function must take into account other significant factors expressing the costs and environmental impacts of the life cycle. This is, therefore, an example of optimization using a multiobjective genetic algorithm (Fonseca and Fleming, 1995). Of the possible approaches to the problem, the one chosen here is that of the weighted sum, recalling a typical formulation of conventional multiobjective optimization. The objective function to be minimized is expressed by the weighted sum of three factors:

- The disassembly time TS, expressed by function (14.2), where the summation is extended to all the disassembled elements
- The cost of the entire life cycle C_{LC} (ignoring the term C_{Dis}, the linear function of the first factor TS), or only the cost of end-of-life C_{EoL} expressed by Equations (14.11)
- The environmental impact of the entire life cycle EI_{LC}, or only the impact of end-of-life EI_{EoL} expressed by Equations (14.8)

This ultimately produces:

$$\Psi = \varphi_1 \cdot TS + \varphi_2 \cdot C_{LC} + \varphi_3 \cdot EI_{LC} \tag{14.13}$$

where φ_1, φ_2, and φ_3 are the weight coefficients of three factors ($\varphi_1 + \varphi_2 + \varphi_3 = 1$).

Given that the factors are not homogeneous, they must be normalized before being introduced into Equation (14.13). With this aim, given a solution to be evaluated it is sufficient to compare the value assumed by each factor, with the maximum value assumed by the same factor in all the solutions under examination. A set of three normalized factors is obtained for each of the solutions to be evaluated. The optimal solution is that corresponding to the lowest value of the function Ψ.

It should be noted that in the context of evolutionary multiobjective optimization, the weight coefficients also generally evolve during the process (Gen and Cheng, 2000). However, in the formulation proposed here these coefficients are excluded from the process of evolution, so that they can be used as parameters to freely direct the investigation.

14.5.4 Generation of Disassembly Sequences and Identification of the Optimal Solution

This second tool takes advantage of the same genetic motor as the first. The generation of disassembly sequences and recovery plans is again intrinsic to the very functioning of the genetic algorithm. Each individual of each generation (represented by the vector in Figure 14.3b), although generated in a completely random manner, in fact represents a potential disassembly sequence (through the first two indices of each triplet) and a potential recovery plan (through the third index). Also in this case, if the sequence of the triplets respects the constraint matrices, and thus the individual represents a truly practicable disassembly sequence, the fitness is evaluated by the calculation of the value assumed by the multiobjective function (14.13).

When the algorithm stops, it provides the optimal (or near-optimal) solution, again expressed in the form of Figure 14.3b, consisting of a series of element–operation–destination triplets, ordered in a way that the constraint matrices are respected. Unlike the previous case, here the series of triplets will not stop at just the elements to be disassembled, but will include them all. It is not certain, however, that the optimal recovery plan will indicate the complete disassembly of the system. As mentioned before, the tool is also designed to define the optimal level of disassembly. Elements that need not be disassembled will, therefore, be identified by the value assumed by the operation index o_k, which will assume the value corresponding to the null operation. For these elements, the third index (s) will always have a value of 1, indicating that the nondisassembled components can only be disposed of as waste.

14.6 Simulations and Analysis of Results

Implementing both of the proposed tools requires the use of appropriate software. In the context of the practical experience reported here, in order to provide a complete description of the implementation phase and of the practical use of the tools, the prototypes were developed using MATLAB® (The MathWorks, Inc., Natick, MA) and tested in a series of simulations on various mechanical systems, characterized by different architecture typologies. To lighten the calculation, the investigation was conducted in the two-dimensional field and on assemblies consisting of a maximum number of elements limited to 10 to 15.

On the basis of the results obtained, the first prototype was found to be an efficient tool for identifying selective disassembly sequences, always arriving at solutions describing feasible sequences, and often at absolute optimal sequences. The second prototype also provides operationally feasible solutions and further guides the optimal planning of disassembly and recovery, by varying the weights given to the economic and environmental aspects in the definition of the multiobjective function. Two significant cases, one for each prototype, are described in detail below. These show the analysis of the same assembly described in Figure 14.4, consisting of 11 elements (6 components and 5 screw fasteners).

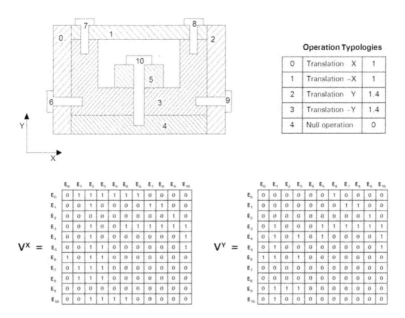

FIGURE 14.4 Case study: Mechanical system.

Since the problem is limited to only two dimensions, the possible typologies of disassembly operations o_k in this case are only the four translations along X, $-$X, Y, and $-$Y, plus the null operation. As reported in the table of Figure 14.4, a value of 1.4 was associated with the vertical translations Y and $-$Y as execution times normalized with respect to horizontal translations. The corresponding constraint matrices were compiled for each of the four possible directions of translation (the figure shows those relative to the two axes, X and Y).

14.6.1 Prototype 1: Selective Disassembly

Simulations where the first prototype was applied on different mechanical assemblies generally gave excellent results. In the case under examination (Figure 14.4), the elements were first characterized through the compilation of the vector (14.3), which associates the relevant typology with each element. Different target elements were then selected (target of the disassembly).

Figure 14.5 shows the more significant results, in the case where element 5 was chosen as the target of the disassembly. These results regard five different simulations performed on the assembly, varying the total number of generations defined as the algorithm's stop criterion (100, 150, 200, 250, and 350, respectively). The solutions identified as optimal are reported for each simulation, with each solution consisting of an ordered series of element–operation

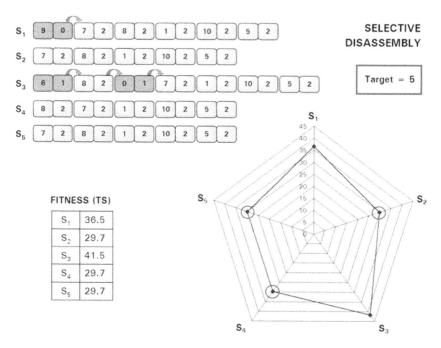

FIGURE 14.5   Selective disassembly: Results of simulation.

couplets. It is clear that, in all cases, the solutions describe feasible disassembly sequences. The simulations S_2, S_4, and S_5 identify the absolute, optimal solution, which, due to the simplicity of the assembly in this case, is easily definable a priori. These simulations all have the same value for the objective function (used as fitness of the algorithm), which coincides with the absolute minimum (normalized disassembly time TS = 29.7).

Simulation S_1 identifies a near-optimal solution, in that it differs from the optimal only in one unnecessary disassembly operation (first couplet, highlighted in the figure), which also results in a change in direction during the disassembly sequence (indicated by curved arrow). Finally, simulation S_3 identifies the least efficient solution, since it foresees the unnecessary disassembly of two elements and three changes in direction, with a considerable increase in the disassembly time (TS = 41.5). The radar graph allows a comparison of the excellence of the results obtained in the simulations, highlighting the three simulations that identified the absolute optimal.

14.6.2   Prototype 2: Partial or Complete Disassembly

The application of the second prototype on the same assembly (Figure 14.4) necessitates additional information for the characterization of the elements (as noted in Section 14.5.1). The compilation of the characterization matrices (14.4) in fact requires, for each element, the typology index together with its constituent material, weight, reusability, and cost and environmental impact of production. Table 14.3 shows the materials and reusability considered for the simulation. The database section includes all the necessary economic and environmental data related to different typologies of materials.

Different investigations were conducted on the same assembly, varying the weight coefficients of the multiobjective function Ψ [function (14.13)]. However, given that it was not possible to compare the results with an optimal solution known a priori, a series of simulations were conducted for each investigation (defined by a specific set of weight coefficients φ), varying the parameters f_1 and f_2 (introduced in Section 14.3.3 to parameterize the algorithm's stop criterion according to a principle of stability of the solutions). Together with a verification of the process of convergence of the algorithm, this investigation allowed the validation of the excellence of the solutions identified as optimal.

Figure 14.6 shows the more significant results from the series of simulations performed with a set of uniform weight coefficients ($\varphi_1 = \varphi_2 = \varphi_3$) and taking the disassembly time, cost of production, and environmental impact of the entire life cycle as factors of the multiobjective function (14.13). On varying the parameters f_1 and f_2 defining the stop condition of the algorithm, the two solutions presented in the figure were identified as optimal disassembly sequences. Each solution is represented by an ordered series of element–operation–destination triplets, which express both the disassembly sequence and the recovery plan. Both solutions suggest conducting a complete disassembly (maximum level). Otherwise, for an element that was not to be removed, the second index of the corresponding triplet would indicate the null operation (table in Figure 14.4).

TABLE 14.3 Partial or complete disassembly: Definition of materials and reusability of elements

ELEMENT	DESCRIPTION	MATERIAL	REUSABILITY (ε_j)
0	Component	ABS	0
1	Component	ABS	0
2	Component	ABS	0
3	Component	Aluminum	0
4	Component	Aluminum	0
5	Component	Steel	1
6	Screw	Steel	0
7	Screw	Steel	0
8	Screw	Steel	0
9	Screw	Steel	0
10	Screw	Steel	1

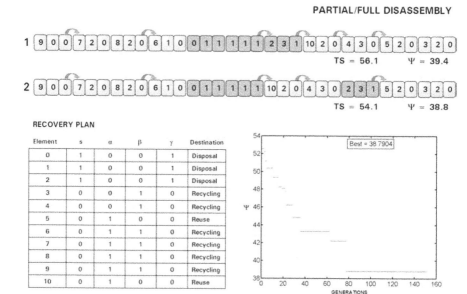

FIGURE 14.6   Partial or complete disassembly: Results of simulation.

Furthermore, the two solutions indicate the same recovery plan, summarized in the figure table, for all the elements of the mechanical system. In this regard, it should be noted that the coefficients of planning and recovery α, β, and γ are evaluated in Equation (14.9) as a function of the third index of each triplet (destination index [s]), and of the element's reusability (Table 14.3).
It is interesting to note that:

- The two solutions differ exclusively in the efficiency of the disassembly sequence, since the second indicates one less change in direction, with a consequent improvement both in the disassembly time TS (−3.5%) and in the final value of the function Ψ (−1.5%).

- The values attributed by the algorithm to the destination index s are entirely coherent with the reusability of the elements and the recyclability properties of the materials. In fact, for elements 0, 1, and 2 (nonreusable, ABS/low recyclable fraction) the algorithm suggests their disposal as waste; for elements 3 and 4 (nonreusable, aluminum/elevated recyclable fraction) it suggests recycling; for elements 5 and 10 (reusable, steel/high recyclable fraction) it suggests reuse; and for elements 6 through 9 (nonreusable, steel/high recyclable fraction) it suggests recycling.

- Despite elements 0, 1, and 2 being destined for disposal (as indicated by the triplets highlighted in the figure), it is nevertheless advantageous to disassemble them in order to recover the other elements.

Finally, Figure 14.6 also shows the graph of the algorithm's convergence toward the optimal solution in the second case ($\Psi_{optimal}$ = 38.8). This graph evidences how, starting from a random population with elevated values of Ψ, the results of the genetic algorithm evolve over successive generations toward solutions with ever-better fitness, converging on the optimal value of Ψ and stabilizing on this.

14.7 Summary

This chapter proposes the definition of a systematic methodology that is able to integrate the different approaches to Disassembly Planning, and to guide in solving different problems using the most effective approach. On the basis of a series of simulations performed on several mechanical assemblies of different typologies, the prototype of the first tool proposed was found to be valid for determining sequences of selective disassembly, defining sequences that are feasible and often coincident with the absolute optimum solution. The prototype of the second tool provided operationally feasible solutions and also guided the optimal planning of disassembly and recovery by varying the weights attributed to economic and environmental aspects in the definition of the multiobjective function.

Furthermore both tools lend themselves not only to the characteristic aims of Disassembly Planning but also to use as aids in Design for Disassembly. The first can be used in planning servicing operations when one or more specific elements must be removed, where it would aid design interventions directed at optimizing the system to facilitate servicing operations (Design for Serviceability). This tool for selective disassembly planning could also be particularly useful for characterizing the difficulty of disassembling a component; that is a basic issue in the method of disassembly depth analysis proposed in Chapter 13.

Because of its particular characteristic of being based on an analysis that can be extended over the entire life cycle, the second prototype can be used as a tool for optimizing the system not only for design specifically oriented toward favoring a product's end-of-life phase (Design for Recovery), but also for a more attentive design intervention concentrating on the product's overall environmental performance. In fact, it evaluates the optimal level of disassembly of a product and the most appropriate final destination for the disassembled components, correlating these with several properties of the components that characterize their behavior over the entire life cycle.

14.8 References

Bonneville, F., Perrard, C., and Henrioud, J.M., A genetic algorithm to generate and evaluate assembly plans, in *Proceedings of INRIA/IEEE Symposium on Emerging Technologies and Factory Automation*, Paris, 1995, 231–239.

Caccia, C. and Pozzetti, A., Genetic algorithm for disassembly strategy definition, in *Proceedings of SPIE Conference on Environmentally Conscious Manufacturing*, Boston, 2000, 68–77.

De Jong, K.A., An analysis of the behaviour of a class of genetic adaptive systems, Ph.D. dissertation, University of Michigan, Ann Arbor, MI, 1975.

Dini, G. et al., Generation of optimized assembly sequences using genetic algorithms, *Annals of the CIRP*, 48(1), 17–20, 1999.

Dini, G., Failli, F., and Santochi, M., A disassembly planning software system for the optimization of recycling processes, *Production Planning and Control*, 12(1), 2–12, 2001.

Dowie, T. and Kelly, P., Estimation of Disassembly Times, Report DDR/TR15, Manchester Metropolitan University, Manchester, UK, 1994.

Fonseca, C.M. and Fleming, P.J., An overview of evolutionary algorithms in multiobjective optimisation, *Evolutionary Computation*, 3(1), 1–16, 1995.

Gen, M. and Cheng, R., *Genetic Algorithms and Engineering Optimisation*, John Wiley & Sons, New York, 2000.

Goldberg, D.E., *Genetic Algorithms in Search, Optimisation and Machine Learning*, Addison-Wesley, Reading, MA, 1989.

Gungor, A. and Gupta, S.M., An evaluation methodology for disassembly processes, *Computers and Industrial Engineering*, 33, 329–332, 1997.

Gungor, A. and Gupta, S.M., Disassembly sequence planning for complete disassembly in product recovery, in *Proceedings of Northeast Decision Sciences Institute Conference*, Boston, 1998, 250–252.

Gungor, A. and Gupta, S.M., Issues in environmentally conscious manufacturing and product recovery: A survey, *Computers and Industrial Engineering*, 36, 811–853, 1999.

Holland, J.H., *Adaptation in Natural and Artificial Systems*, University of Michigan Press, Ann Arbor, MI, 1975.

Homem De Mello, L.S. and Sanderson, A.C., A correct and complete algorithm for the generation of mechanical assembly sequences, *IEEE Transactions on Robotics and Automation*, 7(2), 228–240, 1991.

Hulin, M., An optimal stop criterion for genetic algorithms: A Bayesian approach, in *Proceedings of ICGA International Conference on Genetic Algorithms*, Michigan State University, East Lansing, MI, 1997, 135–143.

Jovane, F. et al., A key issue in product life cycle: Disassembly, *Annals of the CIRP*, 42(2), 651–658, 1993.

Kroll, E. and Carver, B.S., Disassembly analysis through time estimation and other metrics, *Robotics and Computer Integrated Manufacturing*, 15(3), 191–200, 1999.

Kuo, T. C., Zhang, H.C., and Huang, S.H., Disassembly analysis for electromechanical products: A graph-based heuristic approach, *International Journal of Production Research*, 38(5), 993–1007, 2000.

Lambert, A.J.D., Optimal disassembly of complex products, *International Journal of Production Research*, 35(9), 2509–2523, 1997.

Lambert, A.J.D., Disassembly sequencing: A survey, *International Journal of Production Research*, 41(16), 3721–3759, 2003.

Li, J.R., Khoo, L.P., and Tor, S.B., A novel representation scheme for disassembly sequence planning, *International Journal of Advanced Manufacturing Technology*, 20(8), 621–630, 2002.

Mitchell, M., *An Introduction to Genetic Algorithms*, MIT Press, Cambridge, MA, 1998.

Moyer, L.K. and Gupta, S.M., Environmental concerns and recycling/disassembly effects in the electronics industry, *Journal of Electronics Manufacturing*, 7(1), 1–22, 1997.

Navin-Chandra, D., ReStar: A design tool for environmental recovery analysis, in *Proceedings of ICED International Conference on Engineering Design*, The Hauge, The Netherlands, 1993, 780–787.

Navin-Chandra, D., The recovery problem in product design, *Journal of Engineering Design*, 5(1), 67–87, 1994.

Oman, S. and Cunningham, P., Using case retrieval to seed genetic algorithms, *International Journal of Computational Intelligence and Applications*, 1(1), 71–82, 2001.

O'Shea, B., Grewal, S.S., and Kaebernick, H., State of the art literature survey on disassembly planning, *Concurrent Engineering*, 6(4), 345–357, 1998.

Pnueli, Y. and Zussman, E., Evaluating the end-of-life value of a product and improving it by redesign, *International Journal of Production Research*, 35(4), 921–942, 1997.

Seo, K.K., Park, J.H., and Jang, D.S., Optimal disassembly sequence using genetic algorithms considering economic and environmental aspects, *International Journal of Advanced Manufacturing Technology*, 18(5), 371–380, 2001.

Sodhi, R., Sonnenberg, M., and Das, S., Evaluating the unfastening effort in design for disassembly and serviceability, *Journal of Engineering Design*, 15(1), 69–90, 2004.

Srinivasan, H. and Gadh, R., A non-interfering selective disassembly sequence for components with geometric constraints, *IIE Transactions*, 34, 349–361, 2002.

Srinivasan, H., Shyamsundar, N., and Gadh, R., A framework for virtual disassembly analysis, *Journal of Intelligent Manufacturing*, 8, 277–295, 1997.

Subramani, A.K. and Dewhurst, P., Automatic generation of product disassembly sequences, *Annals of the CIRP*, 40(1), 115–118, 1991.

Tang, Y. et al., Disassembly modelling, planning, and application: A review, in *Proceedings of IEEE International Conference on Robotics and Automation*, San Francisco, 2000, 2197–2202.

Ulrich, K.T. and Eppinger, S.D., *Product Design and Development*, 2nd ed., McGraw-Hill, New York, 2000.

Vujosevic, R. et al., Simulation, animation, and analysis of design disassembly for maintainability analysis, *International Journal of Production Research*, 33(11), 2999–3022, 1995.

Woo, T.C. and Dutta, D., Automatic disassembly and total ordering in three dimensions, *Journal of Engineering for Industry*, 113, 207–213, 1991.

Yokota, K. and Brough, D.R., Assembly/disassembly sequence planning, *Assembly Automation*, 12(3), 31–38, 1992.

Zhang, H.C. and Kuo, T.C., A graph-based disassembly sequence planning for end-of-life product recycling, in *Proceedings of IEEE/CPMT International Electronics Manufacturing Technology Symposium*, Austin, TX, 1997, 140–151.

Zussman, E., Kriwet, A., and Seliger, G., Disassembly-oriented methodology to support design for recycling, *Annals of the CIRP*, 43(1), 9–14, 1994.

Chapter 15

Product Recovery Cycles Planning and Cost–Benefit Analysis of Recovery

Various factors influence the suitability of a product to be recovered at different levels (direct reuse, reuse of parts, recycling). An efficient recovery operation, therefore, requires the correct evaluation of all these factors (predicted lifespan, deterioration in performance, recyclability of materials, separability of components from the constructional system). This type of analysis is thus essential in the design phase in order to plan possible recovery cycles at the end of the product's life, and for an evaluation of the efficiency of the product's architecture and component details. An increase in the volumes recovered can be obtained by modifying certain characteristics of product architecture, or by simply varying certain design choices regarding materials or shape of components, in a way that allows a recovery plan involving greater volume flows. In most cases, however, modifications of this type result in higher production costs. In order to plan economically sustainable recovery operations it is thus necessary to evaluate their economic impact. In particular, it is important to determine the conditions under which an increase in the volumes recovered, which entails an undoubted environmental benefit, is also economically advantageous.

This chapter describes a method to aid designers in making the best choices in order to plan the recovery at reuse levels (direct reuse, reuse of parts) for a product at the end of its useful life. Having outlined an effective design methodology for this purpose, calculation models are introduced here for several reasons: to support the quantification of determinant factors, allowing the definition of the reusable parts of the product and the recovery cycles planning and simulation; to evaluate an indicator that translates the environmental effects of recovery cycles in terms of extension of the product's useful life; and to determine the condition for which an increase in the volumes of recovery, of undoubted environmental benefit, is also advantageous from the economic point of view.

A case study on a commonly used industrial unit, a heat exchanger, explains the implementation of the design tool and its calculation models in comparing different constructional systems of a product and their optimization, confirming its effectiveness as a tool for Design for Recovery approach. In particular, the investigation into the relation between the increase in recovery flows and the resulting effects on production costs shows that the economic advantage of modifying the architecture or the choice of materials to achieve an improvement in the recoverability of product components is linked to the possibility of programming an adequate number of recovery cycles.

The fundamental issues in this chapter were previously introduced in Chapters 9 and 11.

15.1 Approach to the Recovery Problem

As stated previously, one of the main objectives of Design for Environment is to optimize the distribution of the flows of resources over the product's whole life cycle; this can be backed by different strategies, as described in Chapters 8 and 9. Of particular importance are those concerning the recovery of components and materials at the end of the product's working life (disassembly, selection, inspection, direct reuse of components, recycling material in primary or external production cycles), in that they are directed at closing the resource flow cycles (Figure 15.1).

In general, the costs of recovery (disassembly, testing, remanufacturing) become prohibitive with increasing depth of the recovery operation. The gains,

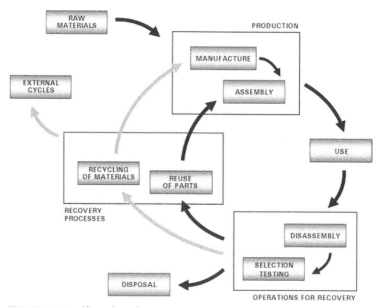

FIGURE 15.1   Life cycle and recovery strategies.

on the other hand, tend to stabilize after a certain depth of recovery is reached; refer to Chapter 9, Figure 9.5. In order to plan economically sustainable recovery operations it is therefore necessary to evaluate their economic impact.

The implementation of the strategies for recovery at the end-of-life translate into design approaches that target disassembly, recovery, and recycling. These design approaches, known as Design for Disassembly and Recycling (Chapter 9, Section 9.3.3), must be complemented by an analysis that evaluates the not-insignificant life cycle costs (Chen et al., 1994; Giudice et al., 1999b).

The suitability and effectiveness of the recovery operations at the end of a product's life (disassembly of parts, selection and inspection of their condition, reuse of complete components, recycling materials in the primary production cycle or in external cycles) are strictly dependent on several determinant factors (see Chapter 11) that characterize the constructional system:

- The detachability of components (depending on the layout of the unit and on the geometry and the junctions between parts)
- The property of component durability, which conditions the process of declining performance over time, and consequently determines their suitability for reuse
- The appropriateness and suitability of recycling material from parts that are not reusable

On the basis of the experience of manufacturers operating in various sectors (e.g., in the industrial plant sector), for some product typologies it seems that resources are generally used in an inefficient way—some parts of a system might still be in working order at the end of its conventional useful life and, therefore, could potentially guarantee further durability, but are nevertheless thrown out with the entire product. This has focused attention on the interesting margins to which component reuse could be applied (see the closed cycle highlighted in Figure 15.1). In light of these considerations, the method proposed here is able to translate certain determining design choices into opportunities for component recovery and reuse by means of a mathematical model that simulates possible recovery cycles at the end of the product's working life and quantifies the benefit in environmental terms. This approach allows product constructional systems to be compared and optimized, in terms of components separability and the distribution of their physical life in the unit, for an effective and sustainable recovery of resources.

15.2 Method for Recovery Cycles Planning

The problem under examination and the stated objective suggest the use of a design methodology such as that schematized in Figure 15.2.

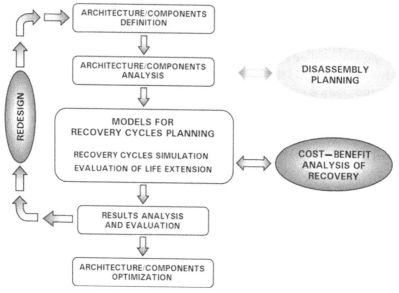

FIGURE 15.2   Design methodology.

The first phase, defining the overall architecture of the product and its components using conventional design tools, consists of identifying the main components of the system, their spatial and functional arrangement, their jointing systems, and the overall geometry and materials of each component. This is followed by a phase in which the conventional constructional system so defined is analyzed in relation to the recovery strategies. This analysis seeks to interpret the system in terms of component modules, highlighting the property of durability of performance in time and the junction restraints (reversible or irreversible) that condition the possibility of separating a component from the rest of the unit (see Chapter 11).

With regard to this last point, a qualitative analysis based on simple observation of the product's architecture and junction systems can be complemented by an analytical approach to the problem of evaluating the separability of each component, or module, using appropriate approaches to the analysis of disassembly plans. This allows the design method to be integrated with the tools of Design and Planning for Disassembly (described in Chapters 13 and 14), which are directed at optimizing the disassembly procedure by reducing the associated costs.

The data obtained in the system analysis is then elaborated using the calculation models proposed here that define and simulate the possible recovery cycles, and evaluate their environmental effectiveness by means of an indicator that quantifies the improvement of exploiting the resources used, in terms of extending the product's useful life. This approach must be complemented by an analysis that evaluates the not-insignificant life cycle costs related to the recovery strategies. The analytical formulation of the problem allows the proposed tool to be integrated with others for the cost–benefit analysis of recovery processes (as highlighted in Figure 15.2), such as those already proposed by the authors (Giudice et al., 1999b). This specific problem will be discussed subsequently.

The final phase consists of evaluating the results obtained with applied calculation models, which can provide suggestions for a correct redesign of the system to obtain a definition that reconciles the requirements of conventional design (functionality, costs) with environmental requirements.

15.3   Calculation Models for Recovery Cycles Planning

The calculation models proposed here, developed as an aid to the design of product recovery cycles, analyze the architecture and components data, translating them into the suitability of each component for reuse and calculating the outcome of the possible recovery cycles at the end of the product's ordinary working life. They quantify the advantages in terms of the recovery value of resources used in production, and consequent environmental protection.

15.3.1   Basic Procedure for Implementing the Calculation Models

The implementation of the calculation models, in the context of the design methodology represented in Figure 15.2, is based on a procedure summarized in Figure 15.3.

The data supplied by the analysis of the architecture and components are initially elaborated to identify the determinant factors for component recovery (i.e., the property of each component that determines the possibility and suitability of its recovery and reuse). As has been shown, these factors were identified as the predicted duration of functional efficiency of each component, and as their separability from the rest of the system. It is then possible to evaluate the reusability of the components (i.e., the possibility of reusing them one or more times) and consequently to plan all the possible recovery cycles. The final phase of the calculation procedure translates the recovery plans into improved use of the resources, introducing an indicator of the extension of the product's useful life.

15.3.2   Determinant Factors for Recovery

The analysis of the overall system, conceived according to conventional design criteria, allows the identification of n main components C_i, each characterized by weight W_i. Having identified the components, which may be grouped

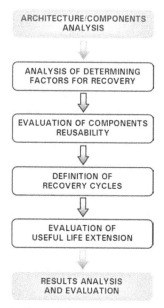

FIGURE 15.3   Basic procedure of calculation model.

in construction modules, two factors determining their suitability for recovery and reuse must be defined: durability and separability.

The durability D_i of the i-th component is defined as the ratio between the predictable duration of component PD_i and the duration of the product's conventional working life T:

$$D_i = int\left(\frac{PD_i}{T}\right) \qquad (15.1)$$

The term PD_i depends on design choices and is quantifiable on the basis of modern design methods, which makes it possible to ensure the complete efficiency of the component in time (Chapter 10), and on the basis of information obtained from the past behavior of preexisting components. The duration T is the time predicted for the normal use of the entire product without the substitution of any part.

The separability of the i-th component expresses the possibility of disassembling and separating the component from the unit in order to recover it. This depends not only on the layout of the architecture and on the junction constraints, but also on the relation between the component in question and other components characterized by the same durability. In recovery terms, a component is separable if it can be disassembled from the unit or if it belongs to a group of components characterized by the same durability and, therefore, constituting a single recoverable subunit, which can be disassembled.

To better define the separability, from Equation (15.1) it is possible to find the number of potential reuses R_i^* of each i-th component on the basis of its durability:

$$R_i^* = D_i - 1 \qquad (15.2)$$

Consequently, the i-th component potentially participates in the j-th recovery cycle if the condition $R_i^* \geq j$ is true. By verifying conditions of this type, it is possible to identify the group of components which, for durability, could be included in each potential recovery cycle.

The separability S_{ij} of the i-th component at the j-th recovery cycle is thus defined as follows: S_{ij} assumes a value 1 if it is possible to disassemble the single i-th component or the group of components potentially reusable at the j-th recovery cycle to which the i-th component belongs; otherwise it is equal to 0.

15.3.3   Determinant Factor Matrices and Recovery Vectors

If m^* indicates the overall number of potential recovery cycles ($m^* = \max(R_i^*)$), it is then possible to define the m^* matrices of the determinant factors for each j-th recovery M_j^{FR}, which compiles the durability and separability factors of all the components:

$$M_j^{FR} = \left[R_i^* \; S_{ij} \right]_{i=1,2,\ldots,n} \qquad (15.3)$$

and the corresponding recovery vectors V_j^R

$$V_j^R = \left[V_{ij}^R \right]_{i=1,2,\ldots,n} \qquad (15.4)$$

which compile the n terms V_{ij}^R defined thus:

$$V_{ij}^R = R_i^* \cdot S_{ij} \qquad (15.5)$$

For the i-th component to be effectively reusable at the j-th recovery cycle, the condition $V_{ij}^R \geq j$ must be true (because $R_i^* \geq j$ and $S_{ij} = 1$ must be true).

15.3.4   Effective Component Reusability

The effective reusability of the i-th component can thus be expressed in two ways. The relative reusability r_{ij}, which expresses the reusability of the i-th component at the j-th recovery cycle, is defined on the basis of condition $V_{ij}^R \geq j$ as follows:

$$r_{ij} = \begin{cases} 0 \text{ if } V_{ij}^R < j \\ 1 \text{ if } V_{ij}^R \geq j \end{cases} \tag{15.6}$$

and, therefore, assumes a value 1 if the i-th component is reusable at the j-th recovery cycle; otherwise it is equal to 0.

The absolute reusability R_i, which expresses the total number of recovery cycles that the i-th component can undergo after its first conventional use, is then given by:

$$R_i = \sum_{j=1}^{m} r_{ij} \tag{15.7}$$

where m is the maximum number of foreseeable recovery cycles for the entire unit, a function of the m^* and of r_{ij} terms. If $R_i = 0$, the i-th component is not reusable; if $R_i \geq 1$, the component can be reused an additional R_i times.

15.3.5   Recovery Fractions

The recovery fraction $\Phi^W{}_j$, which expresses the fraction of the total weight of the product reusable at the j-th recovery cycle, is expressed by:

$$\Phi_j^w = \frac{\sum_{i=1}^{n} r_{ij} \cdot W_i}{\sum_{i=1}^{n} W_i} \tag{15.8}$$

These terms define the quantity of material in play in each recovery cycle, and, therefore, already give a quantitative indication of the efficacy and the benefits of the recovery operations. To better characterize these in terms of environmental impact, however, it is appropriate to express them according to the environmental impact associated with the production of each component. This is possible using Life Cycle Assessment (LCA) techniques and the environmental indicators available in the literature (see Chapter 4 and Chapters 11 through 14 of Part III). These indicators are used to quantify the impact involved in the production of the more common materials and in standard manufacturing processes. Indicating the environmental impact of i-th component production by $EI_{Prod\,i}$, this can be expressed as:

$$EI_{Prod_i} = EI_{Mat_i} + EI_{Mfct_i} = ei_{Mat_i} \cdot W_i + \sum_{k=1}^{h_i} ei_{Pr_k\ i} \cdot \eta_{Pr_k\ i} \tag{15.9}$$

where:

- $ei_{Mat\,i}$ is the unitary eco-indicator for the material making up the i-th component.
- W_i is the weight of the i-th component.
- $ei_{pr_k\,i}$ is the eco-indicator that can be associated with the k-th manufacturing process of the i-th component's production cycle, per unit of main process parameter.
- $\eta_{pr_k\,i}$ is the main process parameter of the k-th manufacturing process of the i-th component's production cycle (e.g., in the case of a cutting process, $\eta_{pr_k\,i}$ is equal to the length of the cut).
- h_i is the total number of manufacturing processes of the i-th component's production cycle.

This expression of $EI_{Prod\,i}$ is similar to those previously introduced in Chapters 12 and 14. It differs exclusively in the generalization of the impact due to the manufacturing processes.

The recovery fraction Φ^{EI}_j, which expresses the fraction of the product weight reusable at the j-th recovery cycle in terms of environmental impact, is expressed by:

$$\Phi^{EI}_j = \frac{\sum_{i=1}^{n} r_{ij} \cdot EI_{Prod_i}}{\sum_{i=1}^{n} EI_{Prod_i}} \tag{15.10}$$

Functions (15.8) and (15.10) express the fraction of the product weight reusable and the fraction of the product "environmental weight" reusable, respectively. They define the recovery cycles in that they are functions of the components that participate in each recovery cycle.

15.3.6   Extension of Useful Life

Useful life, meant as the period of time for which the entire product or part of it is used in the activity that is was designed for, can be considered as an indicator of resource savings. Extending the useful life corresponds to a better use of the resources involved in the production phase, and to a safeguard against the excessive dumping of waste, as was discussed in Chapters 8 and 9. Already formulated by the authors (Giudice et al., 1999a) on the basis of previous suggestions (Navin-Chandra, 1991), this Useful Life indicator (UL) provides a global vision of the product's entire life, beyond the limits of a single life cycle, and is introduced here as a complement to the calculation models proposed.

If at the end of its conventional use (of duration T) the product is not recovered in any way or reused as a whole, then its useful life is equal to time T, during which it performs the function related to that single use. If, instead, the whole product is reused or if part of it is reused, the useful life is extended since all or parts of the resources employed continue to perform the function.

Referring to precedent formulations (Navin-Chandra, 1991; Giudice et al., 1999a), the Useful Life (UL) is expressed as:

$$UL = \left[1 + nr + \sum_{j=1}^{m} \Phi^{j} \right] \cdot T \qquad (15.11)$$

where the number of integral reuses is indicated by nr and the number of recoveries by m. This assumes that at the conclusion of each use subsequent to integral reuses a constant recovery fraction Φ is recovered and again introduced into the same cycle, hypothesizing a constant duration of all the uses equal to T, and there are no losses in recovery process.

In reality, the number of reusable components coming from the original unit decrease with each successive recovery, so the recovery fraction is not constant, and the Useful Life is:

$$UL = \left[1 + nr + \sum_{j=1}^{m} \prod_{k=1}^{j} \Phi_{k} \right] \cdot T \qquad (15.12)$$

where Φ_j is the recovery fraction at the j-th recovery cycle.

From Equation (15.12) one obtains the Extension of Useful Life (EUL), which quantifies the overall extension of the life of the original components within the same product's life cycle:

$$EUL = \left[nr + \sum_{j=1}^{m} \prod_{k=1}^{j} \Phi_{k} \right] \cdot T \qquad (15.13)$$

From Equation (15.13) it is simple to define a final indicator that quantifies the extension of useful life after the generic q-th recovery cycle:

$$EUL_{q} = \left[nr + \sum_{j=1}^{q} \prod_{k=1}^{j} \Phi_{k} \right] \cdot T \qquad (15.14)$$

According to the type of investigation required, the recovery fractions expressed by functions (15.8) or (15.10) are introduced into functions (15.12), (15.13), and (15.14). In the first case, it is only necessary to evaluate the weights W_i of each i-th component. In the second case, functions (15.12), (15.13), and (15.14) express Useful Life (UL) and Extension of Useful Life (EUL) in terms of "life of the environmental impact," and it is necessary to evaluate the terms $EI_{Prod\,i}$ expressed by function (15.9).

15.4 Case Study: Analysis and Optimization of Heat Exchanger Constructional Systems

In this section the model proposed for planning of recovery cycles is applied to a specific design case, the analysis and optimization of the constructional system of heat exchangers (Fraas, 1989). This specific product seems particularly interesting in relation to the problem in question in that it is characterized by construction standards developed according to principles of modular architecture, which make the product highly suitable for the recovery of certain components at the end of its working life. Manufacturers' experience has shown wide margins for the application of recovery operations at the end of the product's life.

15.4.1 Construction Standards of Heat Exchangers

The construction of heat exchangers is codified and standardized by clear design guidelines. The definition of the architecture is directed by its subdivision into three main construction modules (two heads and the central body), each of which can be chosen from a range of different design alternatives (TEMA, 1988). The various possible combinations in the choice of modules give rise to a wide variety of architectures.

15.4.2 Operations for Recovery at the End of Working Life

For some typologies of exchangers and the uses they are destined for, nonhomogeneous deterioration has been found in the metallurgic properties of the steel used. Consequently, some longer-lasting components would be reusable if they were separable from the constructional unit and if they were considered perfectly efficient from the point of view of their mechanical and thermodynamic properties.

An investigation into the potentiality of the component recovery strategies was conducted with particular regard to two typologies of heat exchangers, having the common parameters of overall volume (shell side capacity 3500 L, overall length 7–8 m), thermal pressure working conditions (about 20 bar/400°C shell side, about 40 bar/300°C tube side), and anticipated use in the industrial plant (shell side fluid VPS-BTMS, tube side fluid R-CRUDE). The two architectures differ in that the first has a detachable shell, coded

CFU (Figure 15.4, above), while the second is of the floating head type, coded AES (Figure 15.4, below). In both cases, after about six years of use metallographic analysis showed considerable deterioration in the metallurgical properties of the central module shell, good condition of the tubes (potentially reusable after rigorous cleaning), and excellent condition of the remaining components (front- and rear-end modules). In analogous tests, after another six years of use, the reused tubes showed deterioration, while the good condition of the remaining components suggested their suitability for still longer use and ultimate reuse. Figure 15.5 shows the distribution of component durability (conventional use time T was fixed at six years), and the consequent architecture modularization more functional for recovery.

The data obtained thus confirm the suitability of applying the recovery cycle analysis method proposed here in order to evaluate the efficacy of the tool in relation to two different design operations:

- Comparison between different constructional system typologies to evaluate the recovery potential
- Evaluation of design choices aimed at optimizing a system typology

15.4.3   Application of the Calculation Models

The calculation models were applied following the procedure outlined in Figure 15.3. Analysis of the architectures under examination allowed the

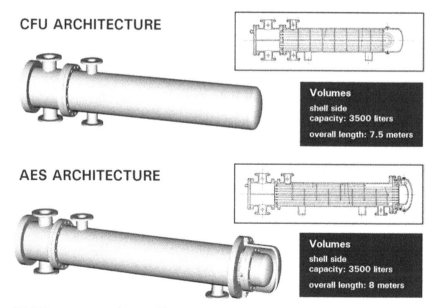

FIGURE 15.4   CFU and AES architecture typologies.

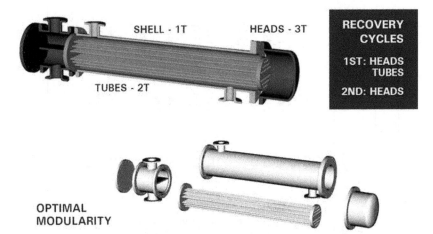

FIGURE 15.5 Distribution of component's durability and optimal modularity.

identification of the main components and constructional modules. The recovery determinant factors were defined, making matrices of type (15.3); this allowed the definition of two possible recovery cycles and the components or groups of components that can be included on the basis of their properties of durability and separability.

In particular, the predicted duration of each component was that found for the preexisting components on the basis of the metallographic testing data reported above. The duration of conventional use T was fixed at six years, according to the indications of the same data. Component separability was identified on the basis of a simple analysis of the typologies of junction between the components, according to whether they were reversible or irreversible. The recovery fractions were calculated in terms of environmental impact, as in function (15.10), because of their greater significance from the environmental viewpoint. The environmental indicators evaluated according to the Eco-indicator 95 method (Chapter 4, Table 4.3) were used in the calculation of $EI_{Prod\,i}$ expressed by function (15.9).

Finally, expressions (15.13) and (15.14) were applied for the calculation of indicators of Extension of Useful Life (EUL), giving results that will be discussed below.

15.4.4 Analysis and Evaluation of Results

The applications presented here to evaluate the efficacy of the proposed model followed two different types of design operation, one of comparison and the other of optimization. In the first case, the aim was to compare the two architecture typologies CFU and AES (some of the data for the two products are reported in Tables 15.1 and 15.2) to evaluate their different recovery potentialities.

TABLE 15.1   Data on the CFU architecture

	Quantity	Material		Weight (Kg)	Weight(Tot) (Kg)	Ei_{mat} (Mpt)	Ei_{mat} (Mpt)	Ei_{mat}(Tot) (Mpt)
CENTRAL MODULE + REAR-END HEAD MODULE								
SHELL	1	FE410.1KW		3953	3953	4.3	16998	16998
SHELL NOZZLE	2	ASTM A105/A106		44	88	4.1	180	361
SHELL COVER	1	FE410.1KW		341	341	4.3	1466	1466
SHELL FLANGE	1	ASTM A105		574	574	4.1	2353	2353
**SUPPORT SADDLE	2	FE410.1KW		101	202	4.3	434	869
					5158			22,047
FRONT-END STATIONARY HEAD MODULE								
*STATIONARY HEAD SHELL	1	FE510.1KW		312	312	4.3	1342	1342
*PASS PARTITION	2	FE510.1KW		91	182	4.3	391	783
*STATIONARY HEAD NOZZLE	2	ASTM A105/A106		44	88	4.1	180	361
*STATIONARY TUBESHEET	1	ASTM A105		2206	2206	4.3	9486	9486
**CHANNEL COVER	1	ASTM A105		1697	1697	4.3	7297	7297
					4485			19,268
TUBES MODULE								
*TUBES	1	FE35.2		7327	7327	4.1	30041	30041
*LONGITUDINAL BAFFLE	2	FE510.1KW		760	1520	4.3	3268	6536
*TRANSVERSE BAFFLE	12	FE430		49	588	4.3	211	2528
					9435			39,105
					19,078			80,420

TABLE 15.2 Data on the AES architecture

	Quantity	Material	Weight (Kg)	Weight(Tot) (Kg)	Ei$_{ma}$ (Mpt)	Ei$_{mat}$ (Mpt)	Ei$_{mat}$(Tot) (Mpt)
CENTRAL MODULE							
SHELL	1	FE410.1KW	3614	3614	4.3	15540	15540
SHELL NOZZLE	2	ASTM A105/A106	44	88	4.1	180	361
SHELL FLANGE (REAR HEAD END)	1	ASTM A105	664	664	4.1	2722	2722
SHELL FLANGE (STAT HEAD END)	1	ASTM A105	574	574	4.1	2353	2353
**SUPPORT SADDLE	2	FE410.1KW	101	202	4.3	434	869
				5142			21,845
REAR-END HEAD MODULE							
**SHELL COVER	1	FE410.1KW	680	680	4.3	2924	2924
**SHELL COVER FLANGE	1	ASTM A105	276	276	4.1	1132	1132
**FLOATING HEAD COVER	1	FE510.1KW	275	275	4.3	1183	1183
**FLOATING HEAD FLANGE	1	ASTM A105	387	387	4.1	1587	1587
**BACKING DEVICE	1	ASTM A105	403	403	4.1	1652	1652
				2021			8477
FRONT-END STATIONARY HEAD MODULE							
**STATIONARY HEAD SHELL	1	FE510.1KW	312	312	4.3	1342	1342
**PASS PARTITION	2	FE510.1KW	91	182	4.3	391	783
**STATIONARY HEAD NOZZLE	2	ASTM A105/A106	44	88	4.1	180	361
**STAT HEAD FLANGE	1	ASTM A105	561	561	4.1	2300	2300
**STAT HEAD FLANGE COVER SIDE	1	ASTM A105	528	528	4.1	2165	2165
**CHANNEL COVER	1	ASTM A105	1697	1697	4.3	7297	7297
				3368			14,247
TUBES MODULE							
*TUBES	1	FE35.2	8793	8793	4.1	36051	36051
*STATIONARY TUBESHEET	1	FE510.1KW	715	715	4.3	3075	3075
*FLOATING TUBESHEET	1	FE510.1KW	515	515	4.3	2215	2215
*TRANSVERSE BUFFLE	14	FE430	38	532	4.3	163	2288
				10,555			43,628
				21,086			88,197

15.4.4.1 Comparison CFU- and AES-Type Architectures

The results of the first type of investigation can be summarized as follows:

- The more complex AES architecture uses more resources and has a greater production impact (overall weight 21,086 kg, impact of materials used 88,197 mPt) compared to the CFU architecture (weight 19,078 kg, impact 80,420 mPt).

- Durability and separability of the components allow a first recovery cycle of essentially the same modules for both AES and CFU, but permit a second and final recovery cycle only in the case of AES. (In Tables 15.1 and 15.2, components with a single asterisk are recovered only in the first cycle, while those with two asterisks are also included in the second recovery cycle.)

- AES architecture gives greater recovery fractions than CFU, as shown by the graph in Figure 15.6. The fraction of the first recovery cycle Φ_j^{EI} is greater than 3.4%, despite the fact that the first recovery covers the same main modules in both cases. This indicates a better distribution of the volumes in AES, with a constructional system that is more functional for recovery (because, unlike in CFU, the rear-end module is not integrated with the central module). The fraction of the second recovery cycle is distinctly greater (+163%), showing a more efficient recovery for AES which, because of its better modularity, allows the second reuse of the entire front-end stationary head module.

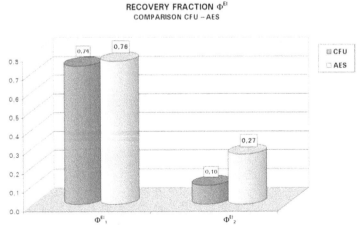

FIGURE 15.6   Recovery fraction (CFU–AES).

- As a direct consequence of the better recovery fractions, the Extension of Useful Life EUL^{EI} values for AES (EUL^{EI} expresses EUL in terms of "life of the environmental impact") are greater than those of CFU (Figure 15.7), slightly greater after the first recovery cycle (+3.4%) and substantially so after the second and final recovery cycle (+19%).

- In conclusion, therefore, the AES construction typology is characterized by a more complex architecture but has better modularity and is more functional for recovery than the CFU typology. This leads, on the one hand, to the use of more resources and greater production costs, while on the other hand it offers greater suitability for the recovery of these resources.

15.4.4.2 Optimization of CFU-Type Architecture

The second design approach had the aim of evaluating the efficacy of the design tool proposed as an aid to the optimization of the CFU-type architecture. From the comparison with the AES architecture above, it is clear that it would be appropriate to make the tube bundle separable from the front-end stationary head module, exactly as in AES, with a reversible flanged coupling that connects the central shell, tube plate, and the front-end shell. The introduction of this modification, which allows a second recovery of the front-end stationary head module, is therefore proposed as a design choice that is directed at a greater efficiency of the CFU recovery cycles. This results in the optimized architecture CFU* shown in Figure 15.8, with the main data reported in Table 15.3.

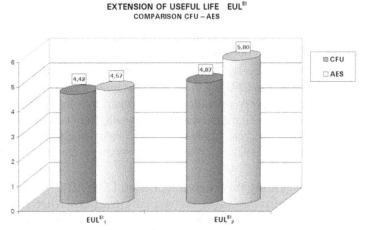

EXTENSION OF USEFUL LIFE EUL^{EI}
COMPARISON CFU – AES

FIGURE 15.7 Extension of useful life (CFU–AES).

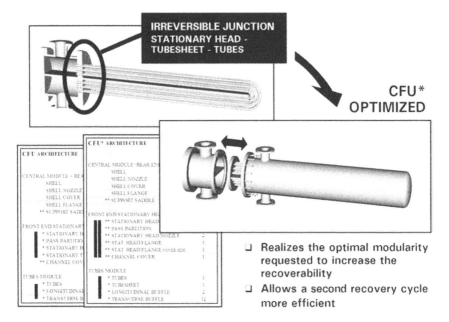

FIGURE 15.8   Optimized architecture CFU*.

The results of the comparison between the two architectures, conventional and optimized, can be summarized as follows:

- CFU* is slightly heavier (+0.2%) but also shows a decrease in the overall impact of the materials used (–0.13%) due to a different percentage of materials characterized by different environmental impacts (although the differences are minimal).

- Durability and separability of the components allow the first recovery cycle to cover essentially the same construction modules for both CFU and CFU*, but the second and final recovery cycle is possible only for CFU* (as indicated by the asterisks in Tables 15.1 and 15.3).

- In the first recovery cycle, CFU* shows a slight decrease in the recovery fraction compared to CFU (–1%), as reported in Figure 15.9. Thus, halting at the first recovery operation would mean that the optimization was inefficient. This slight decrease is, however, fully compensated in the second recovery cycle, made more effective by a better modularity, with a consequent +92% increase in the recovery fraction of CFU* compared to CFU.

- The results regarding the Extension of Useful Life EUL^{EI} (reported in Figure 15.10) mirror those of the recovery fractions. Therefore, the comparison between CFU* and CFU again shows a slight decline in EUL^{EI} after the first recovery cycle (–1%) and a substantial increase

TABLE 15.3 Data on the CFU* architecture

	Quantity	Material	Weight (Kg)	Weight (Tot) (Kg)	Ei_{mat} (Mpt)	Ei_{mat} (Mpt)	Ei_{mat} (Tot) (Mpt)
CENTRAL MODULE + REAR-END HEAD MODULE							
SHELL	1	FE410.1KW	3953	3953	4.3	16998	16998
SHELL NOZZLE	2	ASTM A105/A106	44	88	4.1	180	361
SHELL COVER	1	FE410.1KW	341	341	4.3	1466	1466
SHELL FLANGE	1	ASTM A105	718	718	4.1	2942	2942
** SUPPORT SADDLE	2	FE410.1KW	101	202	4.3	434	868
				5302			22,635
FRONT-END STATIONARY HEAD MODULE							
** STATIONARY HEAD SHELL	1	FE510.1KW	312	312	4.3	1342	1342
** PASS PARTITION	2	FE510.1KW	91	182	4.3	391	783
** STATIONARY HEAD NOZZLE	2	ASTM A105/A106	44	88	4.1	180	361
** STAT. HEAD FLANGE	1	ASTM A105	701	701	4.1	2875	2875
** STAT. HEAD FLANGE COVER SID	1	ASTM A105	528	528	4.1	2165	2165
**CHANNEL COVER	1	ASTM A105	1697	1697	4.3	7297	7297
				3508			14,822
TUBES MODULE							
* TUBES	1	FE35.2	7327	7327	4.1	30041	30041
* TUBESHEET	1	FE510.1KW	873	873	4.3	3752	3752
* LONGITUDINAL BUFFLE	2	FE510.1KW	760	1520	4.3	3268	6536
* TRANSVERSE BUFFLE	12	FE430	49	588	4.3	211	2528
				10,308			42,857
				19,117			80,314

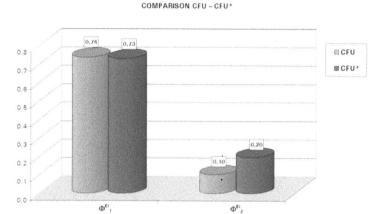

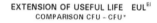

FIGURE 15.9   Recovery fraction (CFU–CFU*).

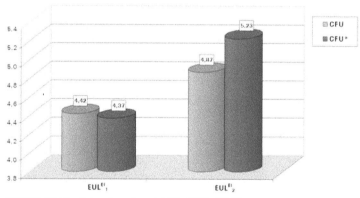

FIGURE 15.10   Extension of useful life (CFU–CFU*).

after the second and final recovery (+7.4%). This demonstrates the importance of using an indicator that extends the evaluation over all the recovery cycles for a complete analysis of the efficiency of the recovery operation.

- In conclusion, the optimization proposed for the CFU architecture is appreciable only with an evaluation of the entire potentiality of recovery.

15.4.4.3   *Final Overview of Results*

The graphs in Figures 15.11 and 15.12 show, respectively, the greater efficiency of the AES architecture compared to the other two examined, in terms of the

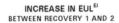

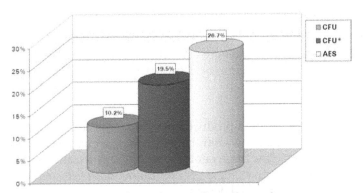

FIGURE 15.11 **Increase in EUL between first and second recovery.**

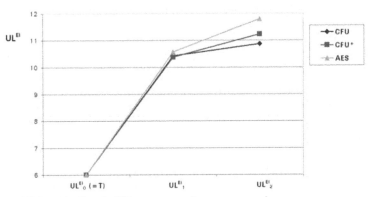

FIGURE 15.12 **Trend of UL over successive recovery cycles.**

increases of EUL^{EI} after the first and second recovery, and the trend of the UL^{EI} over successive recovery cycles for the three architectures (similar to EUL^{EI}, UL^{EI} also expresses UL in terms of "life of the environmental impact").

15.5 Cost–Benefit Analysis of Recovery Cycles

The recovery planning methodology shown in Figure 15.2 and discussed in Section 15.2, can be summarized as follows:

- Definition of the general architecture of the product and component details using tools of traditional design

- Analysis of constructional system previously defined, in relation to the recovery strategies
- Elaboration of the data by calculation models, which define the possible recovery cycles and evaluate their environmental efficacy
- Evaluation of the results given by the calculation models used and definition of suggestions for a correct redesign of the architecture and components

At this point the redesigned product again undergoes the prior processes of analysis and evaluation, now complemented by an appropriate cost–benefit analysis that can evaluate the economic cost of the redesigned product, with the final aim of best reconciling the environmental performance with the requirements of conventional design.

15.5.1 Calculation Models for Cost–Benefit Analysis of Recovery

The cost–benefit analysis model is based on the distinction made between two different production systems (Giudice et al., 1999b):

- First production, in which the product is manufactured from virgin raw materials only
- Second production, which integrates the volumes of virgin raw materials with volumes of recovered materials to manufacture a product of second production with used and remanufactured components

The cost of first production C^I_{Prod} is defined as the sum of the terms relevant to the acquisition of the materials, the production of the parts, and their assembly:

$$C^I_{Prod} = \sum_{i=1}^{n} \left(c_i \cdot W_i + C_{Mfct_i} \right) + \sum_{v=1}^{nv} C_{Ass_v} \qquad (15.15)$$

where:

- c_i is the unitary cost for the material making up the i-th component.
- $C_{Mfct\,i}$ is the manufacturing cost of the i-th component.
- nv is the total number of junctions required in the structure.
- $C_{Ass\,v}$ is the cost of making the v-th assembly junction.

The cost of second production $C^{II}_{Prod_j}$, after the generic j-th recovery cycle, is expressed by:

$$C^{II}_{Prod_j} = \sum_{i=1}^{n} \left(1 - r_{ij}\right) \cdot \left(c_i \cdot W_i + C_{Mfct_i}\right) + \sum_{v=1}^{nv} C_{Ass_v} \qquad (15.16)$$

where r_{ij} is the reusability of the i-th component at the j-th recovery cycle, as defined in Equation (15.6).

The Recovery of Production Cost $RC_{Prod\,j}$ after the j-th recovery cycle can be defined by:

$$RC_{Prod_j} = \frac{C^{I}_{Prod} - C^{II}_{Prod_j}}{C^{I}_{Prod}} = \frac{\displaystyle\sum_{i=1}^{n} r_{ij} \cdot \left(c_i \cdot W_i + C_{Mfct_i}\right)}{\displaystyle\sum_{i=1}^{n} \left(c_i \cdot W_i + C_{Mfct_i}\right) + \sum_{v=1}^{nv} C_{Ass_v}} \qquad (15.17)$$

At this point it is of particular interest to determine the condition for which an increase in the volumes of recovery [i.e., an increase in the recovery fractions as defined in Equations (15.8) or (15.10)], which is of undoubted environmental advantage, is also advantageous from the economic point of view. Such an increase can be obtained by modifying the constructional system to allow recovery cycles that involve greater flows of the volumes making up the product. In the majority of cases, however, modifications of this type lead to an increase not only in the recovery of production cost as defined in Equation (15.17), but also in the production costs themselves. Nevertheless, considering the possibility of reiterating the recovery over time, it can happen that the increase in the production costs is in some way compensated for over time.

The model presented for an investigation of this type was inspired by a model used for a different purpose (Matthews and Lave, 1995), already reformulated by the authors (Giudice et al., 1999b). On the basis of the definitions of first and second production suggested above, the cost of the entire life of the product C_{LC} can be defined, applying the following simplified ideal hypotheses:

- Recovery is reiterated over time, in a regime where, for all the recovery cycles programmed, the volumes involved in the different levels of recovery are always the same. Thus, the recovery fraction is constant for all recovery cycles, and recovery of production cost as defined in Equation (15.17) is constant and expressed as RC_{Prod}.
- The product is sold and recovered in a sufficiently short period of time that a null rate of interest can be assumed.

- The cost of disposing of the unrecoverable parts and the cost of disposal at the end of the product's life are ignored.

$$C_{LC} = C_{Prod}^{I} + \sum_{j=1}^{m}\left(C_{Prod_j}^{II} + C_{Rec}\right) = \left(1 + m - m \cdot RC_{Prod}\right) \cdot C_{Prod}^{I} + m \cdot C_{Rec} \quad (15.18)$$

In Equation (15.18) C_{Rec} is a generic term of recovery cost, assumed constant for each recovery cycle. Now assume that modifications are made to the system, so that they cause an increase in the Recovery of Production Cost RC_{Prod}. If Δ_{RC} indicates the increase in RC_{Prod}, and Δ_C indicates the corresponding increase in C_{Prod}^{I},

$$\begin{cases} RC_{PROD} \rightarrow RC_{PROD}^{*} = RC_{PROD} + \Delta_{RC} \\ C_{PROD}^{I} \rightarrow C_{PROD}^{I*} = C_{PROD}^{I*} + \Delta_C \end{cases}$$

As the last hypothesis, the variation of the recovery costs are negligible and therefore $C_{Rec} \cong C_{Rec}^{*}$. In fact, C_{Rec}^{*} changes because it is dependent on the components that participate in the improved recovery cycle. On the basis of Equation (15.18) it is possible to obtain the new expression of the cost of the entire life C_{LC}^{*}:

$$C_{LC}^{*} = \left(1 + m - m \cdot RC_{Prod} - m\Delta_{RC}\right) \cdot \Delta_C - m \cdot \Delta_{RC} \cdot C_{Prod}^{I} + C_{LC} \quad (15.19)$$

As long as the increase in the recovery of production cost leads to a condition of economic advantage over the entire useful life predicted for the product (even though it causes an increase in the production costs), it is necessary that $C_{LC}^{*} \leq C_{LC}$, which provides the following condition:

$$\frac{\Delta_{RC} \cdot \left(\Delta_C + C_{Prod}^{I}\right)}{\Delta_C} \geq \frac{1 + m \cdot \left(1 - RC_{Prod}\right)}{m}$$

$$\Rightarrow \Psi\left(\Delta_{RC}, \Delta_C, C_{Prod}^{I}\right) \geq \Omega\left(m, RC_{Prod}\right) \quad (15.20)$$

Expression (15.20) represents the condition of economic advantage of the increase in recovery flows. The function Ω decreases as the number of recovery cycles m increases. This confirms what was expected regarding the fact that the economic advantage of modifications to the constructional system [which lead to an increase in Equation (15.17)] is linked to the possibility of planning an elevated number of recovery cycles. Having evaluated the increase in production cost Δ_C caused by any modifications to the constructional system, expression (15.20) can be used to determine the minimum number of recovery cycles m that must be programmed to compensate the increase in Δ_C. In contrast, evaluating the number of possible recovery cycles m, from expression (15.20) it is possible to obtain the maximum limiting value that Δ_C can assume for the condition of economic convenience to be achieved.

15.5.2 Case Study: Implementation of Cost–Benefit Analysis Models

The redesign of architecture CFU, which results in an optimized architecture CFU*, allows a second recovery of the front-end stationary head module and is therefore directed at a greater efficiency of the second recovery cycle. The application of the models for cost–benefit analysis of recovery, particularly the condition (15.20), makes it possible to relate the economic convenience of the constructional system optimization to the increase Δ_C in production cost C^I_{Prod} resulting from this optimization, and to m_{AF} the number of recoveries possible after the first. The typology of the product does not respect the conditions "sold and recovered in a sufficiently short period of time," but the ideal hypothesis that a null rate of interest can be assumed is applied all the same, to obtain an indicative evaluation.

In the graph in Figure 15.13 the economic advantage of the constructional system optimization is represented for different values of Δ_C (expressed as percentage values of the conventional production cost of the system), by the conditions under which the straight lines Ψ are above the curve Ω, which is independent of Δ_C. Clearly, upon increasing Δ_C the point of parity moves toward higher values of m_{AF}. In particular, in the case where only one recovery cycle can be programmed after the first, as in the case under examination, the condition of economic advantage is respected as long as the increase in Δ_C

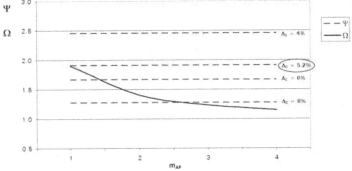

FIGURE 15.13 Cost–benefit analysis of recovery.

remains below the limiting value $\Delta_{C\,lim} = 5.2\%$ of the production cost of the conventional constructional system. If Δ_C exceeds this limit, at least two more recovery cycles (involving the same volumes) after the first would be necessary to ensure economic advantage.

15.6 Acknowledgments

The main contents of this chapter were previously published (Giudice, F., La Rosa, G., and Risitano, A., Product recovery-cycles design: Extension of useful life, in *Feature Based Product Life-cycle Modelling*, Soenen, R. and Olling, G., Eds., Kluwer Academic Publishers, Dordrecht, The Netherlands, 2003, 165–185; Giudice, F., La Rosa, G., and Risitano, A., Optimisation and cost-benefit analysis of product recovery-cycles, in *Proceeding of ICED01 International Conference on Engineering Design: Design Applications in Industry and Education—Design Methods for Performance and Sustainability*. Culley, S. et al., Eds., Professional Engineering Publishing, London, 2001, 629–636,) and are reproduced with permission from Springer and John Wiley and Sons Ltd, respectively.

15.7 Summary

This chapter proposes a method to aid designers in making the best choices in order to plan recovery cycles for a product at the end of its working life. Having outlined an effective design methodology for this purpose, calculation models were presented to support the definition of the reusable parts of the system and the recovery cycles planning; to evaluate an indicator that translates the environmental effects of recovery cycles in terms of extension of the product's useful life; and to determine the condition for which an increase in the volumes of recovery, of undoubted environmental benefit, is also advantageous from the economic point of view. Applied as a tool of analysis and optimization of the suitability for recovery of modules characterizing the architecture of heat exchangers, the design method and calculation models confirm this tool's potential as an aid to plan the best recovery cycles of the product, to optimize redesign choices, and to investigate the relation between system optimization and the resulting effects on costs.

15.8 References

Chen, R.W., Navin-Chandra, D., and Prinz, F.B., Product design for recyclability: A cost–benefit analysis model and its application, *IEEE Transaction on Components, Packaging, and Manufacturing Technology*, 17(4), 502–507, 1994.

Fraas, A.P., *Heat Exchanger Design,* 2nd ed., Wiley & Sons, London, 1989.

Giudice, F., La Rosa, G., and Risitano, A., Indicators for environmentally conscious product design, in *Proceedings of EcoDesign '99: 1st International Symposium on Environmentally Conscious Design and Inverse Manufacturing,* Tokyo, 1999a, 71–76.

Giudice, F., La Rosa, G., and Risitano, A., Models and indicators for the cost–benefit analysis of a green product, in *Proceedings of EcoDesign '99: 1st International Symposium on Environmentally Conscious Design and Inverse Manufacturing,* Tokyo, 1999b, 77–82.

Matthews, H.S. and Lave, L.B., Price setting for green design, in *Proceedings of IEEE International Symposium on Electronics and the Environment,* Orlando, FL, 1995, 304–309.

Navin-Chandra, D., Design for environmentability, in *Proceedings of ASME Design Theory and Methodology Conference,* Miami, FL, 1991, DE-31, 119–125.

TEMA, *Standards of the Tubular Exchanger Manufacturers Association,* 7th ed., Tubular Exchanger Manufacturers Association, New York, 1988.

Chapter 16

Methodological Framework and Analysis Models for Simulation of the Product Life Cycle

---

Product analysis undertaken from the perspective of Life Cycle Design requires the definition of procedures and tools to steer the designer towards the best choices in relation to the product's performance—conventional and environmental—over its whole life cycle. Design choices can be directly linked to product performance using models of product behavior. With these models it is possible to simulate the life cycle in relation to phenomena of deterioration in performance of materials (due to external factors or loading conditions).

This chapter presents a methodological framework and the models for such analysis to allow the simulation of product life cycle at the design stage. The case study reported shows how the simulation method, complemented by analytical tools, can be applied to associate each set of design choices under examination with a broad spectrum of information: the durability and criticality of components and the system, as a function of the period of use; possible faults and the consequent servicing costs; the residual life of components, their possible reuse at the end-of-life, and consequent extension of the system's useful life; and the environmental impact of the whole product life cycle.

The fundamental issues in this chapter were previously introduced in Chapters 3, 9, 10, and 15.

16.1 Simulation and the Life Cycle Approach

Life Cycle Design consists of a design intervention that incorporates all the phases of the product's life cycle (development, production, use, maintenance and repair, retirement, recovery) in the entire design process, from the phase of concept definition to that of detailed design development (Chapter 3, Section 3.1.1). The traditional evolutionary approach is based on the feedback of information flows used to improve the design intervention; a powerful alternative can consist of already predicting the consequences of design choices on the product's life cycle during the solution synthesis phase itself. This is equivalent to simulating the life cycle of the product in the early phases of the design process (Chapter 3, Section 3.2.4).

Therefore, it could be extremely helpful to define a methodological framework and relative analytical tools in support of the design process, allowing the management of design choices (both at the level of product layout and that of the specifications of components), in relation to the product's conventional (functionality, reliability, cost) and environmental performance over its life. When variations in the design choices are proposed, such as reorganization of the architecture or modification of component geometries and materials, these tools must allow the simulation of the resulting life cycle so that it is possible to compare the various alternatives and identify the solutions that best realize appropriate performance goals. With particular regard to the simulation of the final phases of the life cycle (i.e., use and retirement), the objective performances to be optimized must be related not only to the design choices but also to factors of deterioration that can alter the behavior of the product and its components over time (Chapter 10).

16.2 Approach to the Problem and Methodological Framework

With these aims, this chapter proposes a method of simulating product behavior over its life cycle and the relative analytical tools to obtain, for every possible set of design choices:

- Indications of the duration, safety, and criticality of single components and of the overall system, on varying the time of use
- Indications of possible failures and evaluations of the resulting servicing costs
- Indications of the residual life of components, evaluation of their possible reuse, and quantification of the resulting potential extension of the system's useful life
- Evaluation of the environmental impact associated with the main phases of the life cycle

The main difficulty consists of correlating design parameters with product performance in its life cycle after production. This requires not only a modeling of the life cycle, but also an appropriate modeling of the product as a basis for simulating its behavior in response to different design choices.

The importance of life cycle modeling in the process of product development, in relation to the environmental impact, has already been demonstrated (Zust and Caduff, 1997). Also, some approaches to life cycle simulation have been outlined with particular reference to modular products (Tomiyama et al., 1997) and in more general terms (Kato et al., 2001); see also Chapter 3, Section 3.2.4. The first deals with approaches defining life cycle models according to elementary activities that aid the inventory phase of Life Cycle Assessment, in accordance with the indications of ISO14040 standards. The second proposes simulation schemes that provide particular information on the product's environmental impact (quantifying the flow of materials discharged or translating the impact of end-of-life into economic terms) without detailing the correlation with the main design parameters. Furthermore, while introducing the temporal variable governing the simulation, these schemes rarely study in depth this concept in relation to the phenomena of performance decay and to the consequent effect on the system's efficiency (Hata et al., 1997).

As summarized in the scheme in Figure 16.1, the development of the method proposed here makes use of certain tools opportunely correlated with indices expressing the decay of performance over time:

- A model of the constructional system, based on the behavior of functional subgroups

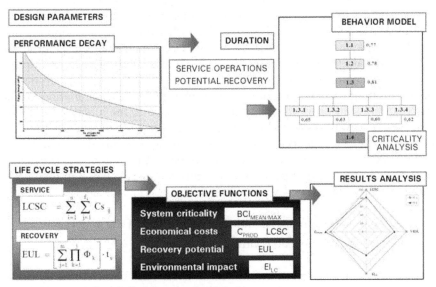

FIGURE 16.1   Methodological framework and tools.

- Certain significant functions that express the system's performance in relation to the possible strategies of improving the life cycle under examination, which are the optimization of the useful life (through servicing operations on the system) and the recovery of resources at end-of-life (through the reuse and recycling of components)

With the support of these tools it is possible to outline a simulation procedure allowing the evaluation, already in the design phase, of the product's possible behavior in the intermediate and final phases of the life cycle (use and end-of-life), according to the main design choices and the duration of use. This behavior must be evaluated using objective functions that are relevant to the aims of optimal product design. They must quantify the behavior of the product in relation to three main aspects:

- Level of functional efficiency and safety of the constructional system

- Costs of the product in relation to the main phases of the life cycle

- Environmental impact of the whole life cycle and the recovery potential of the resources used

16.3 Product Model and Analysis Tools

Again referring to the scheme in Figure 16.1, the main points can be considered in more detail. To correlate the set of design choices and product performance, the methodological approach proposed here is based on advanced Failure Mode and Effect Analysis (FMEA), scheduling the development of a product behavior model based on its function rather than on its structure (Eubanks et al., 1996; Eubanks et al., 1997). This is based on the definition of the main elementary functionalities of the constructional system, on the set of determining variables required for the behavior under examination to take place (initial state) and final conditions reached after the function has taken place (final state), and of the performance characteristics regulating the behavior, which can generally be expressed using mathematical models.

With regard to the evaluation of variations in the system's performance over time, in the method proposed here this can be calculated in relation to different typologies of decay phenomena:

- Independent from load conditions (e.g., aging of the materials)

- Dependent on the load conditions (e.g., fatigue)

Taking into consideration a defined set of materials characterized by decay curves, it is possible to evaluate the indices that provide information on the duration of components (Index of Duration—DI) and on safety (Dynamic Criticality Factor —DCF) over the life cycle. On these bases, the simulation of the system's functionalities allows the identification of possible failures and their classification in terms of their effect on components. It is then possible to derive the criticality of the system through analysis of the configuration of the model (blocks in a series or parallel), the danger of a failure, and the residual performance level.

This information directly influences possible strategies for improving the environmental performance of the life cycle considered here and also discussed in Chapter 9: optimization of the useful life (through servicing operations on the system) and the recovery of resources at end-of-life (through the reuse and recycling of components). In this regard, reference is made to the calculation models already available for both the evaluation of the economic impact of servicing systems during their useful life (Gershenson and Ishii, 1993) and for the planning of recovery cycles at end-of-life (Chapter 15). The indicators proposed by these models are, respectively, the Life Cycle Service Cost (LCSC) and the Extension of Useful Life (EUL).

16.3.1 Model of System Behavior

As mentioned above, design choices and product performance are correlated using a model of the system's behavior based on advanced FMEA (Eubanks et al., 1996; Eubanks et al., 1997). This model defines the main elementary functionalities of the constructional system, the determining variables required for the behavior under examination to take place, the final state after the function has taken place, and the performance conditions regulating the behavior. Figure 16.2 shows the reference scheme for a system based on a main functionality, broken down into elementary behaviors in series and in parallel. BHV_s is the s-th behavior in series; if it is constituted by np_s behavior in parallel, BHV_{sp} is the p-th behavior in parallel constituting the s-th behavior in series. Each elementary behavior is defined by:

- Components directly involved in the behavior; n_s is the number of components involved in the s-th behavior
- Mathematical models expressing the performance conditions that regulate the behaviors; nv_s is the total number of performance conditions for s-th behavior, generally consisting of functions linking performance conditions Pf_v to operating conditions, to fixed and variable geometric parameters, and to the properties of the materials (preconditions)
- Performance limits Pf^*_v, which, compared with Pf_v, make it possible to establish whether the behavior takes place correctly (postconditions)

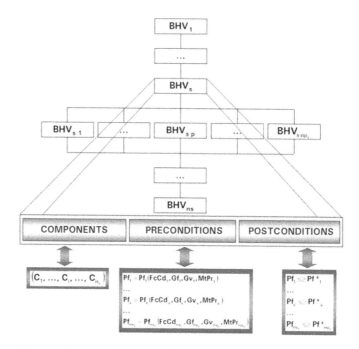

FIGURE 16.2   Model of system behavior: Reference scheme.

16.3.2 Evaluation of Performance Decay

Simulating the system's behavior requires an evaluation of the variations in its behavior over time. While distinguishing between the two different typologies of decay phenomena (independent from and dependent on load conditions), it is generally possible to ascribe the phenomena to material performance diagrams (Figure 16.3) where the time variable t represents real time for the phenomena of the first type, and the real time of use for phenomena of the second type.

Once the materials comprising the system have been chosen and each is characterized by its corresponding decay curve, it is possible to evaluate the indices providing information on the duration of the components and on the level of safety over the time of use, and thus over the life cycle.

16.3.2.1 Duration Index

With regard to the first aspect, an index of the duration of the component (DI) is introduced. This is defined as the ratio between the estimated physical life of the component t_r, determined by the performance required Pf_r, and the fixed useful life t_u, which is a design requisite:

$$DI = \frac{t_r}{t_u} \tag{16.1}$$

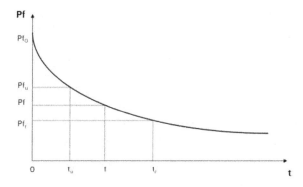

FIGURE 16.3 Decay of performance: Reference diagram.

Having fixed the duration of the useful life t_u, this index DI depends on t_r (i.e., both of the material, since the decay curve varies with this, and of the component geometry, which determines the working conditions and therefore the required performance Pf_r). The value of DI provides indications of the component's operating conditions over the arc of its useful life and allows the quantification of both the need for servicing operations and the possibility of reusing the component. In the case where DI < 1, the need to substitute the component f times over the entire arc of the useful life can be anticipated with:

$$f = \text{int}\left(\frac{1}{DI}\right) \tag{16.2}$$

If 1 < DI < 2, the component can be used only once. In the case where DI > 2, the component can be reused r times, with:

$$r = \text{int}(DI) - 1 \tag{16.3}$$

16.3.2.2 Dynamic Criticality Factor

Considering the variation in a component's level of safety over time, and again referring to Figure 16.3, the Dynamic Criticality Factor (DCF) is introduced:

$$DCF = \frac{Pf_0 - Pf}{Pf_0 - Pf_r} \tag{16.4}$$

where Pf_0 is the component's initial performance, Pf is that corresponding to the generic time t, and Pf_r is again the required performance. This index quantifies the increase in the component's criticality during its use, and for $t = t_u$ expresses the criticality corresponding to the end of use (in this case, the notation DCFu will be used).

If Pf_u indicates the performance level corresponding to the end of useful life, it is seen that:

- If $Pf_u > Pf_r$ then DCFu < 1; therefore, the component is not critical.
- If $Pf_u < Pf_r$ then DCFu > 1; therefore, the component is critical and is the more critical the higher the value assumed by DCFu.

16.3.2.3 Behavior Criticality Index

As above, according to the system model introduced here, each of the system's behaviors is determined by a set of components. Indicating the generic component correlated to the behavior with C_i, and the corresponding value of the dynamic criticality factor at end of use with $DCFu_i$, the Behavior Criticality Index (BCI) is defined as:

$$BCI = c \cdot \max\left(DCFu_i\right) \qquad (16.5)$$

where $c \in [0,1]$ is an additional factor expressing the criticality of the behavior under examination in relation to the functionality of the whole system. If a "nonbehavior" results in the total arrest of the system, this must be classified with a high c factor, given the importance of the failure. The higher the value of BCI, the more critical the behavior under examination, both for the effect of its failure on the system and for the decrease in performance level, quantified by the dynamic criticality factor. The distribution of the values assumed by BCI for each behavior on varying the design parameters allows an analysis of the criticality of the system under examination.

16.3.3 Analysis of Life Cycle Strategies

The values assumed by the indices of performance decay introduced above directly condition the possible strategies for improving the environmental behavior of the life cycle (i.e., the strategies for improvement of resource exploitation, as were introduced in Chapter 9). These include optimization of the useful life (through servicing operations) and recovery of resources at end-of-life (through component reuse and recycling). The following mathematical reference models can be used to quantify these behaviors.

16.3.3.1 Life Cycle Service Cost

The strategies for optimizing the useful life of products include interventions of diagnosis, maintenance, repair, substitution, and any other operations that may be necessary to ensure the correct functioning of the system (Chapter 9, Section 9.2). For an evaluation of the effect that design choices can have in terms of ease of servicing, mathematical models that express the life cycle service costs can be used (Gershenson and Ishii, 1993). Given a component requiring several service operations, the cost of the w-th service operation on the i-th generic component is given by:

$$Cs_{iw} = tl_{iw} \cdot cl_{iw} + c_i \qquad (16.6)$$

where tl_{iw} is the time of intervention, cl_{iw} is the cost of the intervention per unit time, and c_i is the cost of the component or of the material required in the intervention.

Considering as service interventions the substitution of failed components, for a system consisting of n components C_i, the total Life Cycle Service Cost (LCSC) is expressed by:

$$LCSC = \sum_{i=1}^{n} \sum_{w=1}^{f_i} Cs_{iw} \qquad (16.7)$$

where f_i is the number of substitution interventions required for the i-th component, defined by Equation (16.2).

16.3.3.2 Recovery Cycles and Extension of Useful Life

The problem of planning recovery cycles in relation to the duration of the components has already been treated by the authors (Giudice et al., 2003), and some significant results have been reported in Chapter 15, where a calculation model was proposed that takes account of the environmental impact of producing the i-th component, EI_{Prod_i}. This can be expressed in terms of the eco-indicators of the materials and processes, defined by Equation (15.9) of Chapter 15:

$$EI_{Prod_i} = ei_{Mat_i} \cdot W_i + \sum_{k=1}^{h_i} ei_{Pr_k\,i} \cdot \eta_{pr_k\,i} \qquad (16.8)$$

Considering that it is possible to anticipate a number of recovery cycles m, the recovery fraction that can be associated with the j-th cycle (in terms of environmental impact) can be expressed by Equation (15.10) of Chapter 15:

$$\Phi_j^{EI} = \frac{\sum_{i=1}^{n} r_{ij} \cdot EI_{Prod_i}}{\sum_{i=1}^{n} EI_{Prod_i}} \qquad (16.9)$$

where r_{ij} is a binary coefficient that assumes unitary value only in the case where the i-th component is reusable at the j-th cycle. This is, therefore, easily expressed as a function of r, the number of possible component reuses, given by Equation (16.3).

Hypothesizing a constant duration of all the reuses, equal to the fixed duration of the first use t_u, the function quantifying the extension of useful life is defined as (see Chapter 15, Section 15.3.6):

$$EUL = \left[\sum_{j=1}^{m} \prod_{k=1}^{j} \Phi_k^{EI} \right] \cdot t_u \qquad (16.10)$$

This function, which expresses the extension of the life of original components within the life cycle of the product, can be considered an indicator of savings in resources according to their different environmental impacts.

16.4 Definition of Objective Functions

As above, the product's behavior in the life cycle must be evaluated in relation to its functional capacities, to the economic costs, and to the environmental performance. With this aim, and with reference to the mathematical models proposed so far, the following objective functions are proposed:

- The criticality index of the constructional system, identified in the mean or maximum values assumed by the BCI indices expressing, through Equation (16.5), the criticality of each behavior of the system

- The product's costs over the life cycle, which, ignoring the costs of product retirement, can be quantified by the production and servicing costs, the latter expressed by LCSC defined in function (16.7)

- The potential extension of useful life of the resources used in the system through the recovery of components, expressed by EUL defined in function (16.10)

- The environmental impact of the life cycle, quantified by the sum of the impacts associated with the phases of production, use (servicing), and retirement

The last function requires further development of the mathematical models, as discussed below.

These objective functions evaluated for each design choice are suitable for treatment using multiobjective analysis, with the ultimate aim of determining which solution best satisfies the entire spectrum of desired performances.

16.4.1 Environmental Impact of the Life Cycle

The function expressing the environmental impact of the life cycle EI_{LC} is similar to those previously introduced in Chapters 12 and 14. It is defined as the sum of three terms, each relative to the main phases of the cycle: production, use (servicing), and end-of-life.

$$EI_{LC} = EI_{Prod} + EI_{Srv} + EI_{EoL} \tag{16.11}$$

The first term can be expressed as the sum of the environmental impacts associated with the production of the individual components $EI_{Prod\,i}$ [Equation (16.8)]:

$$EI_{Prod} = \sum_{i=1}^{n} \left(ei_{Mat_i} \cdot W_i + \sum_{k=1}^{h_i} ei_{Pr_k\,i} \cdot \eta_{pr_k\,i} \right) \tag{16.12}$$

The environmental impact of use, assimilable to the impact associated with service operations, can be defined on the basis of the servicing costs model LCSC [Equation (16.7)], considering that every substitution of a failed component corresponds to an environmental impact equal to that of the component's production.

$$EI_{Srv} = \sum_{i=1}^{n} f_i \cdot \left(ei_{Mat_i} \cdot W_i + \sum_{k=1}^{h_i} ei_{Pr_k\,i} \cdot \eta_{pr_k\,i} \right) \tag{16.13}$$

The last term of Equation (16.11), relating to the product's end-of-life, is composed of three terms depending on whether the components are discarded or recovered through reuse or recycling:

$$EI_{EoL} = EI_{Dsp} + EI_{Rcl} + EI_{Rs} \tag{16.14}$$

With reference to the parameter of reuse r expressed by Equation (16.3), an additional binary coefficient ρ is introduced. This assumes unitary value only in the case where $r \geqslant 1$; otherwise it is 0.

Incorporating this in the definition of the first term of Equation (16.14), it can be expressed as follows:

$$EI_{Dsp} = \sum_{i=1}^{n} EI_{Dsp_i} = \sum_{i=1}^{n} ei_{Dsp_i} \cdot W_i \cdot (1 + f_i) \cdot (1 - \xi_i) \cdot (1 - \rho_i) \tag{16.15}$$

where $ei_{Dsp\,i}$ is the environmental impact of disposal per unit weight, W_i is the weight of the component, ξ_i is the recyclable fraction of the material forming the component, and f_i is the number of required substitutions during use, expressed by Equation (16.2). Analogously:

$$EI_{Rcl} = \sum_{i=1}^{n} EI_{Rcl_i} = \sum_{i=1}^{n} ei_{Rcl_i} \cdot W_i \cdot \left(1 + f_i\right) \cdot \left(\xi_i\right) \cdot \left(1 - \rho_i\right) \qquad (16.16)$$

where $ei_{Rcl\,i}$ is the environmental impact of the recycling process per unit weight and the other terms are as defined above.

Equations (16.15) and (16.16) are not 0 only for components that cannot be reused ($\rho_i = 0$).

The last term of Equation (16.14) refers to the environmental impact at the end-of-life, due only to reusable components ($\rho_i = 1$), and is expressed by:

$$EI_{Rs} = \sum_{i=1}^{n} EI_{Rs_i} = \sum_{i=1}^{n} \left(-EI_{Prod_i}\right) \cdot r_i \cdot \rho_i \qquad (16.17)$$

where again $EI_{Prod\,i}$ is expressed by Equation (16.8) and r_i by Equation (16.3). This term thus expresses a recovery of environmental impact, as the sum of production impacts associated with reusable components.

As in other previous chapters, all the eco-indicators considered are evaluated according to the Eco-indicator 99 method (Chapter 4, Table 4.3).

16.5 Simulation and Analysis of Results

This simulation procedure follows the development of the models proposed and is summarized in Figure 16.1. It is based on a direct relation between the design parameters that constitute the variables to be optimized, and the objective functions (Section 16.4) that quantify the product's performance in the main phases of the life cycle. This relation is structured using the behavior model of the system (Section 16.3.1) and is expressed analytically, using the indicators of component duration and failure derived from the deterioration of performance over time (Section 16.3.2).

To illustrate the application of the methodology, the results obtained on a mechanical system, the four-speed gear box shown in Figure 16.4, are summarized below.

16.5.1 Behavior Model of the System

From an analysis of the gear box's operation it is possible to define the system model which, in this case, is very simple and can be reduced to elementary

Four-Speed Gear with Clutch

Speed (max)	12 km/h	gear I	
	23 km/h	gear II	
	33 km/h	gear III	
	45 km/h	gear IV	

Power	2.5 CV - 5000 rpm	Shafts wheelbase	77 mm
τ motor	0.2 (motor/clutch transmission)	Total distance to cover	40,000 km

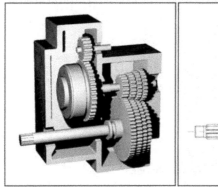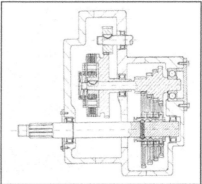

FIGURE 16.4 Case study: Gear box.

behaviors, largely in a series. Only one behavior, that representing the transmission of torque according to the gear engaged, breaks down into four subbehaviors in parallel, corresponding to each gear.

The scheme of the model is partially represented in Figure 16.5, presenting the information associated with behavior 1.1 (transmission of the torque from the motor shaft to the clutch): the components involved, the mathematical models expressing the required performance (in this case the structural strength), and the verification of the final state of the behavior.

16.5.2 Performance Evaluations and Analysis of Criticality

Having developed the behavior model, the analysis continues with the simulation of the functional performance of individual components over the life cycle of the system. On the basis of the design requirements (summarized in Figure 16.4), by varying the design choices (in particular materials and geometric parameters) it is possible to determine the loading conditions of each component and its required duration. In this regard it is necessary to consider certain details. The components comprising the system are subjected to fatigue loading. The gear box is destined for a motor vehicle; if the number of kilometers the vehicle is expected to travel is set as a design datum (vehicle mission), it is possible to determine the number of loading cycles of each component. Thus, all the time variables t introduced above (Section 16.3.2) are to be understood as loading cycles, and the decay of performance is expressed by the resistance to fatigue curves of the materials (Chapter 10, Section 10.2).

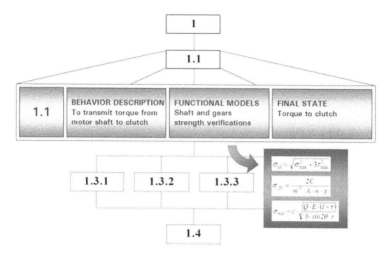

FIGURE 16.5   Case study: Model of system behavior.

Table 16.1 presents some evaluations for a first design alternative I, defined by the choice of materials and the specification of geometric parameters.

For each component, the table gives the chosen material; the number of expected loading cycles t_u related to the fixed mission; the required performance level Pf_r (mechanical strength in MPa); and the number of loading cycles t_r guaranteed by the chosen material, in relation to Pf_r (t_r assumes a value of infinity when Pf_r is below the fatigue limit of the material).

The same table then shows the corresponding values assumed by the indicators introduced in Section 16.3.2. In particular, when the duration index DI [Equation (16.1)] assumes a value of infinity, r assumes 2 as the conventional maximum number of possible reuses. Moreover, given that in this first design alternative the primary shaft is considered a single piece with the corresponding four gear wheels, all five components assume a value of the number of expected substitutions f equal to the maximum value assumed by each single component (reported in brackets), and a value of r equal to the minimum.

On the basis of the values assumed by the factor DCFu for each component, it is then possible to analyze the distribution of the criticality index of the behaviors BCI, evaluating their maximum value (indicating the most critical behavior) and estimating the mean criticality of the entire system (for design alternative I, mean BCI is equal to 0.80).

16.5.3   First Analysis of the Performance in the Life Cycle

Again referring to Table 16.1, from the values assumed by f and r, it is possible to predict the system's poor performance, both in terms of the use phase

TABLE 16.1 Performance evaluation and analysis of criticality: Design alternative I

	MATERIAL	t_u	$Pf_r(\sigma_{fc})$	t_r	DI	f	r	DCFu
Motor Shaft	AISI 1015	8,59E+08	49,2	infinite	infinite	0	2	0,77
Clutch Box	AISI 1015	1,72E+08	110,9	infinite	infinite	0	2	0,91
Motor Gear	AISI 1030	8,59E+08	153,5	infinite	infinite	0	2	0,96
Clutch Spring	AISI 1080	3,20E+06	345,1	3,12E+06	0,98	1	0	1,01
Primary Shaft	AISI 1015	1,72E+08	115,1	infinite	infinite	(0) 5	(2) 0	0,90
Gear 1 — Primary Shaft	AISI 1015	1,72E+08	160,2	3,23E+07	0,19	(5) 5	(0) 0	1,03
Gear 1 — Secondary Shaft	AISI 1030	3,54E+07	195,7	1,23E+07	0,35	2	0	1,05
Gear 2 — Primary Shaft	AISI 1015	1,72E+08	161,9	3,20E+07	0,19	(5) 5	(0) 0	1,04
Gear 2 — Secondary Shaft	AISI 1030	3,54E+07	182,8	3,11E+07	0,88	1	0	1,01
Gear 3 — Primary Shaft	AISI 1015	1,72E+08	156,9	6,15E+07	0,36	(2) 5	(0) 0	1,02
Gear 3 — Secondary Shaft	AISI 1030	3,54E+07	203,9	5,30E+06	0,15	6	0	1,07
Gear 4 — Primary Shaft	AISI 1015	1,72E+08	134,9	infinite	infinite	(0) 5	(2) 0	0,96
Gear 4 — Secondary Shaft	AISI 1015	3,54E+07	144,5	1,00E+08	2,83	0	1	0,96
Secondary Shaft	AISI 1015	3,54E+07	149,9	1,00E+08	2,83	0	1	0,98
Inner Spring	AISI 1080	3,20E+06	363,1	1,07E+06	0,33	2	0	1,04
Mesh Mechanism	AISI 1015	8,85E+06	188,6	1,87E+06	0,21	4	0	1,06

(high number of component failures and substitutions) and the end-of-life phase (little possibility of component reuse). This is confirmed by the values assumed by the functions LCSC and EUL introduced in Section 16.3.3, and evidenced by comparison with design alternative IIa, which predicts better values for f and r, and also for the mean BCI (0.78), while production costs remain substantially the same. This is principally due to a better choice of materials.

The values assumed by LCSC and EUL are further improved in design alternative IIb. This design differs from IIa only in that the primary shaft and gears (Figure 16.6) are no longer considered a single unit but, rather, are able to be disassembled (this modification does not result in variations in the behavior model and criticality of the system). In this case, the failure of a single gear does not require the substitution of the whole shaft–gear unit (improving LCSC), nor does it limit the possibility of reusing other components of the unit (improving EUL). The modification does, however, lead to an increase in the production cost.

All these results are summarized in Figure 16.7, where the values of the functions under examination have been normalized (the reciprocal of EUL is considered to have to minimize all the functions).

16.5.4 Analysis of the Environmental Impact of the Life Cycle

Finally, it is possible to evaluate the environmental impact of the entire life cycle EI_{LC} for the more interesting design alternatives IIa and IIb, using the

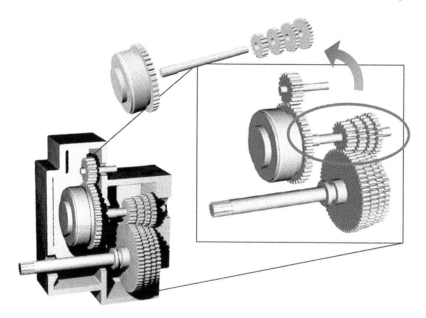

FIGURE 16.6   Design alternative IIb: Primary shaft and gears detachable.

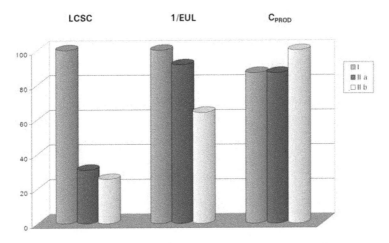

FIGURE 16.7   Comparison between design alternatives I, IIa, and IIb.

mathematical models introduced in Section 16.4.1. Also, in terms of this objective function alternative, IIb is better than IIa (–44%), paralleled by a limited increase in production cost (+15%), as evidenced in Figure 16.8 (EI_{LC} is expressed in mPt, according to Eco-indicator 99 method).

The radar diagram in Figure 16.9 gives an overview of the effectiveness of the two design alternatives considered, in terms of the four most significant objective functions (to be minimized): production cost, servicing costs, the reciprocal of extension of useful life, and the environmental impact of the life cycle.

16.6 Summary

In the context of Life Cycle Design, it is necessary to define analytical procedures and tools allowing the management of design choices (both at the level of product layout and at that of the specifications of single components) in relation to the conventional performances (functionality, safety, cost) and environmental performances of the product over its whole life cycle. In the methodological framework proposed in this chapter, the direct relation between design choices and final performance is obtained through a behavior model of the product. The proposed model allows the simulation of the life cycle in terms of phenomena of decay in performance of the materials.

As shown by the case study reported, the simulation method and analytical tools can be applied to obtain, for each design alternative examined, indications on the duration, safety, and criticality of single components and of the

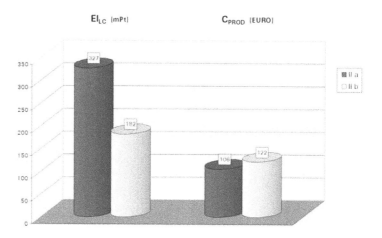

FIGURE 16.8   Comparison between design alternatives IIa and IIb.

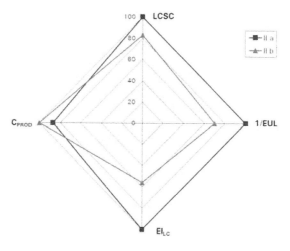

FIGURE 16.9   Comparison between design alternatives IIa and IIb: Radar diagram.

overall system, on varying the time of use; indications of possible expected failures and evaluations of the resulting servicing costs associated with the life cycle; indications of the residual life of components, evaluations of their possible reuse, and quantification of the resulting extension of the system's useful life; and evaluations of the environmental impact associated with the main phases of the life cycle, from production to retirement. By analyzing this broad spectrum of information, evaluated for each design alternative, it is possible to determine, already in the design phase, which solution can best satisfy the extended performance over the life cycle of a product.

16.7 References

Eubanks, C.F., Kmenta, S., and Ishii, K., System behavior modeling as a basis for advanced failure modes and effects analysis, in *Proceedings of the 1996 ASME Computers in Engineering Conference*, Irvine, CA, 1996, 96-DETC/CIE-1340.

Eubanks, C.F., Kmenta, S., and Ishii, K., Advanced failure modes and effects analysis using behavior modeling, in *Proceedings of the 1997 ASME Design Theory and Methodology Conference*, Sacramento, CA, 1997, 97-DETC/DTM-02.

Gershenson, J. and Ishii, K., Life-cycle serviceability design, in *Concurrent Engineering: Automation, Tools and Techniques*, Kusiak, A., Ed., John Wiley & Sons, New York, 1993, 363–384.

Giudice, F., La Rosa, G., and Risitano, A., Product recovery-cycles design: Extension of useful life, in *Feature Based Product Life-Cycle Modelling*, Soenen, R. and Olling, G., Eds., Kluwer Academic Publishers, Dordrecht, The Netherlands, 2003, 165–185.

Hata, T., Kimura, F., and Suzuki, H., Product life cycle design based on deterioration simulation, in *Proceedings of 4th CIRP International Seminar on Life Cycle Engineering*, Berlin, 1997, 59–68.

Kato, S., Hata, T., and Kimura, F., Decision factors of product life cycle strategies, in *Proceedings Supplement of EcoDesign 2001: 2nd International Symposium on Environmentally Conscious Design and Inverse Manufacturing*, Tokyo, 2001, 31–34.

Tomiyama, T., Umeda, Y., and Wallace, D.R., A holistic approach to life cycle design, in *Proceedings of 4th CIRP International Seminar on Life Cycle Engineering*, Berlin, 1997, 92–103.

Zust, R. and Caduff, G., Life-cycle modelling as an instrument for life-cycle engineering, *Annals of the CIRP*, 46(1), 351–354, 1997.

Index